Astrophysical Techniques

Astrophysical Techniques

C R Kitchin

University of Hertfordshire Observatory

Third Edition

Institute of Physics Publishing
Bristol and Philadelphia

British Library Cataloguing-in-Publication Data

A catalogue record for this book is available from the British Library.

ISBN   0 7503 0497 9   hardback
        0 7503 0498 7   paperback

Library of Congress Cataloging-in-Publication Data are available

Consultant Editor: **Professor A J Meadows**

First edition published 1984 by Adam Hilger Ltd and reprinted 1988

Second edition published 1991 by IOP Publishing Ltd under the Adam Hilger imprint and reprinted 1996 under the imprint of Institute of Physics Publishing

Third edition published 1998 by Institute of Physics Publishing, wholly owned by The Institute of Physics, London

Institute of Physics Publishing, Dirac House, Temple Back, Bristol BS1 6BE, UK

US Editorial Office: Institute of Physics Publishing, The Public Ledger Building, Suite 1035, 150 South Independence Mall West, Philadelphia, PA 19106, USA

Typeset in TeX using the IOP Bookmaker Macros
Printed in the UK by J W Arrowsmith Ltd, Bristol

For

Christine

Contents

Preface

The aim of this book is to provide a coherent state-of-the-art account of the instruments and techniques used in astronomy and astrophysics today. While every effort has been made to make it as complete and up-to-date as possible, the author is only too well aware of the many omissions and skimpily treated subjects throughout the work. For some types of instrumentation it is possible to give full details of the instrument in its finally developed form; however, for the 'new astronomies' development is still occurring at a rapid pace and the details will change between the writing and publishing of this edition. For those areas of astronomy therefore, a fairly general guide to the principles behind the techniques is given, and this should enable the reader to follow the detailed designs in the scientific literature.

Another aim has been to try and reduce the recent trend towards fragmentation of astronomical studies. The new techniques which are required for observing in exotic regions of the spectrum bring their own concepts and terminology with them. This can lead to the impression that the underlying processes are quite different, when in fact they are identical but are merely being discussed from differing points of view. Thus, for example, the Airy disc and its rings and the polar diagram of a radio dish do not at first sight look related, but in fact they are simply two different ways of presenting the same data. As far as possible therefore, broad regions of the spectrum are dealt with as a single area, rather than as many smaller disciplines. The underlying unity of all of astronomical observation is also emphasised by the layout of the book; the pattern of detection–imaging–ancillary instruments has been adopted so that one stage of an observation is encountered together with the similar stages required for all other information carriers. This is not an absolutely hard and fast rule, however, and in some places it seemed appropriate to deal with topics out of sequence, either in order to prevent a multiplicity of very small sections, or in order to keep the continuity of the argument going.

The treatment of the topics is at a level appropriate to a science-based undergraduate degree. As far as possible the mathematics and/or physics background which may be needed for a topic is developed or given within that section. In some places it was felt that some astronomy background would be needed as well, so that the significance of the technique under discussion could be properly realised. Although aimed at an undergraduate level, most of the mathematics should be understandable by anyone who has attended a competently taught mathematics course in their final years at school. Thus many amateur astronomers may well find selected aspects of the book of great use and interest. The fragmentation of astronomy, which has already been mentioned, means that there is a third group of people who may find the book useful, and that is professional astronomers themselves. The treatment of the topics in general is at a sufficiently high level, yet in a convenient and accessible form, for those professionals seeking information on techniques in areas of astronomy with which they might not be totally familiar.

I would like to acknowledge the generosity of the Library of the Institute of Astronomy in Cambridge, for allowing me to spend many hours using their facilities whilst preparing this third edition.

Lastly I must pay a tribute to the many astronomers and other scientists whose work is summarised here. It is not possible to list them by name, and to have studded the text with detailed references would have ruined the intention of making the book readable. I would, however, like to take the opportunity afforded by this preface to express my deepest gratitude to them all.

C R Kitchin

Standard symbols

Most of the symbols used in this book are defined when they are first encountered and that definition then applies throughout the rest of the section concerned. Some symbols, however, have acquired such a commonality of use that they have become standard symbols amongst astronomers. Some of these are listed below, and the symbol will not then be separately defined when it is encountered in the text.

amu	Atomic mass unit $= 1.6605 \times 10^{-27}$ kg
c	Velocity of light *in vacuo* $= 2.9979 \times 10^8$ m s^{-1}
e	Charge on the electron $= 1.6022 \times 10^{-19}$ C
e^-	Symbol for an electron
e^+	Symbol for a positron
eV	Electron volt $= 1.6022 \times 10^{-19}$ J
G	Gravitational constant $= 6.670 \times 10^{-11}$ N m^2 kg^{-2}
h	Planck's constant $= 6.6262 \times 10^{-34}$ J s
k	Boltzmann's constant $= 1.3806 \times 10^{-23}$ J K^{-1}
m_e	Mass of the electron $= 9.1096 \times 10^{-31}$ kg
n	Symbol for a neutron
p$^+$	Symbol for a proton
U, B, V	Magnitudes through the standard UBV photometric system
$^{12}_6$C	Symbol for nuclei (normal carbon isotope given as the example). The superscript is the atomic mass (in amu) to the nearest whole number and the subscript is the atomic number
γ	Symbol for a gamma ray
ε_0	Permittivity of free space $= 8.85 \times 10^{-12}$ F m^{-1}
λ	Symbol for wavelength
μ	Symbol for refractive index
μ_0	Permeability of a vacuum $= 4\pi \times 10^{-7}$ H m^{-1}
ν	Symbol for frequency

Chapter 1

Detectors

1.1 OPTICAL AND INFRARED DETECTION

Introduction

In this and the immediately succeeding sections, the emphasis is upon the detection of the radiation or other information carrier, and upon the instruments and techniques used to facilitate and optimise that detection. There is inevitably some overlap with other sections and chapters, and some material might arguably be more logically discussed in a different order from the one chosen. In particular in this section, telescopes are included as a necessary adjunct to the detectors themselves. The theory of the formation of an image of a point source by a telescope, which is all that is required for simple detection, takes us most of the way through the theory of imaging extended sources. Both theories are therefore discussed together even though the latter should perhaps be in Chapter 2. There are many other examples such as x-ray spectroscopy and polarimetry which appear in section 1.3 instead of sections 4.2 and 5.2. In general the author has tried to follow the pattern detection–imaging–ancillary techniques throughout the book, but has dealt with items out of this order when it seemed more natural to do so.

The optical region is taken to include the near infrared and ultraviolet, and thus roughly covers the region from 10 nm to 100 μm. The techniques and physical processes employed for investigations over this region bear at least a generic resemblance to each other and so may be conveniently discussed together.

Detectors

Detector types

In the optical region detectors fall into two main groups, thermal and quantum (or photon) detectors. Both these types are incoherent, that is to say, only the

1

amplitude of the electromagnetic wave is detected, the phase information is lost. Coherent detectors are common at long wavelengths (section 1.2), where the signal is mixed with that from a local oscillator (heterodyne principle). Only recently have optical heterodyne techniques been developed in the laboratory, and these have yet to be applied to astronomy. We may therefore safely regard all optical detectors as incoherent in practice.

In quantum detectors, the individual photons of the optical signal interact directly with the electrons of the detector. Sometimes individual detections are monitored (photon counting), at other times the detections are integrated to give an analogue output like that of the thermal detectors. Examples of quantum detectors include the eye, photographic emulsion, photomultiplier, photodiode and charge-coupled devices.

Thermal detectors, by contrast, detect radiation through the increase in temperature that its absorption causes in the sensing element. They are generally less sensitive and slower in response than quantum detectors, but have a very much broader spectral response. Examples include thermocouples, pyroelectric detectors, bolometers and Golay cells.

The eye

This is undoubtedly the most fundamental of the detectors to a human astronomer although it is a very long way from being the simplest. It is now rarely used for

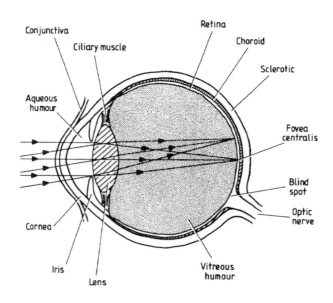

Figure 1.1.1. Optical paths in the eye.

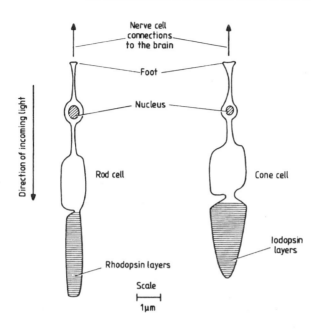

Figure 1.1.2. Retinal receptor cells.

primary detection, although there are still a few applications in which it performs comparably with or possibly even better than other detection systems. Examples of this could be very close double-star work and planetary observation, in which useful work can be undertaken even though the eye may have been superseded for a few specific observations by, say, speckle interferometry and planetary space probes. Much more commonly it is necessary to find and/or guide on objects visually while they are being studied with other detectors. Thus there is some importance in understanding how the eye and especially its defects can influence these processes.

It is vital to realise that a simple consideration of the eye on basic optical principles will be misleading. The eye and brain act together in the visual process, and this may give better or worse results than, say, an optically equivalent camera system depending upon the circumstances. Thus the image on the retina is inverted and suffers from chromatic aberration (see p. 51) but these effects are compensated for by the brain. Conversely, high contrast objects or structures near the limit of resolution, such as planetary transits of the sun and the Martian 'canals', may be interpreted falsely.

The optical principles of the eye are shown in figure 1.1.1 and are too well known to be worth discussing in detail here. The receptors in the retina (figure 1.1.2) are of two types; cones for colour detection, and rods for black and white reception at higher sensitivity. The light passes through the outer layers of nerve

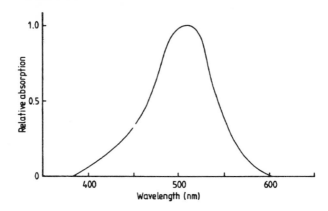

Figure 1.1.3. Rhodopsin absorption curve.

connections before arriving at these receptors. In the rods a pigment known as rhodopsin or, from its colour, visual purple, absorbs the radiation. It is a complex protein with a molecular weight of about 40 000 amu, and its absorption curve is shown in figure 1.1.3. It is arranged within the rods in layers about 20 nm thick and 500 nm wide, and may comprise up to 35% of the dry weight of the cell. Under the influence of light a small fragment of it will split off. This fragment, or chromophore, is a vitamin A derivative called retinal or retinaldehyde and has a molecular weight of 286 amu. One of its double bonds changes from a *cis* to a *trans* configuration (figure 1.1.4) within a picosecond of the absorption of the photon. The portion left behind is a colourless protein called opsin. The instant of visual excitation occurs at some stage during the splitting of the rhodopsin molecule, but its precise mechanism is not yet understood. The reaction causes a change in the permeability of the receptor cell's membrane to sodium ions and thus a consequent change in the electrical potential of the cell. The change in potential then propagates through the nerve cells to the brain. The rhodopsin molecule is slowly regenerated. The response of the cones is probably due to a similar mechanism. A pigment known as iodopsin is found in cones and this

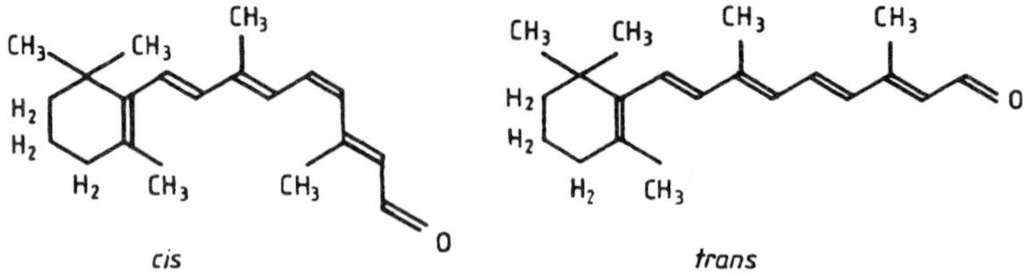

Figure 1.1.4. Transformation of retinaldehyde from *cis* to *trans* configuration by absorption of a photon near 500 nm.

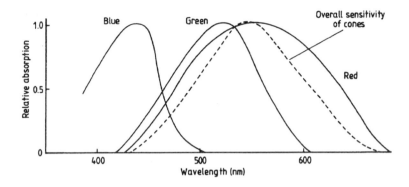

Figure 1.1.5. Absorption curves of pigments involved in cone vision.

also contains the retinaldehyde group. Cone cells, however, are of three varieties with differing spectral sensitivities, and this appears to arise from the presence of much smaller quantities of other pigments. The absorption curves of these other pigments are shown in figure 1.1.5, together with the overall sensitivity curve of the cones for comparison.

In bright light much of the rhodopsin in the rods is split into opsin and retinaldehyde, and their sensitivity is therefore much reduced. Vision is then provided primarily by the cones, although their sensitivity is only about 1% of the maximum for the rods. The three varieties of cones combine their effects to produce colour vision. At low light levels only the rods are triggered by the radiation and vision is then in black and white. The overall spectral sensitivities of the rods and cones differ (figures 1.1.3 and 1.1.5) with that of the rods peaking at about 510 nm and that of the cones at 550 nm. This shift in sensitivity is called the Purkinje effect. It can be a problem for double-star observers since it may cause a hot and bright star to have its magnitude underestimated in comparison with a cooler and fainter star, and vice versa. Upon entering a dark observatory from a brightly illuminated room, the rhodopsin in the rods slowly re-forms over about half an hour, their sensitivity therefore improves concurrently. Thus we have the well known phenomenon of dark adaption, whereby far more can be distinguished after a few minutes in the dark than can be detected initially. If sensitive vision is important for an observation, then, for optimum results, bright lights should be avoided for half an hour before it is due to be made. Most observatories are only faintly illuminated and usually with red light in order to try and minimise any loss of dark adaption.

Astronomical observation is generally due to rod vision. Usually with a dark-adapted eye, between one and ten photons are required to trigger an individual rod. However, several rods must be triggered in order to result in a pulse being sent to the brain. This arises because many rods are connected to a single nerve fibre. Most cones are also multiply connected, although a few, particularly those in the *fovea centralis*, have a one to one relationship with nerve fibres. The total number of rods is about 10^8 with about 6×10^6 cones and about 10^6 nerve fibres, so that upwards of a hundred receptors can be connected to a single nerve. In addition there are many cross connections between groups of receptors. The cones are concentrated towards the *fovea centralis*, and this is the region of most acute vision. Hence the rods are more plentiful towards the periphery of the field of view, and so the phenomenon called averted vision, whereby a faint object only becomes visible when *not* looked at directly, arises. Its image is falling onto a region of the retina richer in rods when the eye is averted from its direct line of sight. The combination of differing receptor sensitivities, change in pigment concentration, and aperture adjustment by the iris mean that the eye is usable over a range of illuminations differing by a factor of 10^9 or 10^{10}.

The Rayleigh limit (see p. 48) of resolution of the eye is about twenty seconds of arc when the iris has its maximum diameter of five to seven millimetres. But for two separate images to be distinguished, they must be separated on the retina by at least one unexcited receptor cell. So that even for images on the *fovea centralis* the actual resolution is between one or two minutes of arc. This is much better than elsewhere on the retina since the *fovea centralis* is populated almost exclusively by small tightly packed cones, many of which are singly connected to the nerve fibres. Its slightly depressed shape may also cause it to act as a diverging lens producing a slightly magnified image in that region. Away from the *fovea centralis* the multiple connection of the rods, which may be as high as a thousand receptors to a single nerve fibre, degrades the resolution far beyond this figure. Other aberrations and variations between individuals in their retinal structure means that the average resolution of the human eye lies between five and ten minutes of arc for point sources. Linear sources such as an illuminated grating can be resolved down to one minute of arc fairly commonly. The effect of the granularity of the retina is minimised in images by rapid oscillations of the eye through a few tens of seconds of arc with a frequency of a few hertz, so that several receptors are involved in the detection when averaged over a short time.

With areas of high contrast, the brighter area is generally seen as too large, a phenomenon which is known as irradiation. We may understand this as arising from stimulated responses of unexcited receptors due to their cross-connections with excited receptors. We may also understand eye fatigue which occurs when staring fixedly at a source (for example when guiding on a star) as due to depletion of the sensitive pigment in those few cells covered by the image. Averting the eye very slightly will then focus the image onto different cells so reducing the problem. Alternatively the eye can be rested momentarily by looking away from the eyepiece to allow the cells to recover.

The response of vision to changes in illumination is logarithmic. That is to say, if two sources, A and B are observed to differ in brightness by a certain amount, and a third source, C, appears midway in brightness between them, then the energy from C will differ from that from A by the same *factor* as that of B differs from C. In other words, if we use L to denote the perceived luminosity and E to denote the actual energy of the sources, we have if

$$\tfrac{1}{2}(L_A + L_B) = L_C \tag{1.1.1}$$

then

$$E_A/E_C = E_C/E_B. \tag{1.1.2}$$

This phenomenon is the reason for the magnitude scale used by astronomers to measure stellar brightnesses (section 3.1), since the scale had its origins in the eye estimates of stellar luminosities by the ancient astronomers. The faintest stars visible to the dark-adapted naked eye from a good observing site on a good clear moonless night are of about magnitude six. This corresponds to the detection of about 3×10^{-15} W. Special circumstances or especially sensitive vision may enable this limit to be improved upon by some one to one and a half stellar magnitudes. Conversely the normal ageing processes in the eye mean that the retina of a 60-year old person receives only about 30% of the amount of light seen by a person half that age. Eye diseases and problems, such as cataracts, may reduce this much further. Thus observers should expect a reduction in their ability to perceive very faint objects as time goes by.

Photomultipliers

Semiconductors and the photoelectric effect
Semiconductors.   The photomultiplier and several of the other detectors considered later in this section derive their properties from the behaviour of semiconductors. Thus some discussion of the relevant aspects of these materials is a necessary prerequisite to a full understanding of the detectors of this type.

Conduction in a solid may be understood by considering the electron energy levels. For a single isolated atom they are unperturbed and are the normal atomic energy levels. As two such isolated atoms approach each other their interaction causes the levels to split (figure 1.1.6). If further atoms approach, then the levels also develop further splits, so that for N atoms in close proximity to each other, each original level is split into N sublevels (figure 1.1.7). Within a solid therefore, each level becomes a pseudo-continuous band of permitted energies, since the individual sublevels all overlap each other. The energy level diagram for a solid then has the appearance shown in figure 1.1.8. The innermost electrons remain bound to their nuclei, while the outermost electrons interact to bind the atoms together. They occupy an energy band called the valence band.

In order to conduct electricity through such a solid, the electrons must be able to move. From figure 1.1.8 we may see that free movement could occur within the valence and higher bands. However, if the original atomic level which became the valence band upon the formation of the solid was fully occupied, then all the sublevels within the valence band will still be fully occupied. But if

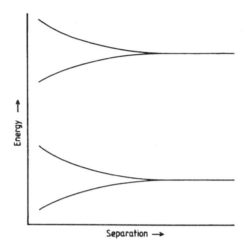

Figure 1.1.6. Schematic diagram of the splitting of two of the energy levels of an atom due to its proximity to another similar atom.

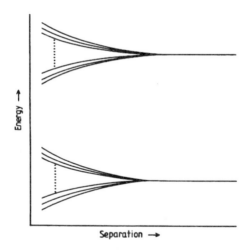

Figure 1.1.7. Schematic diagram of the splitting of two of the energy levels of an atom due to its proximity to many similar atoms.

any given electron is to move under the influence of an electric potential, then its energy must increase. This it cannot do, since all the sublevels are occupied, and so there is no vacant level available for it at this higher energy. Thus the electron cannot move after all. Under these conditions we have an electrical insulator. If the material is to be a conductor, we can now see that there must be vacant sublevels which the conduction electron can enter. There are two ways in which

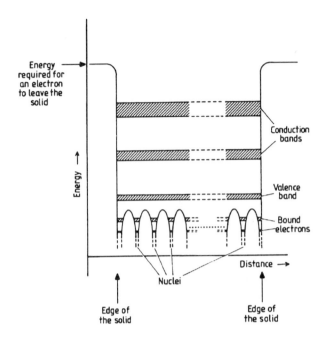

Figure 1.1.8. Schematic energy level diagram of a solid.

such empty sublevels may become available. Either the valence band is unfilled, for example when the original atomic level had only a single electron in an s subshell, or one of the higher energy bands becomes sufficiently broadened to overlap the valence band. In the latter case at a temperature of absolute zero, all the sublevels of both bands will be filled up to some energy which is called the Fermi level. Higher sublevels will be unoccupied. As the temperature rises some electrons will be excited to some of these higher sublevels, but will still leave room for conduction electrons.

A third type of behaviour occurs when the valence and higher bands do not actually overlap, but have only a small separation. Thermal excitation may then be sufficient to push a few electrons into some of the higher bands. An electric potential can now cause the electrons in either the valence or the higher band to move. The material is known as a semiconductor since its conductance is generally better than that of an insulator but considerably worse than that of a true conductor. The higher energy bands are usually known as the conduction bands. A pure substance will have equal numbers of electrons in its conduction bands and of spaces in its valence band. However, an imbalance can be induced by the presence of differing atomic species. If the valence band is full, and one atom is replaced by another which has a larger number of valence electrons (a donor atom), then the excess electron(s) usually occupy new levels in the gap between the valence and conduction bands. From there they may more easily be excited into the conduction band. The semiconductor is then an n-type since any current within it will largely be carried by the (negative) electrons in the conduction band. In the other case when an atom is replaced with one which has fewer valence electrons (an acceptor atom), spaces will become available within the valence band and an electric current will be carried by the movement of electrons within this band. Since the movement of an electron within the valence band is exactly equivalent to the movement of a positively charged hole in the opposite direction, this type of semiconductor is called p-type, and its currents are thought of as being transported by the positive holes in the valence band.

The photoelectric effect.   The principle of the photoelectric effect is well known; a photon with a wavelength less than the limit for the material is absorbed by the material and an electron is emitted. The energy of the electron is a function of the energy of the photon, while the number of electrons depends upon the intensity of the illumination. In practice the situation is somewhat more complex, particularly when we are interested in specifying the properties of a *good* photoemitter.

The main requirements are that the material should absorb the required radiation efficiently, and that the mean free paths of the released electrons within the material should be greater than that of the photons. The relevance of these two requirements may be most simply understood from looking at the behaviour of metals in which neither condition is fulfilled. Since metals are conductors there are many vacant sublevels near their Fermi levels (see the earlier discussion). After absorption of a photon, an electron is moving rapidly and energetically within the metal. Collisions will occur with other electrons and since these electrons have other nearby sublevels which they may occupy, they can absorb some of the energy from our photoelectron. Thus the photoelectron is slowed by collisions until it may no longer have sufficient energy to escape from the metal even if it does reach the surface. For most metals the mean free path of a photon is about 10 nm and that of the released electron less than 1 nm, thus the eventually emitted number of electrons is very considerably reduced by collisional energy losses. Furthermore the high reflectivity of metals results in only a small fraction of the suitable photons being absorbed, and so the actual number of electrons emitted is only a very small proportion of those potentially available.

Thus a good photoemitter must have low energy loss mechanisms for its released electrons while they are within its confines. The loss mechanism in metals (collision) can be eliminated by the use of semiconductors or insulators. Then the photoelectron cannot lose significant amounts of energy to the valence electrons because there are no vacant levels for the latter to occupy. Neither can it lose much energy to the conduction electrons since there are very few of these around. In insulators and semiconductors, the two important energy

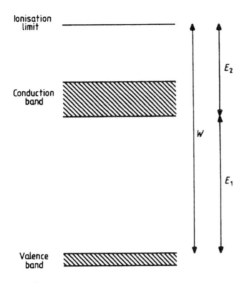

Figure 1.1.9. Schematic partial Grotrian diagram of a good photoemitter.

loss mechanisms are pair production and sound production. If the photoelectron is energetic enough, then it may collisionally excite other valence electrons into the conduction band thus producing *pairs* of electrons and holes. This process may be eliminated by requiring that E_1 the minimum energy to excite a valence electron into the conduction band of the material (figure 1.1.9) is larger than E_2, the excess energy available to the photoelectron. Sound waves or phonons are the vibrations of the atoms in the material and can be produced by collisions between the photoelectrons and the atoms especially at discontinuities in the crystal lattice etc. Only one per cent or so of the electron's energy will be lost at each collision because the atom is so much more massive than the electron. However, the mean free path between such collisions is only one or two nanometres so that this energy loss mechanism becomes very significant when the photoelectron originates deep in the material. The losses may be reduced by cooling the material since this then reduces the available number of quantised vibration levels in the crystal and also increases the electron's mean free path.

The minimum energy of a photon if it is to be able to produce photoemission is known as the work function and is the difference between the ionisation level and the top of the valence band (figure 1.1.9). Its value may, however, be increased by some or all of the energy loss mechanisms mentioned above. In practical photoemitters pair production is particularly important at photon energies above the minimum and may reduce considerably the expected flux as the wavelength decreases. The work function is also strongly dependent upon the surface properties of the material; surface defects, oxidation products, impurities etc can cause it to vary widely even among different samples of the same substance. The work function may be reduced if the material is strongly p-type and is at an elevated temperature. Vacant levels in the valence band may then be populated by thermally excited electrons bringing them nearer to the ionisation level and so reducing the energy required to let them escape. Since most practical photoemitters are strongly p-type, this is an important process and confers sensitivity at longer wavelengths than the nominal cut-off point. The long-wave sensitivity, however, is very variable since the degree to which the substance is p-type is very strongly dependent upon the presence of impurities and is very sensitive to small changes in the composition of the material.

In photomultipliers, once a photoelectron has escaped from the photoemitter, it is accelerated by an electric potential until it strikes a second electron emitter. The primary electron's energy then goes into pair production and secondary electrons are emitted from the substance in a manner analogous to photoelectron emission. Since typically 1 eV (1.6×10^{-19} J) of energy is required for pair production and the primary's energy can reach 100 eV or more by the time of impact, several secondary electron emissions can result from a single primary electron. This multiplication of electrons cannot, however, be increased without bound, since if the primary's energy is too high it will penetrate too far into the material for the secondary electrons to escape. The upper limit, however, is a factor of one hundred or more and is not usually a restriction in practical devices in which multiplications by factors of ten or less are used. The properties of a good secondary electron emitter are the same as those of a good photoemitter except that its optical absorption and reflection is no longer important.

Electrons can be emitted from either the primary or secondary electron emitters through processes other than the required ones, and these electrons then contribute to the noise of the system. The most important such processes are thermionic emission and radioactivity. Thermionic emission is electron escape due to the thermal motions of the electrons. Its rate is governed by the position of the Fermi level. Most practical electron emitters are strongly p-type semiconductors whose Fermi levels are close to their valence bands, thus any further reduction in thermionic emission can only be achieved by cooling. The Fermi level is strongly influenced by impurities and changes in composition so that the rate of thermionic emission can also vary widely from one sample to another. Radioactivity, which for this purpose includes the effects of cosmic rays, can produce electrons in several ways. They may originate in the primary or secondary electron emitter through their normal interactions, or from direct radioactive emission (β rays), or from interactions with other parts of the device. Little can be done to reduce this nuisance except to use uncontaminated materials, to shield the device and to arrange the accelerating electric fields so that electrons which do not originate from one of the electron emitters are deflected away from subsequent stages in the detector. Some discrimination between pulses originating from genuine photons and those due to other processes is possible

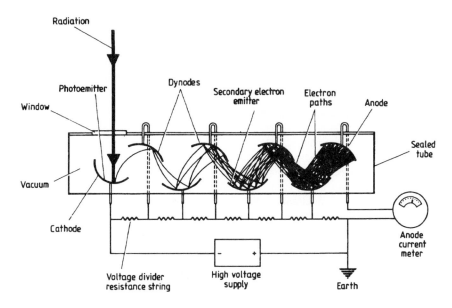

Figure 1.1.10. Schematic arrangement for a photomultiplier.

when the detector is used in a photon counting (see p. 296) mode. Pulses outside the probable size range of those originating from photons can then be disregarded.

Photomultipliers. The main application of the photoelectric effect to radiation detection is in the electron multiplier phototube, or photomultiplier as it is more commonly known. The basic components and construction are shown in figure 1.1.10. The photoemitter is coated onto the cathode and this is at a negative potential of some 1000 V with respect to earth. The secondary emitter is coated onto dynodes which are successively more positive than the cathode by 100 V or so for each stage. The various electrodes are shaped and positioned so that the electrons are channelled towards the correct next electrode in the sequence after each interaction. The final signal pulse may contain 10^6 electrons for each incoming photon, and after arriving at the anode it may be further amplified and detected in any of the usual ways. This large intrinsic amplification of the photomultiplier is one of the major advantages of the device.

The commonest photoelectric and secondary emitter materials in current use are listed in table 1.1.1.

The electrode may be opaque as illustrated in figure 1.1.10 so that the electrons emerge from the illuminated face, or it may be semitransparent, when they emerge from the opposite face. In the latter case some change to the spectral sensitivity occurs due to the absorption in the material, but there are the

Table 1.1.1. Photoelectron emitting substances.

Substance	Long wavelength cut-off point (nm)
Sodium chloride (NaCl)	150
Potassium bromide (KBr)	155
Rubidium iodide (RbI)	185
Cuprous chloride (CuCl)	190
Caesium iodide (CsI)	200
Copper/beryllium (Cu/Be)	200
Copper iodide (CuI)	210
Rubidium telluride ($RbTe_2$)	300
Caesium telluride (Cs_2Te)	350
Caesium antimonide ($Cs_{2.9}Sb$)	600–700
Bi-alkali ((K_2Cs)Sb)	670
Tri-alkali ((Cs) Na_2KSb)	850
Gallium arsenide (GaAs (Cs))	1000
Silver/oxygen/caesium (Ag/Cs_2O)	1000–1100

Secondary electron emitting substances

Beryllium oxide (BeO (Cs))
Caesium antimonide (Cs_3Sb)
Gallium phosphide (GaP (Cs))
Magnesium oxide (MgO (Cs))
Potassium chloride (KCl)
Silver/magnesium (Ag/Mg)

compensating advantages of higher quantum efficiency and convenience since the cathodes may be coated on the inside of the window. Semitransparent cathodes are usually only about 30 nm thick so that it is very easy for small thickness changes to occur and to alter their sensitivities. Their usual application is for long-wave radiation detection.

The dynodes need to be as close to each other and to the cathode as possible. This is to ensure that a high proportion of the electrons from the previous stage are captured and to reduce their transit times and dispersion so that the final pulse is as brief as possible. Since the electron emitting materials are generally coated onto the electrodes after construction of the photomultiplier and while it is in a vacuum just prior to its encapsulation, there is a conflict between the requirements of uniform coatings and of small electrode gaps. Thus a photomultiplier can be optimised towards uniform response or short pulse times but not in general towards both. The appropriate type of photomultiplier must therefore be selected for the application in mind. The electric fields between the dynodes can be arranged so that the electrons are focused onto their next stage. An electron produced at a given point on the cathode is then always mapped to the same points on the dynodes. If the system is unfocused then there is some degree of randomness in the mapping of the electron paths. The latter arrangement gives a more uniform response through statistical averaging in the later stages of the dynode chain, but the pulse lengths are increased from the 10 ns or so of a focused system because of the varying path lengths of the electrons.

The micro-channel plate (figure 1.3.4) can also be used at optical wavelengths. With an array of anodes to collect the clouds of electrons emerging from the plate, it provides an imaging detector with a high degree of intrinsic amplification. Such devices are often referred to as MAMA detectors (Multi-Anode Micro-channel Array).

For detection from the visual out to 30 μm or so, a solid state photomultiplier can be used. This is closely related to the avalanche photodiode (see below). It uses a layer of arsenic-doped silicon on a layer of undoped silicon. A potential difference is applied across the device. An incident photon produces an electron–hole pair in the doped silicon layer. The electron drifts under the potential difference towards the undoped layer. As it approaches the latter, the increasing potential difference accelerates it until it can ionise further atoms. The resulting avalanche of electrons is collected by an electrode on the far side of the undoped silicon layer. A gain of 10^4 or so is possible with this device.

Problems. One of the major problems encountered in using photomultipliers will already have become apparent. That is, the non-uniformity of the response. Not only can nominally identical devices differ in their responses, but the response can vary over the sensitive area of a single device. The variations in the sensitivity of the cathode and the first few dynodes dominate this effect through their amplification by the later stages of the dynode chain. In astronomical photometers (section 3.2) the use of a Fabry lens to provide a stationary extended image so that the effect of the variations is minimised is almost universal. The change in the long-wave sensitivity with temperature and impurities etc has great significance for wide band photometry. The V-filter long-wave cut-off (Chapter 3), for example, is provided by the photomultiplier response and not by the filter. Similarly the standard U and B filters have red 'leaks' which will distort the measurements if there is any long-wave sensitivity to the detector. Other variations in the response will occur for oblique illumination since the effective cathode thickness is then altered. Polarised light can also be a problem since there is some dependence of the response upon the direction of the plane of polarisation. Thus when a photomultiplier is used as a part of a polarimeter (section 5.3), a depolariser must also normally be used as well.

Noise in the signal arises from many sources. The amplification of each pulse can vary by as much as a factor of ten through the sensitivity variations and changes in the number of secondary electrons lost between each stage. The final registration of the signal can be by analogue means and then the pulse strength variation is an important noise source. Alternatively individual pulses may be counted and then it is less significant. Indeed when using pulse counting even the effects of other noise sources can be reduced by using a discriminator to eliminate the very large and very small pulses which do not originate in primary photon interactions. Unfortunately, however, pulse counting is limited in its usefulness to faint sources otherwise the individual pulses start to overlap.

Thermionic emission from the cathode and the first few dynodes can be important and is usually the major source of the 'dark current'; that is, the signal detected when no light is falling onto the cathode. Its effect may generally be reduced by cooling and by offsetting the final signal so that it is only the variations in the thermionic emission which contribute to the noise. Another source of the dark current is the ions from the residual gas inside the photomultiplier. These can originate from interactions between the gas and radioactive particles or cosmic rays, or from collision with the electrons moving between the dynodes, and they can drift towards the cathode or dynodes to produce spurious pulses. Also there may be leakage from the dynodes to the anode and glow discharges from the dynodes and anode.

Cosmic rays and other sources of energetic particles and gamma rays contribute to the dark current in several ways. Their direct interactions with the cathode or dynodes early in the chain can expel electrons which are then amplified in the normal manner. Alternatively electrons or ions may be produced from the residual gas or from other components of the structure of the photomultiplier. The most important interaction, however, is generally Čerenkov radiation produced in the window or directly in the cathode. Such Čerenkov pulses can be up to a hundred times more intense than a normal pulse. They can thus be easily discounted when using the photomultiplier in a pulse counting mode, but make a significant contribution to the signal when simply its overall strength is measured.

Many other phenomena can change a photomultiplier's response. Magnetic and electric fields can directly affect the electron paths, thus changing the collecting efficiency of the dynodes and anode. This is particularly important for the astronomer since the Earth's magnetic field will vary in its effect as the photomultiplier changes its position in space when mounted on a telescope which is tracking a star. The use of a 'mu-metal' shield or something similar is then vital. Conversely magnetic fields may be applied deliberately to improve the noise level of the photomultiplier by restricting the effective area of the photocathode. Humidity can change the behaviour through its effect upon the string current (see p. 18) or by causing glow discharges etc. Other equipment which is operating nearby can induce transient currents or change the level of the high tension voltage supply. Heavy equipment such as dome drive motors etc is particularly important in this respect. The response of the photomultiplier changes with temperature as we have already seen, and the tubes are usually cooled in operation to reduce the dark current. However, the cooling from ambient temperature is a lengthy process; five to ten hours not being an uncommon time needed for the temperature to stabilise. Thus an impatient or forgetful astronomer can also be a source of error by using the photomultiplier before it has cooled down adequately. Yet another problem source is aging. That is, the long term change in performance due to such processes as chemical degradation of the electron emitters, softening of the vacuum through leakage, migration of the chemicals within the structure through their evaporation and recondensation and so on. The last effect may be considerably worsened if the tube is stored or operated at elevated temperatures, or is illuminated by very intense radiation. Lastly in this by no means complete list of ills to which photomultipliers are prone, the tube must be mounted securely and in a light-tight box (apart from the entrance aperture of course). The mounting must hold the tube rigidly in position but must not apply significant stress to the tube at any of the temperatures which it may encounter. The mounting must also be non-magnetic, non-conducting and chemically inert. The electrical connections are usually made via a ceramic or Teflon socket, and Teflon is also a suitable material for holding the photomultiplier in position (see figure 3.2.6 for example). The photometer must be light-tight apart from its entrance aperture (section 3.2) or stray lights in the observatory may add to the noise and non-uniformity problems.

Operation. Extreme care in the use of photomultipliers is necessary if photometry to 1% (0.01 magnitudes) accuracy or better is to be achieved. The foregoing catalogue of problems can also be taken in reverse to indicate the precautions which must be taken during the operation of the device. Photometers (section 3.2) must be designed to eliminate or to reduce to acceptable levels all of the possible faults.

In addition to the problems of the photomultiplier itself there is the high tension voltage supply. This drives currents through both the resistor chain and the photomultiplier. The former is called the string current and the latter the anode current. Thus as the anode current changes in response to the changing illumination of the photomultiplier, so the string current must change also. This then affects the accelerating potentials within the photomultiplier and so the device becomes non-linear in its response. As a general rule-of-thumb, if $x\%$ accuracy is desired in the measurements, then the string current must exceed the anode current by a factor of $100/x$. The high tension voltage supply must obviously also be stabilised or its variations will add to the noise. Generally it must be stable to a factor of ten better than the required signal stability.

On another practical note, cooling or warming rates for photomultipliers should be held to one degree per minute or less. Otherwise the seals around the electrical contacts or the window may crack under the stress of differential expansion and the vacuum inside the device will be destroyed.

Table 1.1.2. Classification scheme for type of detector.

Sensitive parameter	Detector names	Class
Voltage	Photovoltaic cells	Quantum
	Photoelectromagnetic cells	Quantum
	Thermocouples	Thermal
	Pyroelectric detectors	Thermal
Resistance	Bolometer	Thermal
	Photoconductive cell	Quantum
	Phototransistor	Quantum
Charge	Charge-coupled device (CCD)	Quantum
	Charge injection device (CID)	Quantum
Electron excitation	Photographic emulsion	Quantum
Electron emission	Gas cell	Quantum
	Photomultiplier	Quantum
	Television	Quantum
	Image intensifier	Quantum
Pressure	Golay cell	Thermal
Chemical composition	Eye	Quantum

A detector breviary

Two of the four most ubiquitous detectors used in astronomy, the eye and the photomultiplier, have now been discussed in some detail. The third major type of detector is the photographic emulsion and is dealt with in section 2.2 and the fourth is the CCD (see below). There are many other types of detectors but before going on to discuss these, they need placing in some sort of logical framework or the reader is likely to become more confused rather than more enlightened by this section! We may idealise any detector as simply a device wherein some measurable property changes in response to the effects of electromagnetic radiation. We may then classify the types of detector according to the property that is changing, and the scheme used in this work is shown in table 1.1.2.

Other properties may be sensitive to radiation but fail to form the basis of a useful detector. For example forests can catch fire, and some human skins turn brown under the influence of solar radiation, but these are extravagant or slow ways of detecting the sun! Yet other properties may become the basis of detectors in the future. Such possibilities might include the initiation of stimulated radiation from excited atomic states as in the laser and the changeover from superconducting to normally conducting regimes in a material.

Before resuming discussion of the detector types listed earlier, we need to establish the definitions of some of the criteria used to assess and compare detectors. The most important of these are listed in table 1.1.3.

Table 1.1.3. Some criteria for assessment and comparison of detectors.

QE (quantum efficiency)	Ratio of the actual number of photons which are detected to the number of incident photons.
DQE (detective quantum efficiency)	Square of the ratio of the output signal/noise ratio to the input signal/noise ratio.
τ (time constant)	This has various precise definitions. Probably the most widespread is that τ is the time required for the detector output to approach to within $(1 - e^{-1})$ of its final value after a change in the illumination; i.e. the time required for about 63% of the final change to have occurred.
Dark noise	The output from the detector when it is unilluminated. It is usually measured as a root-mean-square voltage or current.
NEP (noise equivalent power or minimum detectable power)	The radiative flux as an input, which gives an output signal-to-noise ratio of unity. It can be defined for monochromatic or black body radiation, and is usually measured in watts.
D (detectivity)	Reciprocal of NEP. The signal-to-noise ratio for incident radiation of unit intensity.
D^* (normalised detectivity)	The detectivity normalised by multiplying by the square root of the detector area, and by the electrical bandwidth. It is usually pronounced 'dee star'.

$$D^* = \frac{(a\Delta f)^{1/2}}{\text{NEP}}. \qquad (1.1.3)$$

	The units cm Hz$^{1/2}$ W^{-1} are commonly used and it then represents the signal-to-noise ratio when 1 W of radiation is incident on a detector with an area of 1 cm^2, and the electrical bandwidth is 1 Hz.
R (responsivity)	Detector output for unit intensity input. Units are usually volts per watt or amps per watt.
Dynamic range	Ratio of the saturation output to the dark signal. Sometimes only defined over the region of linear response.
Spectral response	The change in output signal as a function of changes in the wavelength of the input signal.
λ_m (peak wavelength)	The wavelength for which the detectivity is a maximum.
λ_c (cut-off wavelength)	There are various definitions. Among the commonest are: wavelength(s) at which the detectivity falls to zero; wavelength(s) at which the detectivity falls to 1% of its peak value; wavelength(s) at which D^* has fallen to half its peak value.

For purposes of comparison D^* is generally the most useful parameter. For photomultipliers in the visible region it is around 10^{15} to 10^{16}. Figures for the eye and for photographic emulsion are not directly obtainable, but values of 10^{12} to 10^{14} can perhaps be used for both to give an idea of their relative performances.

Charge-coupled devices (CCD)

Although only invented about three decades ago, these devices have already become as important a class of detector as the other three main types for astronomical work. Only their cost seems likely to restrict their use in the foreseeable future. Their popularity arises from their ability to integrate the detection over long intervals, their dynamic range (10^5), high quantum efficiency, and from the ease with which arrays can be formed to give two-dimensional imaging. In fact, CCDs can only be formed as an array, a single unit is of little use by itself. For this reason they are dealt with here rather than in Chapter 2. In astronomy, CCDs are almost synonymous with detectors of radiation. However, they do have other important applications, especially as computer memories. Their structure is somewhat different for these other applications and the reader should bear this in mind when reading about the devices from non-astronomical sources.

The basic detection mechanism is related to the photoelectric effect. Light incident on a semiconductor (usually silicon) produces electron–hole pairs as we have already seen. These electrons are then trapped in potential wells produced by numerous small electrodes. There they accumulate until their total number is read out by charge coupling the detecting electrodes to a single read-out electrode.

An individual unit of a CCD is shown in figure 1.1.11. The electrode is

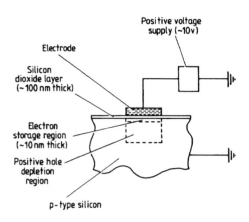

Figure 1.1.11. Basic unit of a CCD.

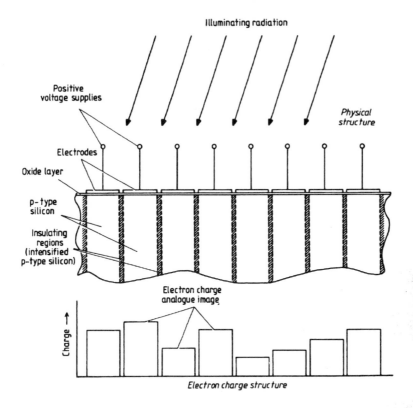

Figure 1.1.12. Array of CCD basic units.

insulated from the semiconductor by a thin oxide layer. In other words, the device is related to the metal oxide–silicon or MOS transistors. It is held at a small positive voltage which is sufficient to drive the positive holes in the p-type silicon away from its vicinity and to attract the minority carriers, the electrons, into a thin layer immediately beneath it. The electron–hole pairs produced in this depletion region by the incident radiation are then separated out and these electrons also accumulate in the storage region. Thus an electron charge is formed whose magnitude is a function of the intensity of the illuminating radiation. In effect, the unit is a radiation-driven capacitor.

Now if several such electrodes are formed on a single silicon chip and the depletion regions are insulated from each other by zones of very high p-type doping, then each will develop a charge which is proportional to its illuminating intensity. Thus we have a spatially digitised electric analogue of the original optical image (figure 1.1.12). All that remains is to be able to retrieve this electron image in some usable form. This is accomplished by charge coupling. Imagine an array of electrodes such as those we have already

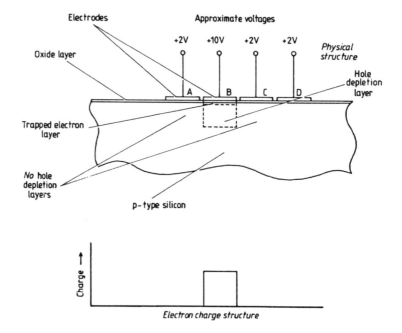

Figure 1.1.13. Active electron charge trapping in CCD.

seen in figure 1.1.11, but without their insulating separating layers. Then if one such electrode acquired a charge, it would diffuse across to the nearby electrodes. However, if the voltage of the electrodes on either side of the one containing the charge were reduced, then their hole depletion regions would disappear and the charge would once again be contained within two p-type insulating regions (figure 1.1.13). This time, however, the insulating regions are not permanent but may be changed by varying the electrode voltage. Thus the stored electric charge may be moved physically through the structure of the device by sequentially changing the voltages of the electrodes. Hence in figure 1.1.13, if the voltage on electrode C is changed to about +10 V, then a second hole depletion zone will form adjacent to the first. The stored charge will diffuse across between the two regions until it is shared equally. Now if the voltage on electrode B is gradually reduced to +2 V, its hole depletion zone will gradually disappear and the remaining electrons will transfer across to be stored under electrode C. Thus by cycling the voltages of the electrodes as shown in figure 1.1.14, the electron charge is moved from electrode B to electrode C. With careful design the efficiency of this charge transfer (or coupling) may be made as high as 99.9999%. Furthermore we may obviously continue to move the charge through the structure to electrodes D, E, F, etc by continuing to cycle the voltages in an appropriate fashion. Eventually the charge may be brought to an output electrode from whence its value may be determined by discharging

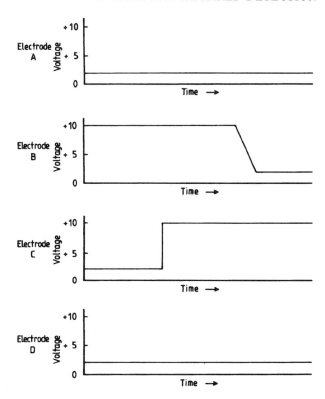

Figure 1.1.14. Voltage changes required to transfer charge from electrode B to electrode C in the array shown in figure 1.1.13.

it through an integrating current meter or some similar device. In the scheme outlined here, the system requires three separate voltage cycles to the electrodes in order to move the charge. Hence it is known as a three-phase CCD; two-phase CCDs are discussed later. Three separate circuits are formed, with each electrode connected to those three before and three after it (figure 1.1.15). The voltage supplies, α, β and γ (figure 1.1.15), follow the cycles shown in figure 1.1.16 in order to move charge packets from the left towards the right. Since only every third electrode holds a charge in this scheme the output follows the pattern shown schematically at the bottom of figure 1.1.16. The order of appearance of an output pulse at the output electrode is directly related to the position of its originating electrode in the array. Thus the original spatial charge pattern and hence the optical image may easily be inferred from the time-varying output.

The complete three-phase CCD is a combination of the detecting and charge transfer systems. Each pixel has three electrodes and is isolated from pixels in adjacent columns by insulating barriers (figure 1.1.17). During an exposure electrodes B are at their full voltage and the electrons from the whole area of

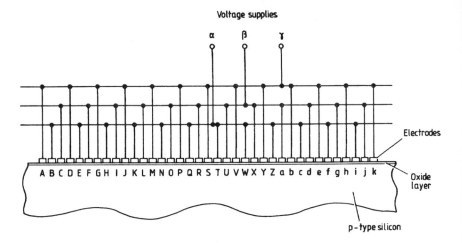

Figure 1.1.15. Connection diagram for a three-phase CCD.

the pixel accumulate beneath them. Electrodes A and C meanwhile are at a reduced voltage and so act to isolate each pixel from its neighbours along the column. When the exposure is completed, the voltages in the three electrode groups are cycled as shown in figure 1.1.16 until the first set of charges reaches the end of the column. At the end of the column a second set of electrodes running orthogonally to the columns (figure 1.1.17) receives the charges into the middle electrode for each column. That electrode is at the full voltage, and its neighbours are at reduced voltages, so that each charge package retains its identity. The charges in the read-out row of electrodes are then cycled to move the charges to the output electrode where they appear as a series of pulses. When the first row of charges has been read out, the voltages on the column electrodes are cycled to bring the second row of charges to the read-out electrodes, and so on until the whole image has been retrieved. With the larger-format CCDs, the read-out process can take some time. In order to improve observing efficiency, some devices therefore have a storage area between the imaging area and the read-out electrodes. This is simply a second CCD which is not exposed to radiation. The image is transferred to the storage area, and while it is being read out from there, the next exposure can commence on the detecting part of the CCD.

A two-phase CCD requires only a single clock, but needs double electrodes to provide directionality to the charge transfer (figure 1.1.18). The second electrode, buried in the oxide layer, provides a deeper well than that under the surface electrode and so the charge accumulates under the former. When voltages cycle between 2 V and 10 V (figure 1.1.19), the stored charge is attracted over to the nearer of the two neighbouring surface electrodes and then accumulates again under the buried electrode. Thus cycling the electrode voltages, which

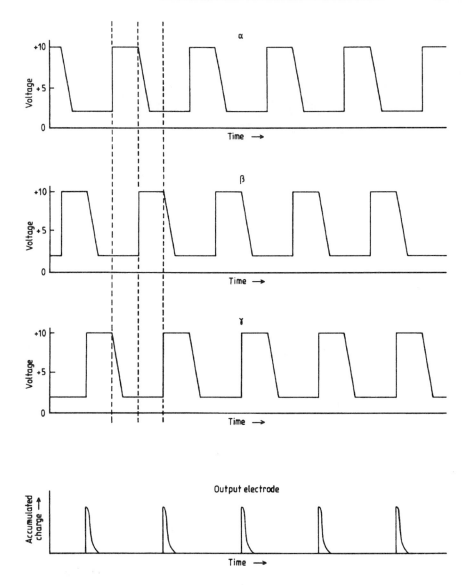

Figure 1.1.16. Voltage and output cycles for a three-phase CCD.

may be done from a single clock, causes the charge packets to move through the structure of the CCD, just as for the three-phase device.

A virtual phase CCD requires just one set of electrodes. Additional wells with a fixed potential are produced by p and n implants directly into the silicon substrate. The active electrode can then be at a higher or lower potential as

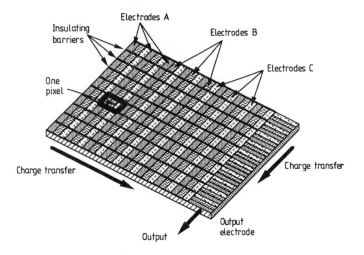

Figure 1.1.17. Schematic structure of a three-phase CCD.

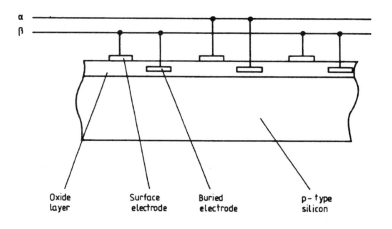

Figure 1.1.18. Physical structure of a two-phase CCD.

required to move the charge through the device. The active electrodes in a virtual phase CCD are physically separate from each other leaving parts of the substrate directly exposed to the incoming radiation. This enhances their sensitivity, especially at short wavelengths.

A combination of photomultiplier/image intensifier (sections 2.1 and 2.3) and CCD, known as electron bombarded CCDs or intensified CCDs (EBCCD or ICCD) has recently been developed. This places a negatively-charged photocathode before the CCD. The photoelectron is accelerated by the voltage difference between the photocathode and the CCD, and hits the CCD at high energy, producing many electron–hole pairs in the CCD for each incident photon.

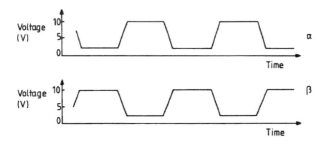

Figure 1.1.19. Voltage cycles for a two-phase CCD.

This might appear to give the device a quantum efficiency of over 100% but it is in effect merely another type of amplifier; the signal-to-noise ratio remains that of the basic device (or worse), and so no additional information is obtained.

The electrodes on or near the surface of a CCD can reflect some of the incident light, thereby reducing the quantum efficiency and changing the spectral response. To overcome this problem several approaches have been adopted. Firstly, the metallic electrodes used in early devices are replaced by transparent polysilicon electrodes. Secondly, the CCD may be illuminated from the back so that the radiation does not have to pass through the electrode structure at all. This, however, requires that the silicon forming the CCD be very thin so that the electrons produced by the incident radiation are collected efficiently. Nonetheless, even with thicknesses of only 10 μm, some additional cross talk (see below) may occur. More importantly, the process of thinning the CCD chips is risky and many devices may be damaged during the operation. Successfully thinned CCDs are therefore very expensive in order to cover the cost of the failures. They are also very fragile and can become warped, but they do have the advantage of being less affected by cosmic rays than thicker CCDs (see below). With a suitable anti-reflection coating, a back-illuminated CCD can now reach a quantum efficiency of 90% in the red and near infrared. Other methods of reducing reflection losses include using extremely thin electrodes, or an open electrode structure (as in a virtual phase CCD) which leaves some of the silicon exposed directly to the radiation. At longer wavelengths, the thinned CCD may become semi-transparent. Not only does this reduce the efficiency because fewer photons are absorbed, but interference fringes may occur between the two faces of the chip. These can become very obtrusive and have to be removed as a part of the data reduction process.

The CCD as so far described suffers from loss of charge during transfer because of imperfections at the interface between the substrate and the insulating oxide layer. The faintest images are worst affected by this since the few electrons which have been accumulated are physically close to the interface. Artificially illuminating the CCD with a low-level uniform background during an exposure will ensure that even the faintest parts of the image have the surface states filled. This offset or 'fat zero' can then be removed later in signal processing. Alternatively, a positively charged layer of n-type silicon between the substrate and the insulating layer forces the charge packets away from the interface, producing a buried-channel CCD. Charge transfer efficiencies for CCDs now approach 99.9999%, leading to typical read-out noise levels of one to two electrons per pixel. The read-out noise can be reduced further by using a non-destructive read-out mechanism and repeating and averaging many read-out sequences. To save time, only the lowest intensity parts of the image are repeatedly read out, with the high intensity parts being skipped. The devices have therefore become known as skipper CCDs.

The spectral sensitivity of CCDs ranges from 400 nm to 1100 nm, with a peak where the quantum efficiency can approach 90% near 750 nm. Short-wave sensitivity can be conferred by coating the device with an appropriate phosphor (see p. 38) to convert the radiation to the sensitive region. The CCD regains sensitivity at very short wavelengths because x-rays are able to penetrate the device's surface electrode structure. For integration times longer than a fraction of a second, it is usually necessary to cool the device in order to reduce its dark signal. Using liquid nitrogen for this purpose, integration times of minutes to hours are easily attained. The dark signal can be further reduced by subtracting a dark frame from the image. The dark frame is in all respects (exposure length, temperature, etc) identical to the main image, excepting that the camera shutter remains closed whilst it is obtained. It therefore just comprises the noise elements present within the detector and its electronics. Because the dark noise itself is noisy, it may be better to use the average of several dark frames obtained under identical conditions.

Two-dimensional arrays can be formed by stacking linear arrays. The operating principle is unchanged, and the correlation of the output with position within the image is only slightly more complex than before. Currently CCDs with $10\,000 \times 10\,000$ pixels are being envisaged, but this is probably approaching the practical limit because of the time taken to read out such large images, and the need for massive computer memories and processing power to cope with them. For typical noise levels and pixel capacities, astronomical CCDs have dynamic ranges of 100 000 to 500 000 (usable range in accurate brightness determination of up to 14.5^m). This therefore compares very favourably with the dynamic range of less than 1000 (brightness range of less than 7.5^m) available from a typical photographic image.

A major problem with CCDs used as astronomical detectors is the noise introduced by cosmic rays. A single cosmic ray particle passing through one of the pixels of the detector can cause a large number of ionisations. The resulting electrons accumulate in the storage region along with those produced by the photons. It is usually possible for the observer to recognise such events in the final image because of the intense 'spike' that is produced. The cosmic ray noise can then be reduced by replacing the signal from the affected pixel by the average of the eight surrounding pixels. However, at the time of writing this correction often has to be done 'by hand' and is a time-consuming process. When two or more images of the same area are available, automatic removal of the cosmic ray spikes is now possible with reasonable success rates.

Another serious defect of CCDs is the variation in background noise between pixels. This takes two forms. There may be a large scale variation of 10–20% over the whole sensitive area, and there may be individual pixels with permanent high background levels (hot spots). The first problem may be largely eliminated during subsequent signal processing (flat fielding), if its effect can be determined by observing a uniform source. The effect of a single hot spot may also be reduced in signal processing by replacing its value with the mean of the four surrounding pixels. However, the hot spots are additionally often poor transferers of charge and there may be other bad pixels. All preceding pixels are then affected as their charge packets pass through the hot spot or bad pixel and a spurious line is introduced into the image. There is little that can be done to correct this last problem, other than to buy a new CCD! Even the 'good' pixels do not have 100% charge transfer efficiency, so that images containing very bright stars show a 'tail' to the star caused by the electrons lost to other pixels as the star's image is read out.

Yet another problem is that of cross talk or blooming. This occurs when an electron strays from its intended pixel to one nearby. It affects rear-illuminated CCDs because the electrons are produced at some distance from the electrodes, and is the reason why such CCDs have to be thinned. It can also occur for any CCD when the accumulating charge approaches the maximum capacity of the pixel. Then the mutual repulsion of the negatively-charged electrons may force some over into adjacent pixels. The capacity of each pixel depends upon its physical size and is around half a million electrons for the typical 15 μm sized pixels used in many astronomical CCDs. Some CCDs have extra electrodes to enable excess charges to be bled away before they spread into nearby pixels. Such electrodes are known as drains, and their effect is often adjustable by means of an anti-blooming circuit. Anti-blooming should not be used if you intend making photometric measurements on the final image, but otherwise it can be very effective in improving the appearance of images containing both bright and faint objects.

The size of the pixels in a CCD can be too large to allow a telescope to operate at its limiting resolution. In such a situation, the images can be 'dithered' to provide improved resolution. Dithering consists simply of obtaining multiple images of the same field of view, with a shift of the position of the image on the CCD between each exposure by a fraction of the size of a pixel. The images are then combined to give a single image with sub-pixel resolution.

One of the major advantages of a CCD over a photographic emulsion is the improvement in the speed of detection. However, widely different estimates of the degree of that improvement may be found in the literature, based upon different ways of assessing performance. At one extreme, there is the ability of a CCD to provide a recognisable image at very short exposures, because of its low noise. Based upon this measure, the CCD is perhaps 1000 times faster than photographic emulsion. At the other extreme, one may use the time taken to reach the mid-point of the dynamic range. Because of the much greater dynamic range of the CCD, this measure suggests CCDs are about 5 to 10 times faster than photography. The most sensible measure of the increase in speed, however, is based upon the time required to reach the same signal-to noise ratio in the images (i.e. to obtain the same information). This latter measure suggests, that in their most sensitive wavelength ranges (around 750 nm for CCDs and around 450 nm for photographic emulsion), CCDs are 20 to 50 times faster than photographic emulsion.

Recently, charge coupling has been used to provide read-out of the signals from photoconductive infrared arrays. The CCD array also allows the signal to be integrated before it is output. Current devices have array sizes up to 128×128.

Other types of detectors

Detectors other than the four major types may still find use in particular applications or have historical interest. Some of these devices are briefly surveyed below.

Photovoltaic cells

These are also known as photodiodes and barrier junction detectors. They rely upon the properties of a p–n junction in a semiconductor. The energy level diagram of such a junction is shown in figure 1.1.20. Electrons in the conduction band of the n-type material are at a higher energy than the holes in the valence band of the p-type material. Electrons therefore diffuse across the junction and produce a potential difference across it. Equilibrium occurs when the potential difference is sufficient to halt the electron flow. The two Fermi levels are then coincident, and the potential across the junction is equal to their original difference. The n-type material is positive and the p-type negative, and we have a simple p–n diode. Now if light of sufficiently short wavelength falls onto such a junction then it can generate electron–hole pairs in both the p- and the n-type materials. The electrons in the conduction band in the p region will be attracted towards the n region by the intrinsic potential difference across the junction, and they will be quite free to flow in that direction. The holes in the valence band of the p-type material will be opposed by the potential across the junction and so will not move. In the n-type region the electrons will be similarly trapped while the holes will be pushed across the junction. Thus a current is generated by the illuminating radiation and this may simply be monitored and used as a measure of the light intensity.

The response of a p–n junction to radiation is shown in figure 1.1.21. It can be operated under three different regimes. The simplest, which is labelled B

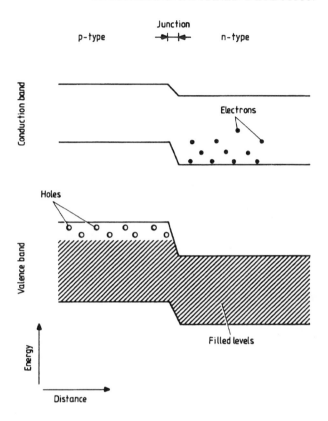

Figure 1.1.20. Schematic energy level diagram of a p–n junction at the instant of its formation.

in figure 1.1.21, has the junction short circuited through a low-impedance meter. The current through the circuit is the measure of the illumination. In regime C the junction is connected to a high impedance so that the current is very close to zero, and it is the change in voltage which is the measure of the illumination. Finally in regime A the junction is back biased. That is, an opposing voltage is applied to the junction. The voltage across a load resistor in series with the junction then measures the radiation intensity.

The construction of a typical photovoltaic cell is shown in figure 1.1.22. The material in most widespread use for the p and n semiconductors is silicon which has been doped with appropriate impurities. The solar power cells found on most artificial satellites are of this type. The silicon-based cells have a peak sensitivity near 900 nm and cut-off wavelengths near 400 and 1100 nm. Their quantum efficiency can be up to 50% near their peak sensitivity and D^* can be up to 10^{12}.

Indium arsenide, indium selenide, indium antimonide, gallium arsenide

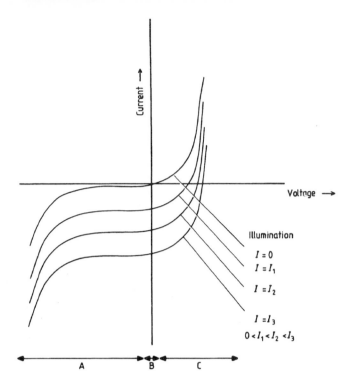

Figure 1.1.21. Schematic V/I curves for a p–n junction under different levels of illumination.

and indium gallium arsenide can all be formed into photodiodes. They are particularly useful in the infrared where indium antimonide out-performs bolometers (see below) for wavelengths up to 10 μm. Arrays of such diodes are now available to give direct imaging in the infrared. For example the infrared camera (IRCAM) on the UK infrared telescope (UKIRT) uses an array of 58 × 62 indium antimonide diodes bonded to an array of silicon MOSFET devices to give direct read-out from each pixel.

If a p–n junction is reverse-biased to more than half its breakdown voltage an avalanche photodiode (APD) results. The original electron–hole pair produced by the absorption of a photon will be accelerated by the applied field sufficiently to cause further pair production through inelastic collisions. These secondary electrons and holes can in their turn produce further ionisations and so on (cf Geiger and proportional counters, pp 109, 110). Eventually an avalanche of carriers is created, leading to an intrinsic gain in the signal by a factor of one hundred or more. The typical voltage used is 200 V, and the avalanche is quenched (because otherwise the current would continue to flow once started) by connecting a resistor of several hundred kΩ in series with the diode. Then

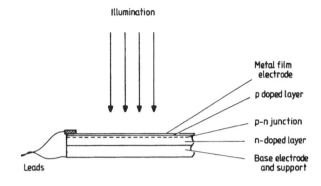

Figure 1.1.22. Cross section through a p–n junction photovoltaic detector.

as the current starts to flow, the voltage across the diode is reduced and the breakdown is quenched. A major drawback for APDs is that the multiplication factor depends very sensitively upon the bias. The power supply typically has to be stable to a factor of ten better than the desired accuracy of the output.

Another type of junction photovoltaic cell uses a p-type semiconductor in contact with a metal. A potential difference develops over the junction as the electrons diffuse across from the metal. The sensitivity of devices of this form of construction, however, is very low, and they are slow in their response. Selenium is the most commonly used material and their major applications occur in the light meters of cameras.

Photoelectromagnetic detectors

These are generally of little importance in astronomy since their sensitivity is low. Some use is made of indium antimonide based devices, however, in the 1 to 10 mm region. They consist of a single layer of an intrinsic semiconductor in a magnetic field. The magnetic field serves to separate the electron–hole pairs in a manner similar to the way in which the junction potential separates them in a photodiode. The voltage which is produced by radiation can then be used as a measure of its intensity.

Thermocouples

As is well known from school physics, two dissimilar metals in contact can develop a potential difference across their junction. This is called the Seebeck effect. It arises from the differing Fermi levels of the two metals. Electrons flow from the one with the higher level until the potential across the junction is sufficient to halt the transfer. Now the Fermi level at temperatures above absolute zero is the energy level which has a 50% chance of being occupied. As the temperature is increased, more and more of the electrons become thermally excited. However, since the holes will be densely packed near the top of the unexcited electrons while the excited electrons can be much more widely dispersed among the upper levels, the position of the Fermi level will change with temperature. The change in the Fermi level will not necessarily be identical for two different metals, and so at a junction the difference between the two Fermi levels will vary with temperature. Thus the Seebeck potential at the junction will change in order to continue to produce a balance. In a thermocouple therefore two dissimilar metals are joined into a circuit which incorporates a galvanometer. When one junction is at a different temperature from the other, their Seebeck potentials differ and a current flows around the circuit.

A practical thermocouple for radiation detection is made from two metals in the form of wires which are twisted together and blackened to improve their absorption. The other junction is kept in contact with something with a large thermal inertia so that it is kept at a constant temperature. The sensitive junction absorbs the radiation and so heats up slightly. The junctions are then at different temperatures so that a current flows as described above. A sensitive galvanometer can detect this current and it can be used to measure the intensity of the radiation. Several thermocouples are usually connected serially so that their potentials combine. Such an array is called a thermopile.

Practical thermocouples and thermopiles are usually made from antimony and bismuth or from nickel with various admixtures of copper, silicon, chromium, aluminium etc. They are useful wideband detectors, especially for infrared work. Their simplicity of operation and their robustness has led to many applications being found for them in satellite-borne instrumentation, in spite of their relatively low sensitivity (they have values of D^* only up to 10^9 to 10^{10}).

Pyroelectric detectors

These usually have very low values of D^* (typically around 10^8) and so are currently little used in astronomy. They are based upon dielectric materials such as lithium tantalate and triglycine sulphate. These materials have a large temperature dependent electrical polarisation. Thus if incorporated into a capacitor as its dielectric medium, they will induce a spontaneous charge in the capacitor which will be temperature dependent. Illuminating the detector then changes its temperature and so also its charge. The magnitude of the change can be used as a measure of the intensity of the illumination.

Bolometers

A bolometer is simply a device which changes its electrical resistivity in response to heating by illuminating radiation. At its simplest, two strips of the material are used as arms of a Wheatstone bridge. When one is heated by the radiation its resistance changes and so the balance of the bridge alters. Two strips of the material are used to counteract the effect of slower environmental temperature changes, since they will both vary in the same manner under that influence.

Bolometers used in astronomy are of two main varieties, room temperature thermistor bolometers and cooled semiconductor bolometers. The former have found extensive application in rockets and spacecraft because of their simplicity and sturdiness. They consist of mixtures of manganese, nickel and cobalt oxides sintered together and coated with an appropriate absorber for the wavelength which is to be observed. They have a high negative coefficient of resistance. Values of D^* of up to 10^9 over a very wide range of wavelengths can be obtained. Cooled semiconductor bolometers are the most sensitive of the infrared detectors at wavelengths more than 10 μm. D^* values of 10^{12} to 10^{13} or higher are possible using beryllium or gallium-doped germanium and monolithic silicon. With appropriate blackening they can be used out to a wavelength of one millimetre or longer. The bolometer is cooled by liquid helium and this is sometimes held under reduced pressure so that temperatures of less than 3 K can be used. The resistance of the bolometer is measured by monitoring a very small bias current. Arrays of up to 256 × 256 bolometers allow direct imaging at long wavelengths.

Photoconductive cells

Like bolometers, photoconductive cells exhibit a change in conductivity with the intensity of their illumination. The mechanism for that change, however, is quite different in the two cases. For the photoconductive cell it is the absorption of the radiation by the electrons in the valence band of the material and their consequent elevation to the conduction band. The conductivity therefore increases with increasing illumination. There is a cut-off point determined by the minimum energy required to excite a valence electron. A very wide variety of materials may be used, with widely differing sensitivities, cut-off wavelengths, operating temperatures etc. A small sample of the ones in current use is given in the table below as an example.

Material	Operating temperature (K)	Cut-off wavelength (μm)	D^*
Mercury-cadmium-telluride	60	2–12	
Lead sulphide (PbS)	77–300	3.5	10^8–10^9
Indium antimonide (InSb)	77–195	6.5	10^{11}
Gold-doped germanium (Ge(Au))	77	1-9	10^{10}
Mercury-doped germanium (Ge(Hg))	4	3–4	10^{10}
Copper-doped germanium (Ge(Cu))	4	6–30	10^{10}
Boron-doped germanium (Ge(B))	4	120	10^9

Recently photoconductive detectors which do not require the electrons to be excited all the way to the conduction band have been made using alternating layers of gallium arsenide (GaAs) and indium gallium arsenide phosphide (InGaAsP), each layer being only ten or so atoms thick. The detectors are known as QWIPs (quantum well infrared photodetectors). The lower energy required to excite the electron gives the devices a wavelength sensitivity ranging from 1 to 12 μm. It should be possible to fabricate arrays of 1000 × 1000 QWIPs in the fairly near future.

The photoconductive cell is operated like the bolometer by monitoring a small bias current.

Phototransistors

These are of little direct use in astronomy because of their low sensitivity. They find wide application, however, within the apparatus used by astronomers, for example in conjunction with a light-emitting diode (LED) to generate a switching signal. They consist simply of a pnp or npn transistor with the minority current carriers produced by the illumination instead of the normal emitter. Thus the current rises with increasing radiation and provides a measure of its intensity.

The photovoltaic cell discussed earlier when operated in mode A is acting as a phototransistor. It is then sometimes called a photodiode. Arrays of such diodes have found some astronomical applications when allied to image intensifiers (section 2.3).

Charge injection devices (CID)

The detection mechanism of these devices is identical to that of the CCD. The difference between them occurs in their read-out system. With these devices two electrodes are used for each sensitive element and the charge can be switched between them by alternating their voltages. If the voltages of both elements are simultaneously reduced to zero, then the accumulated charge is deposited (injected) into the substrate. The resulting pulse can be detected and its magnitude is a measure of the original charge. A two-dimensional read-out can be attained by connecting all the first electrodes of each sensitive element serially along rows, while all the second electrodes are connected serially along columns. Switching a given row and column to zero volts then causes the charge in the sensitive element at their intersection to be read out, while all other charges are either unaffected or jump to their alternative electrode.

The connections for a CID device are more complex than those for a CCD, but its data are easier to retrieve separately, and can be read out non-destructively. Although still occasionally to be encountered as an astronomical detector, for example using InSb to form a near to medium infrared detector array, the CID is now rare compared with the all-conquering CCD (see earlier).

Photographic emulsion

This is dealt with in detail in section 2.2. Here it is sufficient to point out that the basic mechanism involved in the formation of the latent image is pair production. The electrons excited into the conduction band are trapped at impurities, while the holes are removed chemically. Unsensitised emulsion is blue-sensitive as a consequence of the minimum energy required to excite the valence electrons in silver bromide.

Gas cells

These devices are closely related to the photomultipliers, but do not have a dynode chain. The photoelectron emitting cathode ejects its electrons directly towards the anode. Some degree of amplification is achieved, however, by the presence of a low pressure gas in the tube. Collisions between its atoms and the photoelectrons eject further electrons which add to the anode current. They are now almost totally superseded by photomultipliers.

Television tubes

See section 2.3.

Image intensifiers

See section 2.3.

Golay cell

Quite different from any of the devices so far considered is the Golay cell. It has a very wide wavelength range and was one of the first infrared detectors to be used in astronomy. Essentially it is a pressure thermometer. The radiation illuminates the blackened wall of a sealed chamber. The wall heats up and in turn warms the gas inside the chamber. The resulting pressure rise may then be used to deflect a small mirror, and the mirror's position can be monitored optically.

Ultraviolet detectors

Some of the detectors that we have reviewed are intrinsically sensitive to shortwave radiation, although some modification from their standard optical forms may be required. For example, photomultipliers will require windows which are transparent down to the required wavelengths. Lithium fluoride and sapphire are common materials for such windows. They may also need to exclude the visible region in order to restrict the sensitivity to just the ultraviolet. Filters may be used for this or a photocathode which is insensitive to visible light employed. A photomultiplier of the latter type is often called a solar blind type of device.

Photography can be used down to 200 nm, after which the gelatine absorption becomes important. At shorter wavelengths special ultra-thin emulsions with a very low gelatine content may be used and are called Schumann emulsions.

Another common method of shortwave detection is to use a standard detector for the visual region, and to add a fluorescent or short glow phosphorescent material to convert the radiation to longer wavelengths. Sodium salycylate and tetraphenyl butadiene are the most popular such substances since their emission spectra are well matched to standard photocathodes. Sensitivity down to 60 nm can be achieved in this manner. Additional conversions to even longer wavelengths can be added for matching to CCD and other solid state detectors whose sensitivities peak in the red and infrared. Ruby (Al_2O_3) is suitable for this, and its emission peaks near 700 nm.

Future possibilities

Many of the devices discussed above are in the early stages of their development and considerable improvement in their characteristics and performance may be expected over the next decade or so. Other possibilities such as laser action initiation and switching of superconducting states have already been mentioned.

A possible replacement for the CCD in a few years could be the superconducting tunnel junction detector (SJT). This can operate out to longer wavelengths than the CCD, can detect individual photons, and provide intrinsic spectral resolution of perhaps 500 or 1000 in the visible. Its operating principle is based upon a Josephson junction. This has two superconducting layers separated by a very thin insulating layer. Electrons are able to tunnel across the junction because they have a wave-like behaviour as well as a particle-like behaviour, and so a current may flow across the junction despite the presence of the insulating layer. Within the superconductor, the lowest energy state for the electrons occurs when they link together to form Cooper pairs. The current flowing across the junction due to paired electrons can be suppressed by a magnetic field. The SJT detector therefore comprises a Josephson junction placed within a magnetic field to suppress the current, having an electric field applied across it. A photon absorbed in the superconductor may split one of the Cooper pairs. This requires an energy of a milli-electron-volt or so compared with about an electron-volt for pair production in a CCD. Potentially therefore the SJT can detect photons with wavelengths up to a millimetre. Shorter wavelength photons will split many Cooper pairs, with the number split being dependent upon the energy of the photon, hence the device's intrinsic spectral resolution. The free electrons are able to tunnel across the junction under the influence of the electric field, and produce a detectable burst of current. SJT detectors and arrays made from them are still very much under development at the time of writing, but may become viable for astronomical use early into the twenty-first century.

Another technique which is currently in the early stages of application to astronomy is optical or infrared heterodyning. A suitable laser beam is mixed with the light from the object. The slow beat frequency is at radio frequencies and may be detected using relatively normal radio techniques (section 1.2). Both the amplitude and phase information of the original signal can be preserved with this technique. Very high dispersion spectroscopy seems likely to be its main area of application.

Noise

In the absence of noise any detector would be capable of detecting any source, however faint. Noise, however, is never absent and generally provides the major limitation on detector performance. A minimum signal-to-noise ratio of unity is required for reliable detection. Noise sources in photomultipliers and CCDs have already been mentioned, and the noise for an unilluminated detector (dark signal) is a part of the definitions of DQE, NEP, D^* and dynamic range (see earlier discussion). Now we must look at the nature of detector noise in more detail.

We may usefully separate noise sources into four main types: *intrinsic noise*, i.e. noise originating in the detector, *signal noise*, i.e. noise arising from the character of the incoming signal, particularly its quantum nature, *external noise* such as spurious signals from cosmic rays etc, and *processing noise*, arising from amplifiers etc used to convert the signal from the detector into a usable form. We may generally assume processing noise to be negligible in any good detection system. Likewise external noise sources should be reduced as far as possible by external measures. Thus an infrared detector should be in a cooled enclosure and be allied to a cooled telescope (e.g. the Infrared Astronomy Satellite, IRAS) to reduce thermal emission from its surroundings, or a photomultiplier used in a photon-counting mode should employ a discriminator to eliminate the large pulses arising from Čerenkov radiation from cosmic rays, etc. Thus we are left with intrinsic and signal noise to consider further.

Intrinsic noise

Intrinsic noise in photomultipliers has already been discussed and arises from sources such as variation in photoemission efficiency over the photocathode, thermal electron emission from the photocathode and dynode chain etc. Noise in photographic emulsion, such as chemical fogging, is discussed later (section 2.2), while irradiation etc in the eye has also been covered earlier.

In solid state devices, intrinsic noise comes from four sources.

Thermal noise, also known as Johnson or Nyquist noise (see section 1.2) arises in any resistive material. It is due to the thermal motion of the charge carriers. These motions give rise to a current, whose mean value is zero, but which may have non-zero instantaneous values. The resulting fluctuating voltage is given by equation (1.2.5).

Shot noise occurs in junction devices and is due to variation in the diffusion rates in the neutral zone of the junction because of random thermal motions. The general form of the shot noise current is

$$i = \left(2eI\Delta f + 4eI_0\Delta f\right)^{1/2} \tag{1.1.4}$$

where e is the charge on the electron, Δf is the measurement frequency bandwidth, I is the diode current, and I_0 is the reverse bias or leakage current. When the detector is reverse-biased, this equation simplifies to

$$i = \left(2eI_0\Delta f\right)^{1/2} \tag{1.1.5}$$

g–r noise (generation–recombination) is caused by fluctuations in the rate of generation and recombination of thermal charge carriers, which in turn leads to fluctuations in the device's resistivity. g–r noise has a flat spectrum up to the inverse of the mean carrier lifetime and then decreases roughly with the square of the frequency.

Flicker noise, or $1/f$ noise, occurs when the signal is modulated in time, either because of its intrinsic variations or because it is being 'chopped' (i.e. the source and background or comparison standard are alternately observed). The mechanism of flicker noise is unclear but its amplitude follows an f^{-n} power spectrum where f is the chopping frequency and n lies typically in the region 0.75 to 2.0. This noise source may obviously be minimised by increasing f. Furthermore, operating techniques such as phase-sensitive detection (section 3.2) are commonplace especially for infrared work, when the external noise may be many orders of magnitude larger than the desired signal. Modulation of the signal is therefore desirable or necessary on many occasions. However, the improvement of the detection by faster chopping is usually limited by the response time of the detector. Thus an optimum chopping frequency normally may be found for each type of detector. For example this is about 10 Hz for thermocouples and Golay cells, 100 Hz for bolometers, and 1000 Hz or more for most photoconductive and photovoltaic cells.

The relative contributions of these noise sources are shown in figure 1.1.23.

Signal noise

Noise can be present in the signal for a variety of reasons. One obvious example is background noise. The source under observation will generally be superimposed upon a signal from the sky due to scattered terrestrial light sources, scattered starlight, diffuse galactic emission, zodiacal light, microwave background radiation etc. The usual practice is to reduce the importance of this noise by measuring the background and subtracting it from the main signal. Often the source and its background are observed in quick succession (chopping,

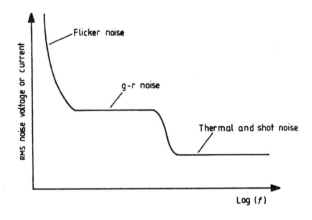

Figure 1.1.23. Relative contributions of various noise sources (schematic).

see *flicker noise* above and section 3.2). Alternatively, there may only be measurements of the background at the beginning and end of an observing run. In either case, some noise will remain, due to fluctuations of the background signal about its mean level.

Noise also arises from the quantum nature of light. At low signal levels photons arrive at the detector sporadically. The probability of arrival is given by a Poisson distribution, and this has a standard deviation of $\sqrt{n}$ (where n is the mean number of photons per unit time). Thus the signal will fluctuate about its mean value. In order to reduce the fluctuations to less than $x\%$, the signal must be integrated for $10^4/(nx^2)$ times the unit time. At high photon densities, photons tend to cluster more than a Poisson distribution would suggest because they are subject to Bose–Einstein statistics. This latter noise source may dominate at radio wavelengths (section 1.2), but is not normally of importance over the optical region.

Digitisation

A further source of noise appears when data is digitised. In practice this applies to all forms of data collection, since even continuously recording instruments have time constants and limits to resolution which mean that they are effectively sampling at discrete time intervals and to the nearest nth decimal place, where n is much less than infinity. Thus a chart recorder with a time constant of 0.01 s and a full scale deflection of 1 V could not pick up a transient of 1 nV lasting 1 μs.

Signals are thus digitised in two ways, signal strength and time. The effect of the first is obvious, there is an uncertainty (i.e. noise) in the measurement corresponding to plus or minus half the measurement resolution. The effect of sampling a time varying signal is more complex. The well known sampling theorem (section 2.1) states that the highest frequency in a signal that can be determined is half the measurement frequency. Thus, if a signal is bandwidth-limited to some frequency, f, then it may be completely determined by sampling at $2f$ or higher frequencies. In fact sampling at higher than twice the limiting (or Nyquist) frequency is a waste of effort; no further information will be gained. However, in a non-bandwidth-limited signal, or a signal containing components above the Nyquist frequency, errors or noise will be introduced into the measurements through the effect of those higher frequencies. In this latter effect, or 'aliasing' as it is known, the beat frequencies between the sampling rate and the higher frequency components of the signal appear as spurious low frequency components of the signal.

Telescopes

The *apologia* for the inclusion of telescopes in the chapter entitled detectors has already been given, but it is perhaps just worth repeating that for astronomers the use of a detector in isolation is almost unknown. Some device to restrict the angular acceptance zone and possibly to increase the radiation flux as well, is almost invariably a necessity. In the optical region this generally implies the use of a telescope, and so they are discussed here as a required ancillary to detectors.

The telescope may justly be regarded as the symbol of the astronomer, for notwithstanding the heroic work done with the naked eye in the past and the more recent advances in 'exotic' regions of the spectrum, our picture of the universe is still primarily based upon optical observations made through telescopes. The optical telescope has passed through four major phases of development, each of which has caused a quantum jump in astronomical knowledge. We may now be near the start of the fifth phase when very large diffraction limited telescopes can be formed from smaller mirrors or segments with active control of their alignment etc. We may hope for a similar rapid development of our knowledge as these instruments come on stream in the next few decades.

Optical theory

Before looking at the designs of telescopes and their properties it is necessary to cover some of the optical principles upon which they are based. It is assumed that the reader is familiar with the basics of the optics of lenses and mirrors; such terminology as focal ratio, magnification, optical axis etc with laws such as Snell's law, the law of reflection and so on, and with equations such as the lens and mirror formulae and the lens maker's formula. If it is required, however, any basic optics book or reasonably comprehensive physics book will suffice to provide this background.

We start by considering the resolution of a lens. That this is limited at all is due to the wave nature of light. As light passes through any opening it is

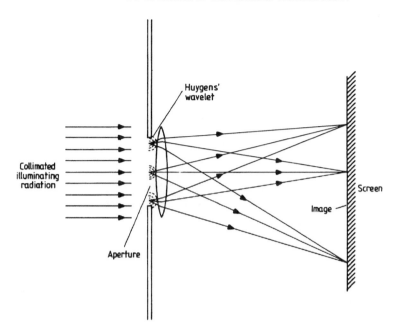

Figure 1.1.24. Fraunhofer diffraction at an aperture.

diffracted and the wavefronts spread out in a shape given by the envelope of the Huygens' secondary wavelets (figure 1.1.24). Huygens' secondary wavelets radiate spherically outwards from all points on a wavefront with the velocity of light in the medium concerned. Three examples are shown in figure 1.1.24. Imaging the wavefront after passage through a slit-shaped aperture produces an image whose structure is shown in figure 1.1.25. The variation in intensity is due to interference between waves originating from different parts of the aperture. The paths taken by such waves to arrive at a given point will differ and so the distances that they have travelled will also differ. The waves will be out of step with each other to a greater or lesser extent depending upon the magnitude of this path difference. When the path difference is a half wavelength or an integer number of wavelengths plus a half wavelength, then the waves will be 180° out of phase with each other and so will cancel out. When the path difference is a whole number of wavelengths, the waves will be in step and will reinforce each other. Other path differences will cause intermediate degrees of cancellation and reinforcement. The central maximum arises from the reinforcement of many waves. While the first minimum occurs when the path difference for waves originating at opposite edges of the aperture is one wavelength, then for every point in one half of the aperture there is a point in the other half such that their path difference is half a wavelength, and all the waves cancel out completely. The intensity at a point within the image of a narrow slit may be obtained from

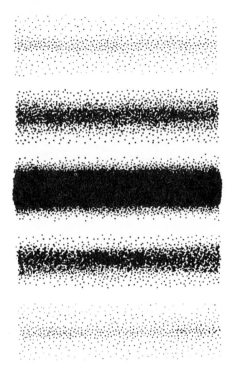

Figure 1.1.25. Image of a narrow slit (negative image).

the result that the diffraction pattern of an aperture is the power spectrum of the Fourier transform of its shape, and is given by

$$I_\theta = I_0 \frac{\sin^2(\pi d \sin\theta/\lambda)}{(\pi d \sin\theta/\lambda)^2} \qquad (1.1.6)$$

where θ is the angle to the normal from the slit, d is the slit width and I_0 and I_θ are the intensities within the image on the normal and at an angle θ to the normal from the slit respectively. With the image focused onto a screen at distance F from the lens, the image structure is as shown in figure 1.1.26 when d is assumed to be large when compared with the wavelength λ. For a rectangular aperture with dimensions $d \times l$, the image intensity is similarly

$$I(\theta, \varphi) = I_0 \frac{\sin^2(\pi d \sin\theta/\lambda)}{(\pi d \sin\theta/\lambda)^2} \frac{\sin^2(\pi l \sin\varphi/\lambda)}{(\pi l \sin\varphi/\lambda)^2} \qquad (1.1.7)$$

where φ is the angle to the normal from the slit measured in the plane containing the side of length l. To obtain the image structure for a circular aperture, which is the case normally of interest in astronomy, we must integrate over the surface of the aperture. The image is then circular with concentric light and dark fringes.

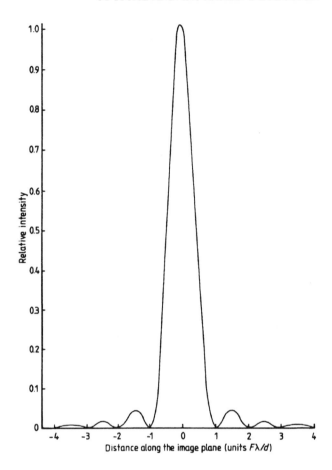

Figure 1.1.26. Cross section through the image of a narrow slit.

The central maximum is known as Airy's disc after the Astronomer Royal who first succeeded in completing the integration. The major difference from the previous case occurs in the position of the fringes.

Consider a circular aperture of radius r, which is illuminated by a beam of light normal to its plane (figure 1.1.27), and consider also the light which is diffracted at an angle θ to the normal to the aperture from a point P, whose cylindrical coordinates with respect to the centre of the aperture C, and the line AB which is drawn through C parallel to the plane containing the incident and diffracted rays, are (φ, ρ). The path difference, Δ, between diffracted rays from P and A is then

$$\Delta = (r - \rho \cos \varphi) \sin \theta \tag{1.1.8}$$

and the phase difference is

$$\frac{2\pi \Delta}{\lambda} = \frac{2\pi}{\lambda}(r - \rho \cos \varphi) \sin \theta. \tag{1.1.9}$$

The elemental area at P is just

$$dA = \rho \, d\varphi \, d\rho. \tag{1.1.10}$$

So the contribution to the electric vector of the radiation in the image plane by the elemental area around P for an angle θ to the normal is proportional to

$$\sin(\omega t + (2\pi/\lambda)(r - \rho \cos \varphi) \sin \theta)\rho \, d\varphi \, d\rho \tag{1.1.11}$$

where $\omega/2\pi$ is the frequency of the radiation, and the net effect is obtained by integrating over the aperture

$$\int_0^{2\pi} \int_0^r \sin\left[\omega t + \left(\frac{2\pi r \sin \theta}{\lambda}\right) - \left(\frac{2\pi \rho \cos \varphi \sin \theta}{\lambda}\right)\right] \rho \, d\rho \, d\varphi$$

$$= \sin\left(\omega t + \frac{2\pi r \sin \theta}{\lambda}\right) \int_0^{2\pi} \int_0^r \rho \cos\left(\frac{2\pi \rho \cos \varphi \sin \theta}{\lambda}\right) d\rho \, d\varphi$$

$$- \cos\left(\omega t + \frac{2\pi r \sin \theta}{\lambda}\right) \int_0^{2\pi} \int_0^r \rho \sin\left(\frac{2\pi \rho \cos \varphi \sin \theta}{\lambda}\right) d\rho \, d\varphi. \tag{1.1.12}$$

The second integral on the right-hand side is zero, which we may see by substituting

$$s = \frac{2\pi \rho \cos \varphi \sin \theta}{\lambda} \tag{1.1.13}$$

so that

$$\cos\left(\omega t + \frac{2\pi r \sin \theta}{\lambda}\right) \int_0^{2\pi} \int_0^r \rho \sin\left(\frac{2\pi \rho \cos \varphi \sin \theta}{\lambda}\right) d\rho \, d\varphi$$

$$= \cos\left(\omega t + \frac{2\pi r \sin \theta}{\lambda}\right)$$

$$\times \int_0^r \rho \int_{s=2\pi\rho \sin\theta/\lambda}^{s=2\pi\rho \sin\theta/\lambda} \frac{-\sin s}{[(4\pi^2 \rho^2 \sin^2 \theta/\lambda^2) - s^2]^{1/2}} \, ds \, d\rho \tag{1.1.14}$$

$$= 0 \tag{1.1.15}$$

since the upper and lower limits of the integration with respect to s are identical. Thus we have the image intensity $I(\theta)$ in the direction θ to the normal

$$I(\theta) \propto \left[\sin\left(\omega t + \frac{2\pi r \sin \theta}{\lambda}\right) \int_0^{2\pi} \int_0^r \rho \cos\left(\frac{2\pi \rho \cos \varphi \sin \theta}{\lambda}\right) d\rho \, d\varphi\right]^2 \tag{1.1.16}$$

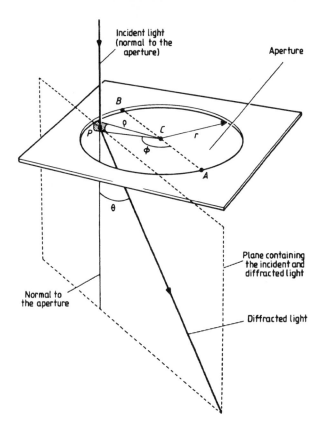

Figure 1.1.27. Diffraction by a circular aperture.

$$\propto \left(r^2 \int_0^{2\pi} \frac{\sin(2m \cos \varphi)}{2m \cos \varphi} \, d\varphi - \frac{1}{2} r^2 \int_0^{2\pi} \frac{\sin^2(m \cos \varphi)}{(m \cos \varphi)^2} \, d\varphi \right)^2 \quad (1.1.17)$$

where

$$m = \frac{\pi r \sin \theta}{\lambda}. \quad (1.1.18)$$

Now

$$\frac{\sin(2m \cos \varphi)}{2m \cos \varphi} = 1 - \frac{(2m \cos \varphi)^2}{3!} + \frac{(2m \cos \varphi)^4}{5!} - \cdots \quad (1.1.19)$$

and

$$\frac{\sin^2(m \cos \varphi)}{(m \cos \varphi)^2} = 1 - \frac{2^3(m \cos \varphi)^2}{4!} + \frac{2^5(m \cos \varphi)^4}{6!} - \cdots \quad (1.1.20)$$

so that

$$I(\theta) \propto \left(r^2 \sum_{n=0}^{\infty} (-1)^n \int_0^{2\pi} \frac{(2m\cos\varphi)^{2n}}{(2n+1)!}\, d\varphi \right.$$
$$\left. -\frac{1}{2}r^2 \sum_{n=0}^{\infty} (-1)^n \int_0^{2\pi} \frac{2^{(2n+1)}(m\cos\varphi)^{2n}}{(2n+2)!}\, d\varphi \right)^2. \qquad (1.1.21)$$

Now

$$\int_0^{2\pi} \cos^{2n}\varphi\, d\varphi = \frac{(2n)!}{2(n!)^2}\pi \qquad (1.1.22)$$

and so

$$I(\theta) \propto \pi^2 r^4 \left[\sum_{n=0}^{\infty} (-1)^n \frac{1}{n+1} \left(\frac{m^n}{n!} \right)^2 \right]^2 \qquad (1.1.23)$$

$$\propto \frac{\pi^2 r^4}{m^2} (J_1(2m))^2 \qquad (1.1.24)$$

where $J_1(2m)$ is the Bessel function of the first kind of order unity. Thus $I(\theta)$ is a maximum or zero accordingly as $J_1(2m)/m$ reaches an extremum or is zero. Its variation is shown in figure 1.1.28. The first few zeros occur for values of m of

$$m = 1.916,\ 3.508,\ 5.087, \ldots \qquad (1.1.25)$$

and so in the image, the dark fringes occur at values of θ given by

$$\sin\theta = \frac{1.916\lambda}{\pi r},\ \frac{3.508\lambda}{\pi r},\ \frac{5.087\lambda}{\pi r}, \ldots \qquad (1.1.26)$$

or for small θ

$$\theta \simeq \frac{1.220\lambda}{d},\ \frac{2.233\lambda}{d},\ \frac{3.238\lambda}{d}, \ldots \qquad (1.1.27)$$

where d is now the diameter of the aperture. The image structure along a central cross section is therefore similar to that of figure 1.1.26, but with the minima expanded outwards to the points given in equation (1.1.27) and the fringes, of course, are circular.

If we now consider two distant point sources separated by an angle α, then two such images will be produced and will be superimposed. There will not be any interference effects between the two images since the sources are mutually incoherent, and so their intensities will simply add together. The combined image will have an appearance akin to that shown in figure 1.1.29. When the centre of the Airy disc of one image is superimposed upon the first minimum of the other image (and vice versa), then we have Rayleigh's criterion for the resolution of a lens. This is the normally accepted measure of the theoretical resolution of a lens. It is given by (from equation 1.1.27)

$$\alpha = \frac{1.220\lambda}{d}. \qquad (1.1.28)$$

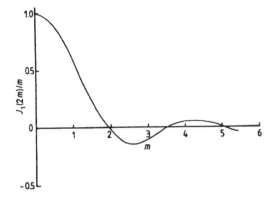

Figure 1.1.28. Variation of $J_1(2m)/m$ with m.

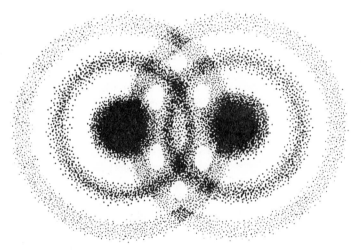

Figure 1.1.29. Image of two distant point sources through a circular aperture (negative image).

It is a convenient measure, but it is quite arbitrary. For sources of equal brightness the image will appear non-circular for separations of about one third of α, while for sources differing appreciably in brightness, the separation may need to be an order of magnitude larger than α for them to be resolved. For the eye, the effective wavelength is 510 nm for faint images so that the resolution of a telescope used visually is given by

$$R = \frac{0.128}{d} \tag{1.1.29}$$

where d is the objective's diameter in metres and R is the angular resolution in seconds of arc. To achieve this resolution in practice, the magnification must be sufficient for the angular separation of the images in the eyepiece to exceed the resolution of the eye (see the earlier discussion). Taking this to be an angle, β, we have the minimum magnification required to realise the Rayleigh limit of a telescope, M_m

$$M_m = \frac{\beta d}{1.220\lambda} \qquad (1.1.30)$$

so that for β equal to three minutes of arc, which is about its average value,

$$M_m = 1300\,d \qquad (1.1.31)$$

where d is again measured in metres. Of course most astronomical telescopes are actually limited in their resolution by the atmosphere. A one metre telescope might reach its Rayleigh resolution on one night a year from a good observing site. On an average night, scintillation (see p. 80 *et seq*) will spread stellar images to about two seconds of arc so that only telescopes smaller than about 0.07 m can regularly attain their diffraction limit. Since telescopes are rarely used visually for serious work such high magnifications as are implied by equation (1.1.31) are hardly ever encountered today. However, some of William Herschel's eyepieces still exist and these, if used on his 1.2 m telescope would have given magnifications of up to 8000 times.

The theoretical considerations of resolution that we have just seen are only applicable if the lens or mirror is of sufficient optical quality that the image is not already degraded beyond this limit. There are many effects which will blur the image and these are known as aberrations. With one exception they can all affect the images produced by either lenses or mirrors. The universal or monochromatic aberrations are known as the Seidel aberrations after Ludwig von Seidel who first analysed them. The exception is chromatic aberration and the related second-order effects of transverse chromatic aberration and secondary colour, and these affect only lenses.

Chromatic aberration arises through the change in the refractive index of glass or other optical material with the wavelength of the illuminating radiation. Some typical values of the refractive index of some commonly used optical glasses are tabulated below.

Glass type	Refractive index at the specified wavelengths				
	361 nm	486 nm	589 nm	656 nm	768 nm
Crown	1.539	1.523	1.517	1.514	1.511
High dispersion crown	1.546	1.527	1.520	1.517	1.514
Light flint	1.614	1.585	1.575	1.571	1.567
Dense flint	1.705	1.664	1.650	1.644	1.638

The degree to which the refractive index varies with wavelength is called the dispersion, and is measured by the constringence, ν

$$\nu = \frac{\mu_{589} - 1}{\mu_{486} - \mu_{656}} \qquad (1.1.32)$$

where μ_{λ} is the refractive index at wavelength λ. The three wavelengths which are chosen for the definition of ν are those of strong Fraunhofer lines: 486 nm—the F line (Hβ); 589 nm—the D lines (Na); 656 nm—the C line (Hα). Thus for the glasses listed earlier, the constringence varies from 57 for the crown glass to 33 for the dense flint (note that the higher the value of the constringence, the *less* that rays of different wavelengths diverge from each other). The effect of the dispersion upon an image is to string it out into a series of different coloured images along the optical axis (figure 1.1.30). Looking at this sequence of images with an eyepiece, then at a particular point along the optical axis, the observed image will consist of a sharp image in the light of one wavelength surrounded by blurred images of varying sizes in the light of all the remaining wavelengths. To the eye, the best image occurs when yellow light is focused since it is less sensitive to the red and blue light. The image size at this point is called the circle of least confusion. The spread of colours along the optical axis is called the longitudinal chromatic aberration, while that along the image plane containing the circle of least confusion is called the transverse chromatic aberration.

Two lenses of different glasses may be combined to reduce the effect of chromatic aberration. Commonly in astronomical refractors, a biconvex crown glass lens is allied to a planoconcave flint glass lens to produce an achromatic doublet. In the infrared, achromats can be formed using barium and strontium fluoride and infrared-transmitting glasses. In the sub-millimetre region (i.e. wavelengths of several hundred microns) crystal quartz and germanium can

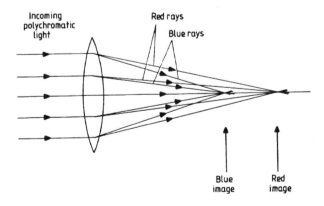

Figure 1.1.30. Chromatic aberration.

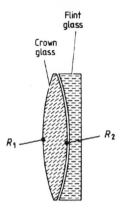

Figure 1.1.31. An achromatic doublet.

be used. The lenses are either cemented together or separated by only a small distance (figure 1.1.31). Despite its name there is still some chromatic aberration remaining in this design of lens since it can only bring two wavelengths to a common focus. If the radii of the curved surfaces are all equal, then the condition for two given wavelengths, λ_1 and λ_2, to have coincident images is

$$2\Delta\mu_C = \Delta\mu_F \qquad (1.1.33)$$

where $\Delta\mu_C$ and $\Delta\mu_F$ are the differences between the refractive indices at λ_1 and λ_2 for the crown glass and the flint glass respectively. More flexibility in design can be attained if the two surfaces of the converging lens have differing radii. The condition for achromatism is then

$$\frac{|R_1| + |R_2|}{|R_1|}\Delta\mu_C = \Delta_F \qquad (1.1.34)$$

where R_2 is the radius of the surface of the crown glass lens which is in contact with the flint lens (NB: the radius of the flint lens surface is almost invariably R_2 as well, in order to facilitate alignment and cementing) and R_1 is the radius of the other surface of the crown glass lens. By a careful selection of λ_1 and λ_2 an achromatic doublet can be constructed to give tolerable images. For example, by achromatising at 486 and 656 nm, the longitudinal chromatic aberration is reduced when compared with a simple lens of the same focal length by a factor of about thirty. Nevertheless since chromatic aberration varies as the square of the diameter of the objective and inversely with its focal length, refractors larger than about 0.25 m still have obtrusively coloured images. More seriously, if filters are used then the focal position will vary with the filter. Similarly the scale on photographic plates will alter with the wavelengths of their sensitive regions. Further lenses may be added to produce apochromats which have three corrected wavelengths and superapochromats with four corrected wavelengths.

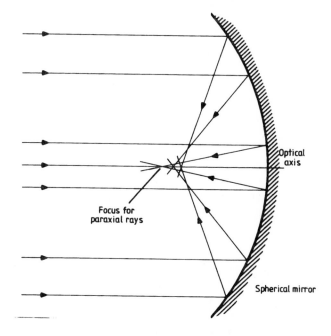

Figure 1.1.32. Spherical aberration.

But such designs are impossibly expensive for telescope objectives of any size, although eyepieces and camera lenses may have eight or ten components and achieve very high levels of correction.

A common and severe aberration of both lenses and mirrors is spherical aberration. In this effect, annuli of the lens or mirror which are of different radii have different focal lengths. It is illustrated in figure 1.1.32 for a spherical mirror. For rays parallel to the optical axis it can be eliminated completely by deepening the sphere to a paraboloidal surface for the mirror. It cannot be eliminated from a simple lens without using aspheric surfaces, but for a given focal length it may be minimised. The shape of a simple lens is measured by the shape factor, q

$$q = \frac{R_2 + R_1}{R_2 - R_1}. \tag{1.1.35}$$

where R_1 is the radius of the first surface of the lens and R_2 is the radius of the second surface of the lens. The spherical aberration of a thin lens then varies with q as shown in figure 1.1.33, with a minimum at $q = +0.6$. The lens is then biconvex with the radius of the surface nearer to the image three times the radius of the other surface. Judicious choice of surface radii in an achromatic doublet can lead to some correction of spherical aberration while still retaining the colour correction.

The deepening of a spherical mirror to a paraboloidal one in order to correct

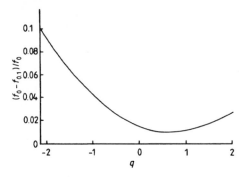

Figure 1.1.33. Spherical aberration in thin lenses. f_x is the focal length for rays parallel to the optical axis, and at a distance x times the paraxial focal length away from it.

for spherical aberration unfortunately introduces a new aberration called coma. This also afflicts mirrors of other shapes, and lenses. It causes the images for objects away from the optical axis to consist of a series of circles which correspond to the various annular zones of the lens or mirror, and which are progressively shifted towards or away from the optical axis (figure 1.1.34). Coma is zero in a system which obeys Abbe's sine condition

$$\frac{\sin \theta}{\sin \varphi} = \frac{\theta_p}{\varphi_p} = \text{constant} \tag{1.1.36}$$

where the angles are defined in figure 1.1.35. A doublet lens can be simultaneously corrected for chromatic and spherical aberrations, and coma within acceptable limits, if the two lenses can be separated. Such a system is called an aplanatic lens. A parabolic mirror can be corrected for coma by adding thin correcting lenses before or after the mirror, as discussed in more detail later in this section under the heading 'telescope designs'. The severity of the coma at a given angular distance from the optical axis is inversely proportional to the square of the focal ratio. Hence its effect can also be reduced by using as large a focal ratio as possible. In Newtonian reflectors (see p. 69) a focal ratio of $f8$ or larger gives acceptable coma for most purposes. At $f3$, coma will limit the useful field of view to about one minute of arc so that prime focus (see p. 67) photography almost always requires the use of a correcting lens to give reasonable fields of view.

Astigmatism is an effect whereby the focal length differs for rays in the plane containing an off-axis object and the optical axis (the tangential plane), in comparison with rays in the plane at right angles to this (the sagittal plane). It decreases more slowly with focal ratio than coma so that it may become the dominant effect for large focal ratios. It is possible to correct astigmatism, but only at the expense of introducing yet another aberration; field curvature. This is simply that the surface containing the sharply focused images is no longer a

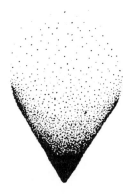

Figure 1.1.34. Shape of the image of a point source due to coma.

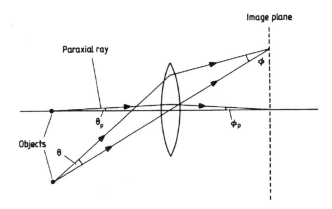

Figure 1.1.35. Parameters for Abbe's sine condition.

flat plane but is curved. A system in which a flat image plane is retained and astigmatism is corrected for at least two radii is termed an anastigmatic system.

The final aberration is distortion, and this is a variation in the magnification over the image plane. An optical system will be free of distortion only if the condition

$$\frac{\tan \theta}{\tan \varphi} = \text{constant} \tag{1.1.37}$$

holds for all values of θ (see figure 1.1.36 for the definition of the angles). Failure of this condition to hold results in pincushion or barrel distortion (figure 1.1.37) accordingly as the magnification increases or decreases with distance from the optical axis. A simple lens is very little affected by distortion and it can frequently be reduced in more complex systems by the judicious placing of stops within the system.

A fault of optical instruments as a whole is vignetting. This is not an

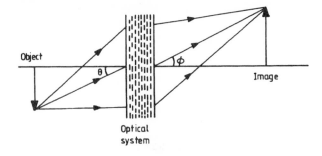

Figure 1.1.36. Terminology for distortion.

attribute of a lens or mirror and so is not included amongst the aberrations. It arises from the uneven illumination of the image plane, usually due to obstruction of the light path by parts of the instrument. Normally it can be avoided by careful design, but it may become important if stops are used in a system to reduce other aberrations.

This long catalogue of faults of optical systems may well have led the reader to wonder if the Rayleigh limit can ever be reached in practice. However, optical designers have a wide variety of variables to play with; refractive indices, dispersion, focal length, mirror surfaces in the form of various conic sections, spacings of the elements, number of elements and so on, so that it is usually possible to produce a system which will give an adequate image for a particular purpose. Other criteria such as cost, weight, production difficulties etc may well prevent the realisation of the system in practice even though it is theoretically

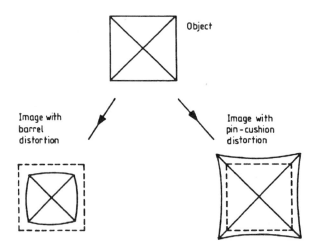

Figure 1.1.37. Distortion.

possible. Multi-purpose systems can usually only be designed to give lower quality results than single-purpose systems. Thus the practical telescope designs which are discussed later in this section are optimised for objects at infinity and if they are used for nearby objects their image quality will deteriorate. The approach to designing an optical system is largely an empirical one. There are two important methods in current use. The older approach requires an analytical expression for each of the aberrations. For example, the third-order approximation for spherical aberration of a lens is

$$\frac{1}{f_x} - \frac{1}{f_p} = \frac{\chi^2}{8f^3\mu(\mu-1)}\left[\frac{\mu+2}{\mu-1}q^2 + 4(\mu+1)\left(\frac{2f}{v}-1\right)q\right.$$
$$\left. + (3\mu+2)(\mu-1)\left(\frac{2f}{v}-1\right)^2 + \frac{\mu^2}{\mu-1}\right] \tag{1.1.38}$$

where f_χ is the focal distance for rays passing through the lens at a distance χ from the optical axis, f_p is the focal distance for paraxial rays from the object (paraxial rays are rays which are always close to the optical axis and which are inclined to it by only small angles), f is the focal length for paraxial rays which are initially parallel to the optical axis and v is the object distance. For precise work it may be necessary to involve fifth-order approximations of the aberrations, and so the calculations rapidly become very cumbersome. The alternative and more modern approach is via ray tracing. The concepts involved are much simpler since the method consists simply of accurately following the path of a selected ray from the object through the system and finding its arrival point on the image plane. Only the basic formulae are required; Snell's law for lenses

$$\sin i = \frac{\mu_1}{\mu_2}\sin r \tag{1.1.39}$$

and the law of reflection for mirrors

$$i = r \tag{1.1.40}$$

where i is the angle of incidence, r is the angle of refraction or reflection as appropriate and μ_1 and μ_2 are the refractive indices of the materials on either side of the interface.

The calculation of i and r for a general ray requires a knowledge of the ray's position and direction in space, and the specification of this is rather more cumbersome. Consider first of all a ray passing through a point P, which is within an optical system (figure 1.1.38). we may completely describe the ray by the coordinates of the point P, together with the angles that the ray makes with the coordinate system axes (figure 1.1.39). we may without any loss of generality set the length of the ray, l, equal to unity, and therefore write

$$\gamma = \cos\theta \tag{1.1.41}$$
$$\delta = \cos\varphi \tag{1.1.42}$$
$$\epsilon = \cos\psi \tag{1.1.43}$$

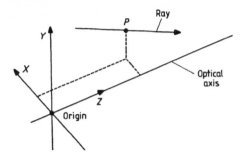

Figure 1.1.38. Ray tracing coordinate system.

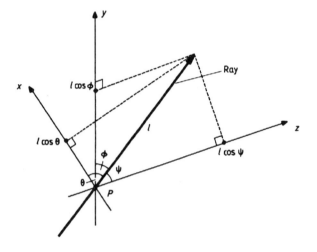

Figure 1.1.39. Ray tracing angular coordinate system.

The angular direction of the ray is thus specified by the vector

$$v = (\gamma, \delta, \epsilon). \tag{1.1.44}$$

The quantities γ, δ and ϵ are commonly referred to as the direction cosines of the ray. If we now consider the point P as a part of an optical surface, then we require the angle of the ray with the normal to that surface in order to obtain i, and thereafter, r. Now we may similarly specify the normal at P by its direction cosines, forming the vector v', say

$$v' = (\gamma', \delta', \epsilon') \tag{1.1.45}$$

and we then have

$$\cos i = \frac{v \cdot v'}{\|v\| \|v'\|} \tag{1.1.46}$$

or

$$i = \cos^{-1}(\gamma\gamma' + \delta\delta' + \epsilon\epsilon'). \tag{1.1.47}$$

The value of r can now be obtained from equations (1.1.39) or (1.1.40). We may again specify its direction by direction cosines (the vector v'', say)

$$v'' = (\gamma'', \delta'', \epsilon'') \tag{1.1.48}$$

and we may find the values of the components as follows. Three simultaneous equations can be obtained; the first from the angle between the incident ray and the refracted or reflected ray, the second from the angle between the normal and the reflected or refracted ray, and the third by requiring the incident ray, the normal, and the refracted or reflected ray to be coplanar, and these are

$$\gamma\gamma'' + \delta\delta'' + \epsilon\epsilon'' = \cos 2r \quad \text{(reflection)} \tag{1.1.49}$$

$$= \cos(i - r) \quad \text{(refraction)} \tag{1.1.50}$$

$$\gamma'\gamma'' + \delta'\delta'' + \epsilon'\epsilon'' = \cos r \tag{1.1.51}$$

$$(\epsilon\delta' - \epsilon'\delta)\gamma'' + (\gamma\epsilon' - \gamma'\epsilon)\delta'' + (\delta\gamma' - \delta'\gamma)\epsilon'' = 0. \tag{1.1.52}$$

After a considerable amount of manipulation one obtains

$$\begin{aligned}
\epsilon' = &\{[(\epsilon\delta' - \epsilon'\delta)\delta' - (\gamma\epsilon' - \gamma'\epsilon)\gamma'](\gamma'\cos\alpha - \gamma\cos r) \\
&- (\gamma'\delta - \gamma\delta')(\epsilon\delta' - \epsilon'\delta)\cos r\} \\
\div &\{[(\epsilon\delta' - \epsilon'\delta)\delta' - (\gamma\epsilon' - \gamma'\epsilon)\gamma'](\gamma'\epsilon - \gamma\epsilon') \\
&- [(\epsilon\delta' - \epsilon'\delta)\epsilon' - (\delta\gamma' - \delta'\gamma)\gamma'](\gamma'\delta - \gamma\delta')\}
\end{aligned} \tag{1.1.53}$$

$$\delta'' = \frac{\gamma'\cos\alpha - \gamma\cos r - (\gamma'\epsilon - \gamma\epsilon')\epsilon''}{(\gamma'\delta - \gamma\delta')} \tag{1.1.54}$$

$$\gamma'' = \frac{\cos\alpha - \delta\delta'' - \epsilon\epsilon''}{\gamma} \tag{1.1.55}$$

where α is equal to $2r$ for reflection, and $(i - r)$ for refraction.

The direction cosines of the normal to the surface are easy to obtain when its centre of curvature is on the optical axis

$$\gamma' = x_1/R \tag{1.1.56}$$

$$\delta' = y_1/R \tag{1.1.57}$$

$$\epsilon' = (z_1 - z_R)/R \tag{1.1.58}$$

where (x_1, y_1, z_1) is the position of P and $(0, 0, z_R)$ is the position of the centre of curvature. If the next surface is at a distance s from the surface under consideration, then the ray will arrive on it at a point (x_2, y_2, z_2) still with the direction cosines γ'', δ'' and ϵ'', where

$$x_2 = (\gamma''s/\epsilon'') + x_1 \tag{1.1.59}$$

$$y_2 = (\delta''s/\epsilon'') + y_1 \tag{1.1.60}$$

$$z_2 = s + z_1. \tag{1.1.61}$$

We may now repeat the calculation again for this surface, and so on. Ray tracing has the advantage that all the aberrations are automatically included by it, but it has the disadvantage that many rays have to be followed in order to build up the structure of the image of a point source at any given place on the image plane, and many images have to be calculated in order to assess the overall performance of the system. Ray tracing, however, is eminently suitable for programming onto a computer, while the analytical approach is not, since it requires frequent value judgements. The approach of a designer, however, to a design problem is similar whichever method is used to assess the system. The initial prototype is set up purely on the basis of what the designer's experience suggests may fulfil the major specifications such as cost, size, weight, resolution, etc. Its performance is then assessed either analytically or by ray tracing. In most cases it will not be good enough, so a slight alteration is made with the intention of improving the performance, and it is reassessed. This process continues until the original prototype has been optimised for its purpose. If the optimum solution is within the specifications then there is no further problem. If it is outside the specifications, even after optimisation, then the whole procedure is repeated starting from a different prototype. The performances of some of the optical systems favoured by astronomers are considered in the next subsection.

Even after a design has been perfected, there remains the not inconsiderable task of physically producing the optical components to within the accuracies specified by the design. The manufacturing steps for both lenses and mirrors are broadly similar, although the details may vary. The surface is roughly shaped by moulding or by diamond milling. More recently, the blanks for large thin mirrors have been cast in a rotating furnace, so that centrifugal acceleration causes the finished blank to be close to its finally required shape. It is then matched to another surface formed in the same material whose shape is its inverse, and the two surfaces are ground together with coarse Carborundum or other grinding powder in between until the required surface begins to approach its specifications. The pits left behind by this coarse grinding stage are removed by a second grinding stage in which finer powder is used. The pits left by this stage are then removed in turn by a third stage using still finer powder, and so on. As many as eight or ten such stages may be necessary. When the grinding pits are reduced to a micron or so in size, the surface may be polished. This process employs a softer powder such as iron oxide or cerium oxide which is embedded in a soft matrix such as pitch. Once the surface has been polished it can be tested for the accuracy of its fit to its specifications. Since in general it will not be within the specifications after the initial polishing, a final stage, which is termed figuring, is necessary. This is simply additional polishing to adjust the surface's shape until it is correct. The magnitude of the changes involved during this stage is only about a micron or two, so that if the alteration that is needed is larger than this, it may be necessary to return to a previous stage in the grinding to obtain a better approximation. Recently, the requirements for non axi-symmetric mirrors for segmented mirror telescopes (see later discussion) and for glancing incidence x-ray telescopes (section 1.3) have led to the development of numerically controlled diamond milling machines which can produce the required shaped and polished surface to an accuracy of 10 nm or better.

The defects in an image which are due to surface imperfections on a mirror will not exceed the Rayleigh limit if the imperfections are less than about one eighth of the wavelength of the radiation for which the mirror is intended. Thus we have the commonly quoted $\lambda/8$ requirement for the maximum allowable deviation of a surface from its specifications. The restriction on lens surfaces is about twice as large since the ray deflection is distributed over two surfaces. However, as we have seen, the Rayleigh limit is an arbitrary one, and for some purposes the fit must be several times better than this limit. This is particularly important when viewing extended objects such as planets, and improvements in the contrast can continue to be obtained even when the fit is accurate to $\lambda/20$ or better by further improvements.

The surface must normally receive its reflecting coating after its production. The vast majority of astronomical mirror surfaces have a thin layer of aluminium evaporated onto them by heating aluminium wires suspended over the mirror in a vacuum chamber. Other metals are occasionally used, especially for ultraviolet work, since the reflectivity of aluminium falls off below 300 nm. The initial reflectivity of an aluminium coating in the visual region is around 90%. This however can fall to 75% or less within a few months as the coating ages. Mirrors therefore have to be re-aluminised at regular intervals. The intervals between re-aluminising can be lengthened by gently cleaning the mirror every month or so. The currently favoured methods of cleaning are rinsing with de-ionised water and/or drifting carbon dioxide snow across the mirror surface. Mirrors coated with a suitably protected silver layer can achieve 99.5% reflectivity in the visible, and are becoming more and more required since some modern telescope designs can have four or five reflections. With ancillary instrumentation the total number of reflections can then reach 10 or more.

Lens surfaces also normally receive a coating after their manufacture, but in this case the purpose is to reduce reflection. Uncoated lenses reflect about 5% of the incident light from each surface so that a system containing, say, five uncoated lenses could lose 40% of its available light through this process. To reduce the reflection losses a thin layer of material covers the lens, for which

$$\mu' = \sqrt{\mu} \qquad (1.1.62)$$

where μ' is the refractive index of the coating and μ is the refractive index of the lens material. The thickness t should be

$$t = \tfrac{1}{4}\lambda/\mu'. \qquad (1.1.63)$$

This gives almost total elimination of reflection at the selected wavelength, but some will still remain at other wavelengths. Lithium fluoride and silicon dioxide are commonly used materials for the coatings.

Mirrors only require the glass as a support for their reflecting film. Thus there is no requirement for it to be transparent, on the other hand it is essential for its thermal expansion coefficient to be low. Thus today's large telescope mirrors are made from materials other than glass. The coefficient of thermal expansion of glass is about $9 \times 10^{-6}\,\mathrm{K}^{-1}$, that of Pyrex about $3 \times 10^{-6}\,\mathrm{K}^{-1}$, and for fused quartz is about $4 \times 10^{-7}\,\mathrm{K}^{-1}$. Amongst the largest telescopes almost the only glass mirror is the one for the 2.5 m Hooker telescope on Mount Wilson. After its construction Pyrex became the favoured material, until the last three decades, when quartz or artificial materials with a similar low coefficient of expansion such as 'Cervit', 'Zerodur', 'ULE', etc but which are easier to manufacture, have been used. A radically different alternative would be to use a material with a very high thermal conductivity, such as a metal. The mirror is then always at a uniform temperature and its surface has no stress distortion. So far such mirrors have only been applied successfully to small telescopes.

Telescope designs

Background

Most serious work with telescopes uses equipment placed directly at the focus of the telescope. But for visual work such as finding and guiding on objects, an eyepiece is necessary. Often it matters little whether the image produced by the eyepiece is of a high quality or not. Ideally, however, the eyepiece should not degrade the image noticeably more than the main optical system. There are an extremely large number of eyepiece designs, whose individual properties can vary widely. For example, one of the earliest eyepiece designs of reasonable quality is the Kellner. This combines an achromat and a simple lens and typically has a field of view of 40° to 50°. The Plössl uses two achromats and has a slightly wider field of view. More recently, the Erfle design employs six or seven components and gives fields of view of 60° to 70°, while the current state-of-the-art is represented by designs such as the Nagler with eight or more components and fields of view up to 85°. Details of these and other designs may generally be found in books on general astronomy or on optics or from the manufacturers. For our purposes, only four aspects of eyepieces are of any concern; light losses, eye relief, exit pupil, and angular field of view.

Light loss occurs through vignetting when parts of the light beam fail to be intercepted by the eyepiece optics, or are obstructed by a part of the structure of the eyepiece, and also through reflection, scattering and absorption by the optical components. The first of these can generally be avoided by careful eyepiece design and selection, while the latter effects can be minimised by antireflection coatings and by keeping the eyepieces clean.

The exit pupil is the image of the objective produced by the eyepiece (figure 1.1.40). All the rays from the object pass through the exit pupil, so that it must be smaller than the pupil of the human eye if all of the light gathered by the

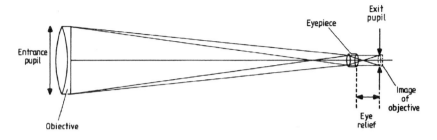

Figure 1.1.40. Exit pupil and eye relief.

objective is to be utilised. Its diameter, E, is given by

$$E = F_e D / F_o \qquad (1.1.64)$$

where D is the objective's diameter, F_e is the focal length of the eyepiece and F_o is the focal length of the objective. Since magnification is given by

$$M = F_o / F_e \qquad (1.1.65)$$

and the diameter of the pupil of the dark-adapted eye is six or seven millimetres, we must therefore have

$$M \gtrsim 170D \qquad (1.1.66)$$

where D is in metres, if the whole of the light from the telescope is to pass into the eye.

The eye relief is the distance from the final lens of the eyepiece to the exit pupil. It should be about six to ten millimetres for comfortable viewing.

The angular field of view is defined by the acceptance angle of the eyepiece, θ'. Usually this is about 40° but it may be up to 90° for wide-angle eyepieces. The angular diameter of the area of sky which is visible when the eye is positioned at the exit pupil, and which is known as the angular field of view, θ, is then just

$$\theta = \theta' / M. \qquad (1.1.67)$$

The brightness of an image viewed through a telescope is generally expected to be greater than when it is viewed directly. However, this is not the case for extended objects. The naked eye brightness of a source is proportional to the eye's pupil diameter squared, while that of the image in a telescope is proportional to the objective diameter squared. If the eye looks at that image, then its angular size is increased by the telescope's magnification. Hence the increased brightness of the image is spread over a greater area. Thus we have

$$R = \frac{\text{brightness through a telescope}}{\text{brightness to the naked eye}} = \frac{D^2}{M^2 P^2} \qquad (1.1.68)$$

where D is the objective's diameter, P is the diameter of the pupil of the eye and M is the magnification. But from equations (1.1.64) and (1.1.65) we have

$$R = 1 \qquad (1.1.69)$$

when the exit pupil diameter is equal to the diameter of the eye's pupil, and

$$R < 1 \qquad (1.1.70)$$

when it is smaller than the eye's pupil.

If the magnification is less than $170D$, then the exit pupil is larger than the pupil of the eye and some of the light gathered by the telescope will be lost. Since the telescope will generally have other light losses due to scattering, imperfect reflection and absorption, the brightness of an extended source is always fainter when viewed through a telescope than when viewed directly with the naked eye. This result is in fact a simple consequence of the second law of thermodynamics; for if it were not the case, then one could have a net energy flow from a cooler to a hotter object. The apparent increase in image brightness when using a telescope arises partly from the increased angular size of the object so that the image on the retina must fall to some extent onto the regions containing more rods (see p. 6) even when looked at directly, and partly from the increased contrast resulting from the exclusion of extraneous light by the optical system. Even with the naked eye, faint extended sources such as M31 can be seen much more easily by looking through a long cardboard tube.

The analysis that we have just seen does not apply to images which are physically smaller than the detecting element. For this situation, the image brightness is proportional to D^2. Again, however, there is an upper limit to the increase in brightness which is imposed when the energy density at the image is equal to that at the source. This limit is never approached in practice since it would require 4π steradians for the angular field of view. Thus stars may be seen through a telescope which are fainter, by a factor called the light grasp, than those visible to the naked eye. The light grasp is simply given by D^2/P^2. Taking $+6$ as the magnitude of the faintest star visible to the naked eye (see section 3.1), the faintest star which may be seen through a telescope has a magnitude, m_1, which is the limiting magnitude for that telescope

$$m_1 = 17 + 5 \log D \qquad (1.1.71)$$

where D is in metres. If the image is magnified to the point where it spreads over more than one detecting element, then we must return to the analysis for the extended sources. For an average eye, this upper limit to the magnification is given by

$$M \simeq 850D \qquad (1.1.72)$$

where D is again in metres.

Designs

Probably the commonest format for large telescopes is the Cassegrain system, although most large telescopes can usually be used in several alternative different modes by interchanging their secondary mirrors. The Cassegrain system is based upon a paraboloidal primary mirror and a convex hyperboloidal secondary mirror (figure 1.1.41). The nearer focus of the conic section which forms the surface of the secondary is coincident with the focus of the primary, and the Cassegrain focus is then at the more distant focus of the secondary mirror's surface. The major advantage of the Cassegrain system lies in its telephoto characteristic; the secondary mirror serves to expand the beam from the primary mirror so that the effective focal length of the whole system is several times that of the primary mirror. A compact and hence rigid and relatively cheap mounting can thus be used to hold the optical components while retaining the advantages of long focal length and large image scale. The Cassegrain design is afflicted with coma and spherical aberration to about the same degree as an equivalent Newtonian telescope (see p. 69), or indeed to just a single parabolic mirror with a focal length equal to the effective focal length of the Cassegrain. The beam expanding effect of the secondary mirror means that Cassegrain telescopes normally work at focal ratios between 12 and 30, even though their primary mirror may be $f3$ or $f4$. Thus the images remain tolerable over a field of view which may be several tenths of a degree across (figure 1.1.42). Astigmatism and field curvature are stronger than in an equivalent Newtonian system, however. Focusing of the final image in a Cassegrain system is customarily accomplished by moving the secondary mirror along the optical axis. The amplification due to the secondary means that it only has to be moved a short distance away from its optimum position in order to move the focal plane considerably. The movement away from the optimum position, however, introduces severe spherical aberration. For the 0.25 m $f4/f16$ system whose images are shown in figure 1.1.42,

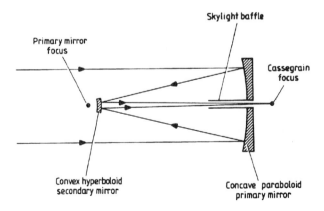

Figure 1.1.41. Cassegrain telescope optical system.

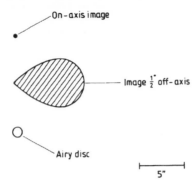

Figure 1.1.42. Images in a 0.25 m $f4/f16$ Cassegrain telescope. (The images were obtained by ray tracing. Since this does not allow for the effects of diffraction, the on-axis image in practice will be the same as the Airy disc, and the $\frac{1}{2}°$ off-axis image will be blurred even further.)

the secondary mirror can only be moved by six millimetres either side of the optimum position before even the on-axis images without diffraction broadening become comparable in size with the Airy disc. Critical work with Cassegrain telescopes should therefore always be undertaken with the secondary mirror at or very near to its optimum position.

A very great improvement to the quality of the images may be obtained if the Cassegrain design is altered slightly to the Ritchey–Chrétien system. The optical arrangement is identical with that shown in figure 1.1.41 except that the primary mirror is deepened to an hyperboloid and a stronger hyperboloid is used for the secondary. With such a design both coma and spherical aberration can be corrected and we have an aplanatic system. The improvement in the images can be seen by comparing figure 1.1.43 with figure 1.1.42. It should be noted,

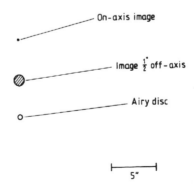

Figure 1.1.43. Ray tracing images in a 0.5 m $f3/f8$ Ritchey–Chrétien telescope.

however, that the improvement is in fact considerably more spectacular since we have a 0.5 m Ritchey–Chrétien and a 0.25 m Cassegrain with the same effective focal lengths. A 0.5 m Cassegrain telescope would have its off-axis image twice the size of that shown in figure 1.1.42 and its Airy disc half the size shown there.

Alternatively, a Cassegrain system can be improved by the addition of correctors just before the focus. The correctors are low- or zero-power lenses whose aberrations oppose those of the main system. There are numerous successful designs for correctors although many of them require aspheric surfaces and/or the use of exotic materials such as fused quartz. Images can be reduced to less than the size of the seeing disc over fields of view of up to one degree, or sometimes over even larger angles.

Another telescope design which is again very closely related to the Cassegrain is termed the Coudé system. It is in effect a very long focal length Cassegrain or Ritchey–Chrétien whose light beam is folded and guided by additional flat mirrors to give a focus whose position is fixed in space irrespective of the telescope position. One way of accomplishing this is shown in figure 1.1.44. After reflection from the secondary, the light is reflected down the hollow declination axis by a diagonal flat mirror, and then down the hollow polar axis by a second diagonal. The light beam then always emerges from the end of the polar axis, whichever part of the sky the telescope may be inspecting. Designs with similar properties can be devised for most other types of mountings although additional flat mirrors may be needed in some cases. With alt–az mountings (see p. 77), the light beam can be directed along the altitude axis to one of the two Nasmyth foci on the side of the mounting. These foci still rotate as the telescope changes its azimuth, but this poses far fewer problems than the changing altitude and attitude of a conventional Cassegrain focus. On large modern telescopes, platforms of considerable size are often constructed at the Nasmyth foci allowing large ancillary instruments to be used. The fixed focus of the Coudé and Nasmyth systems is a very great advantage when bulky items of equipment, such as high dispersion spectrographs, are to be used, since these can be permanently mounted in a nearby separate laboratory and the light brought to them, rather than have to have the equipment mounted on the telescope. The design also has several disadvantages; the field of view rotates as the telescope tracks an object across the sky, and is very tiny due to the large effective focal ratio ($f25$ to $f40$) which is generally required to bring the focus through the axes, finally the additional reflections will cause loss of light.

The simplest of all designs for a telescope, is a telescope used at its prime focus. That is, the primary mirror is used directly to produce the images and the photographic plate or other detector is placed at the top end of the telescope. The largest telescopes have a platform or cage which replaces the secondary mirror and which is large enough for the observer to ride in while he or she operates and guides the telescope from the prime focus position. With smaller instruments too much light would be blocked, so that they must be guided more

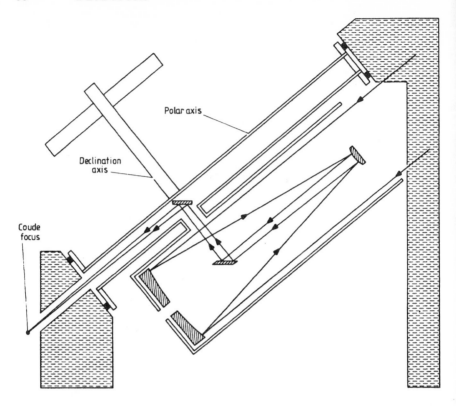

Polar axis

Declination
axis

Coude
focus

Figure 1.1.44. Coudé system for a modified English mounting.

normally using a separate guide telescope. The image quality at the prime focus is usually very poor, even a few seconds of arc away from the optical axis, because the primary mirror's focal ratio may be as short as $f3$ or less in order to reduce the length of the instrument to a minimum. Thus, correcting lenses are invariably essential to give acceptable images and reasonably large fields of view. These are similar to those used for correcting Cassegrain telescopes and are placed immediately before the prime focus.

A system which is almost identical to the use of the telescope at prime focus and which was the first design to be made into a working reflecting telescope, is that due to Newton and hence called the Newtonian telescope. A secondary mirror is used which is a flat diagonal and which is placed just before the prime focus. This reflects the light beam to the side of the telescope from where access to it is relatively easy (figure 1.1.45). The simplicity and cheapness of the design make it very popular as a small telescope for the amateur market, but it is rarely encountered in telescopes larger than about 0.5 m. There are several reasons for its lack of popularity for large telescope designs; the main ones being that it has no advantage over the prime focus position for large telescopes since the

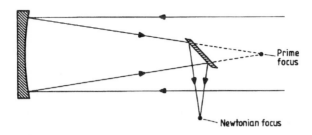

Figure 1.1.45. Newtonian telescope optical system.

equipment used to detect the image blocks out no more light than the secondary mirror, the secondary mirror introduces additional light losses, and the position of the equipment high on the side of the telescope tube causes difficulties of access and counterbalancing. The images in a Newtonian system and at prime focus are very similar and are of poor quality away from the optical axis as shown in figure 1.1.46.

A very great variety of other catoptric (reflecting objective) telescope designs abound, but most have few or no advantages over the two groups of designs which are discussed above. Some find specialist applications, and a few may be built as small telescopes for amateur use because of minor advantages in their production processes, but most such designs will be encountered very rarely.

Of the dioptric (refracting objective) telescopes, only the basic refractor using an achromatic doublet, or very occasionally a triplet, as its objective is in any general use (figure 1.1.47). Apart from the large refractors which were built towards the end of the last century, most refractors are now found as the guide telescopes of larger instruments. Their enclosed tube and relatively firmly mounted optics means that they need little adjustment once aligned with the main telescope.

The one remaining class of optical telescopes is the catadioptric group, of which the Schmidt camera is probably the best known. A catadioptric system uses both lenses and mirrors in its primary light gathering section. Very high degrees of correction of the aberrations can be achieved because of the wide range of variable parameters which become available to the designer in such systems. The Schmidt camera uses a spherical primary mirror so that coma is eliminated by not introducing it into the system in the first place! The resulting spherical aberration is eliminated by a thin correcting lens at the mirror's radius of curvature (figure 1.1.48). The only major remaining aberration is field curvature and the effect of this is eliminated by moulding the photographic plate to the shape of the image surface or through the use of additional correcting lenses (field flatteners). Small amounts of coma and chromatic aberration can be introduced by the correcting lens, but usually this is so thin that these aberrations

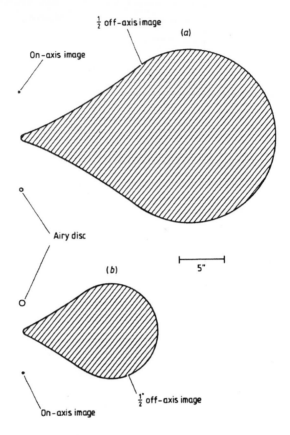

Figure 1.1.46. Images in Newtonian telescopes. (*a*) 1 m *f*4 telescope; (*b*) 0.5 m *f*8 telescope.

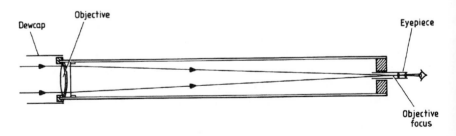

Figure 1.1.47. The astronomical refractor.

are negligible. Diffraction-limited performance over fields of view of several degrees with focal ratios as fast as *f*1.5 or *f*2 is therefore possible. As larger CCDs are becoming available, they are being used on Schmidt cameras. They

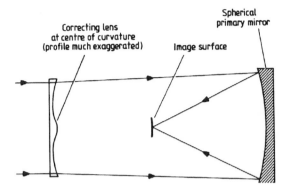

Figure 1.1.48. The Schmidt camera optical system.

are still smaller than photographic plates, and so fields of view are restricted to only a degree or so. However, the high detecting efficiency, the linear response, and the machine-readable output of the CCD means that they may be preferred to the photographic plate for many purposes.

The need to use a lens in the Schmidt design has limited the largest such instruments to an entrance aperture 1.35 m diameter (the Tautenburg Schmidt camera). The design also suffers from having a tube length at least twice its focal length. A number of alternative designs to provide high quality images over fields of view of 5° to 10° have therefore been proposed, though none with any significant sizes have yet been built. For example, the Willstrop three-mirror telescope (figure 1.1.49) is completely achromatic and can potentially be made as large as the largest conventional optical telescope. The primary mirror is close to parabolic, the corrector (secondary mirror) is close to spherical with a turned-down edge, while the tertiary is also close to spherical in shape. It has a 5° field and a tube length one third to one half that of the equivalent Schmidt camera. But its focal surface is curved like that of the Schmidt, and all except

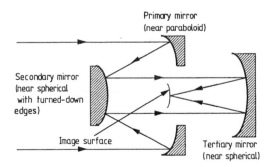

Figure 1.1.49. The Willstrop three-mirror camera optical system.

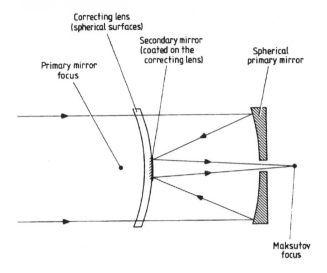

Figure 1.1.50. The Maksutov optical system.

the central 2° of the field of view suffers from vignetting. A 0.5 m version of the design has been successfully built and operated at Cambridge for several years now.

The Schmidt camera cannot be used visually since its focus is inaccessible. There are several modifications of its design, however, which produce an external focus while retaining most of the beneficial properties of the Schmidt. One of the best of these is the Maksutov (figure 1.1.50), which originally also had an inaccessible focus, but which is now a kind of Schmidt-Cassegrain hybrid. All the optical surfaces are spherical, and spherical aberration, astigmatism and coma are almost eliminated, while the chromatic aberration is almost negligible. A similar system is the Schmidt–Cassegrain telescope itself. This uses a thin correcting lens like the Schmidt camera and has a separate secondary mirror. The focus is accessible at the rear of the primary mirror as in the Maksutov system. Schmidt–Cassegrain telescopes are now available commercially in sizes up to 0.4 m diameter. They are produced in large numbers by several firms, at very reasonable costs, for the amateur and education markets. They are also finding increasing use in specific applications, such as site testing, for professional astronomy.

Although catadioptric telescopes approach perfection from the standpoint of image production, their size is limited by the necessity of using a large lens supported only at its edges. Thus the largest such systems currently in use have lens diameters of only 1.35 m.

Although it is not a telescope, there is one further system which deserves mention and that is the coelostat. This comprises two flat mirrors which are driven so that a beam of light from any part of the sky is directed into a fixed direction. They are particularly used in conjunction with solar telescopes whose extremely long focal lengths make them impossible to move. One mirror of the coelostat is mounted on a polar axis and driven at half sidereal rate. The second mirror is mounted and driven to reflect the light into the fixed telescope. Various arrangements and adjustments of the mirrors enable differing declinations to be observed.

No major further direct developments in the optical design of large telescopes seem likely to lead to improved resolution, since such instruments are already limited by the atmosphere. Better resolution therefore requires that the telescope be lifted above the Earth's atmosphere, or that the distortions that it introduces be overcome by a more subtle approach. Telescopes mounted on rockets, satellites, and balloons are discussed later in this section, while the four ways of improving the resolution of Earth-based telescopes: interferometry, speckle interferometry, occultations, and real-time compensation are discussed in sections 2.5, 2.6, 2.7 and later in this section, respectively. A fairly common technique for improving resolution, deconvolution, is discussed in section 2.1. This requires that the precise nature of the degradations of the image is known so that they can be removed. Since the atmosphere's effects are changing in a fairly random manner on a timescale of milliseconds, it is not an appropriate technique for application here, although it should be noted that one approach to speckle interferometry is in fact related to this method.

Now although resolution may not be improved in any straightforward manner, the position is less hopeless for the betterment of the other main function of a telescope, which is light gathering. The upper limit to the size of an individual mirror is probably being approached with the 5 and 6 m telescopes presently in existence. A diameter of 10 m for a metal-on-glass mirror would probably be about the ultimate possible with present day techniques and their foreseeable developments such as a segmented honeycomb construction for the mirror blank or a thin mirror on active supports which compensate for its flexure. The UK currently has plans for two 8 m telescopes using thin mirrors with active supports, and several other instruments of the same class are planned elsewhere.

Greater light-gathering power may therefore be acquired only by the use of several mirrors, or by the use of a single mirror formed from separate segments, or by aperture synthesis. The latter technique is used extensively in radioastronomy (section 2.5), and in principle could be applied at shorter wavelengths. The requirements on the stability of the system of a tenth of its operating wavelength over periods of days or weeks, however, currently limit its practical realisation. The European Southern Observatory's very large telescope will have four 8 m independently mounted mirrors acting as an interferometer and aperture synthesis system. Three apertures in combination can be combined to give results independent of atmospheric delays and optical interferometers based upon this 'closure phase' are currently starting to come on stream (section 2.5). For example, the Cambridge Optical Aperture Synthesis Telescope (COAST) and the Infrared Optical Telescope Array (IOTA). The former uses four 0.5 m sideriostats to feed 0.4 m Cassegrain telescopes and which can have base lines up to 100 m giving milli-arc second resolution in the red and near infrared, while the latter uses three 0.45 m sideriostats with a maximum base line of 38 m.

The construction of a single large mirror from several independent segments is a less severe problem. Several designs are currently operating or are under construction, for example the two 10 m Keck telescopes which are formed from 36 1.8 m hexagonal segments (see figure 1.1.51). The Hobby–Eberly spectroscopic telescope has 91 1.0 m spherical segments forming an 11 m ×10 m mirror. It uses correcting optics to produce an adequate image and is on a fixed mounting (see later discussion). This type of design requires that each segment of the mirror be independently mounted, and that their positions be continuously monitored and adjusted to keep them within a tenth of a wavelength of their correct positions.

Alternatively, the secondary mirror may be segmented as well, and those segments moved to compensate for the variations in the primary mirror segments. The monitoring and control of the segments would be very similar to those required for the remaining option: the multimirror telescope. The first such telescope was built at Mount Hopkins about two decades ago. This telescope was based upon six fairly conventional 1.8 m Cassegrain systems, with mirrors of 'egg-crate' construction to reduce their weight, and arranged in an hexagonal array 6.9 m across. They were supported on a single mounting and fed a common focus. The system was equivalent to a 4.4 m telescope in area, but only cost

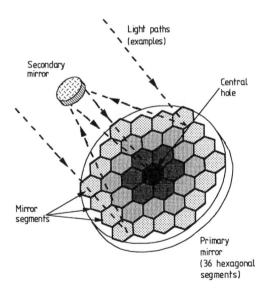

Figure 1.1.51. The optical arrangement of the Keck I and II 10 m telescopes.

about one third as much. Since the cost of a telescope varies as about the cube of its diameter, far larger savings would be possible with the use of more mirrors. The Mount Hopkins telescope, however, experienced problems, especially with the mirror position monitoring system. It has therefore been refitted with a single 6.5 m mirror. An 8 m 'binocular' telescope is under construction in Arizona using honeycomb mirrors and the European Southern Observatory plans four 8 m telescopes which can operate singly or together to give the equivalent collecting area of a 16 m telescope. This system will use very thin mirrors on active supports.

The accuracy and stability of surfaces, alignments, etc required for both the multimirror and the segmented mirror systems is a function of the operating wavelength. For infrared and short microwave work, far less technically demanding limits are therefore necessary and their telescopes are correspondingly easier to construct. Several such large instruments are currently under construction or in the design stage. The limits may also be relaxed if the system is required as a 'light bucket' rather than for imaging. Limits of ten times the operating wavelength may still give acceptable results when a photometer or spectroscope is used on the telescope, providing that close double stars of similar magnitudes are avoided.

A route to potentially very large mirrors is to use a reflecting liquid and to rotate it. The liquid is usually mercury, and its surface naturally takes up a parabolic shape when it is spun at a constant rate. Alternative materials such as gallium and gallium–indium alloy have the advantages over mercury of low toxicity and lower density. They can be supercooled and so remain liquid well below their nominal melting points of 15 to 30°C. Precautions have to be taken to ensure that the spin rate is constant and that vibrations are minimised. The resulting telescope can only observe the zenith or very close to it, and the detector has to be moved to track the motion of the object being observed. With CCDs, the tracking may be accomplished by moving the accumulating charges from one pixel to the next, without the need to move the detector physically. With correcting lenses, fields of view of a degree or so around the zenith may be possible. The use of active optical components for the correcting system may extend that to 20° or more in the future. In space a liquid mirror of potentially almost any size could be formed from a ferromagnetic liquid confined by electromagnetic forces.

Telescopes in space
The most direct way to improve the resolution of a telescope which is limited by atmospheric degradation is to lift it above the atmosphere, or at least above the lower, denser parts of the atmosphere. The two main methods of accomplishing this are to place the telescope onto a balloon-borne platform or onto an artificial satellite. Balloon-borne telescopes, of which the 0.9 m Stratoscope II is the best known but far from the only example, have the advantages of relative

cheapness, and that their images may be recorded by fairly conventional means. Their disadvantages include low maximum altitude (40 km) and short flight duration (a few days).

Many small telescopes have already been launched on satellites. That aboard the international ultraviolet explorer spacecraft (IUE) was a fairly typical example. It used a 0.45 m Ritchey–Chrétien telescope which fed images to a pair of ultraviolet spectrometers. In the infrared, the entire telescope and its ancillary instrumentation may have to be cooled. The ISO spacecraft (Infrared Space Observatory) for example, had a 0.6 m telescope cooled to 3 K with the detectors held at 1.8 K, using 2300 litres of liquid helium. The use of unconventional telescope designs, however, is not uncommon for this type of work, since they may have minor advantages in their weight or volume which become vital when the restrictions imposed by the balloon's lift or launcher's power are taken into account. For example off-axis systems, Gregorian telescopes (in Stratoscope II), and the Wynne camera, which is a sort of inverse Cassegrain system, have all been used recently. Glancing incidence systems which are an extreme form of the off-axis format, are of great importance for short-wave ultraviolet and x-ray observations and are discussed in section 1.3. The Hubble space telescope is a 2.4 m $f2.3/f24$ Ritchey–Chrétien telescope designed to operate from 115 nm to 1 mm, with a variety of instrumentation at its focus. With its corrective optics to compensate for the deficiencies of the primary mirror, and up-dated ancillary instruments, it is now regularly providing diffraction-limited images.

Suggestions for the future include arrays of orbiting optical telescopes with separations of 100 km or more, capable of directly imaging Earth-sized planets around other stars.

Mountings

The functions of a telescope mounting are simple—to hold the optical components in their correct mutual alignment, and to direct the optical axis towards the object to be observed. The problems in successfully accomplishing this to within the accuracy and stability required by astronomers are so great, however, that the cost of the mounting is usually the major item in funding a telescope. We may consider the functions of a mounting under three separate aspects; firstly to support the optical components, secondly to preserve their correct spatial relationships, and thirdly to acquire and hold the object of interest in the field of view.

Mounting the optical components is largely a question of supporting them in a manner which does not strain them and distort their surfaces. Lenses may only be supported at their edges, and it is the impossibility of doing this adequately for large lenses which limits their practicable sizes to a little over a metre. There is little difficulty with small lenses and mirrors since most types of mount may grip them firmly enough to hold them in place without at the same time straining them. However, large mirrors require very careful mounting. They are usually held on a number of mounting points to distribute their weight, and these mounting points, especially those around the edges, may need to be active so that their support changes to compensate for the different directions of the mirror's weight as the telescope moves. The active support can be arranged by systems of pivoted weights or by computer control of the supports. As discussed earlier some recently built telescopes and many of those planned for the future have active supports for the primary mirror which deliberately stress the mirror so that it retains its correct shape whatever the orientation of the telescope or the temperature of the mirror.

The optical components are held in their relative positions by the telescope tube. With most large telescopes, the 'tube' is in fact an open-work structure, but the name is still retained. For small instruments the tube may be made sufficiently rigid that its flexure is negligible. But this becomes impossible as the size increases. The solution then is to design the flexure so that it is identical for both the primary and secondary mirrors. The optical components then remain aligned on the optical axis, but are no longer symmetrically arranged within the mounting. The commonest structure in use which allows this equal degree of flexure is the Serrurier truss (figure 1.1.52). However, this design is not without its own problems. The lower trusses may need to be excessively thin or the upper ones excessively thick in order to provide the required amount of flexure. More seriously their deflection introduces a relative rotation of the optical components which deflects the focus and which may introduce or intensify aberrations. Even so, the majority of the large reflecting telescopes built in the last four decades have used Serrurier truss designs for their tubes.

By far the most usual means of mounting the telescope tube so that it may be pointed at an object and then moved to follow the object's motion across the sky is the equatorial mounting. This is a two-axis mounting, with one axis, the polar axis, aligned parallel with the Earth's rotational axis, and the other, the declination axis, perpendicular to the polar axis. This design has the enormous advantage that only a single constant velocity motor is required to rotate the mounting around the polar axis in order to track an object. It is also very convenient in that angular read-outs on the two axes give the hour angle or right ascension and declination directly. A very wide variety of different arrangements for the basic equatorial mounting exist, but these will not be reviewed here. Books on telescopes, and most general astronomy books list their details by the legion, and should be checked if the reader requires further information.

The alt–az mounting which has motions in altitude and azimuth is the other main two-axis mounting system. Structurally it is a much simpler form than the equatorial and has therefore been adopted for most large radio telescope dishes. Its drawbacks are that the field of view rotates with the telescope motion, and that it needs driving continuously in both axes and with variable speeds in order to track an object. Only relatively recently therefore, since cheap microprocessors became readily available to control the drive speeds, has much use been made of the design for optical telescope mountings. Most of the large telescopes planned

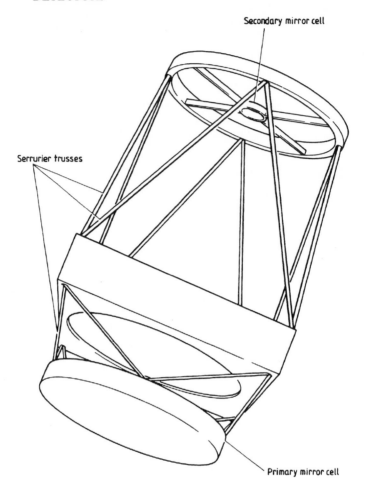

Secondary mirror cell

Serrurier trusses

Primary mirror cell

Figure 1.1.52. Telescope tube construction based on Serrurier trusses.

for the near future, however, will use alt–az mountings.

Several telescopes under construction or being planned at the time of writing will have fixed positions. These include liquid-mirror telescopes which always point near the zenith, and the 11 m Hobby–Eberly telescope (see earlier). Tracking is then accomplished by moving the detector and any correcting optics to follow the motion of the image, or for short exposures, by moving the charges along the pixels of a CCD detector. Zenith-pointing telescopes are generally limited to a degree or two either side of the zenith, but the Hobby–Eberly telescope can point to objects with declinations ranging from $-11°$ to $+71°$. It accomplishes this because it points to a fixed zenith angle of $35°$, but can be rotated in azimuth between observations. There are also a few instruments mounted on altitude–altitude mountings. These use a universal joint (such as a ball and socket) and tip the instrument in two orthogonal directions to enable it to point anywhere in the sky.

With any type of mounting, pointing and tracking accuracies of a second of arc or thereabouts are required, but these are not often achieved for Earth-based telescopes. Thus the observer may have to search for his or her object of interest over an area tens to hundreds of seconds of arc across after initial acquisition, and then guide, either himself or herself, or with an automatic system, to ensure that the telescope tracks the object sufficiently closely for his purposes. With balloon-borne and space telescopes such direct intervention is difficult or impossible. However, the reduction of external disturbances and possibly the loss of weight means that the initial pointing accuracy is higher. For space telescopes tracking is easy since the telescope will simply remain pointing in the required direction once it is fixed, apart from minor perturbations such as light pressure, the solar wind, gravitational anomalies and internal disturbances from the spacecraft. Automatic control systems can therefore usually be relied upon to operate the telescope. In a few cases, however, facilities for transmitting a picture of the field of view after the initial acquisition to the observer on the ground, followed by corrections to the position in order to set onto the correct object, may be necessary.

With terrestrial telescopes the guiding may be undertaken via a second telescope which is attached to the main telescope and aligned with it, or a small fraction of the light from the main telescope may be diverted for the purpose. The latter method may be accomplished by beam splitters, dichroic mirrors, or by intercepting a fraction of the main beam with a small mirror, depending upon the technique which is being used to study the main image. The whole telescope can then be moved using its slow-motion controls to keep the image fixed, or additional optical components can be incorporated into the light path whose movement counteracts the image drift, or the detector can be moved around the image plane to follow the image motion or the charges can be moved within a CCD chip. Since far less mass needs to be moved with the latter methods, their response times can be much faster than that of the first method. The natural development of the second method leads on to the active surface control which is discussed later in this section and which can go some way towards eliminating the effects of scintillation.

If guiding is attempted by the observer then there is little further to add, except to advise a plentiful supply of black coffee in order to avoid going to sleep through the boredom of the operation! Automatic guiding has two major problem areas. The first is sensing the image movement and generating the error signal, while the second is the design of the control system to compensate for this movement. Detecting the image movement has been attempted in three main ways—CCDs, quadrant detectors, and rotating sectors. With a CCD, the guide image is read out at frequent intervals, and any movement of the object being tracked is detected via appropriate software, which also generates the correction signals. In some small CCD systems aimed at the amateur market, it is the main image which is read out at frequent intervals and the resulting multiple images are then shifted into mutual alignment before being added together. This enables poor tracking to be corrected provided that the image shift is small during any single exposure. A quadrant detector is one whose detecting area is divided into four independent sectors, the signal from each of which can be separately accessed. If the image is centred on the intersection of the sectors, then their signals will all be equal. If the image moves, then at least two of the signals will become unbalanced and the required error signal can then be generated. A variant of this system uses a pyramidal 'prism'. The image is normally placed on the vertex, and so is divided into four segments each of which can then be separately detected. In the rotating sector method, a single detector looks at the image. Half of the field of view is blanked off by an occulting disc or sector and this revolves rapidly. The image is normally centred in the field of view and so is bisected by the sector. The detector output is constant so long as the image remains in this position. If the image drifts away from the centre of the field of view, however, the output will then become modulated. The phase relationship between the modulated signal and the position of the sector can then be used to generate the error signal. When guiding on stars by any of these automatic methods it may be advantageous to increase the image size slightly by operating a little way from the focal plane of the telescope, scintillation 'jitter' then becomes less important. The problem of the design of the control system for the telescope to compensate for the image movement should be handed to a competent control engineer. If it is to work adequately, then proper damping, feedback, rates of motion and so on, must be calculated so that the correct response to an error signal occurs without 'hunting' or excessive delays.

Real-time atmospheric compensation

The resolution of ground-based telescopes of more than a fraction of a metre in diameter is limited by the turbulence in the atmosphere. The maximum diameter of a telescope before it becomes seriously affected by atmospheric turbulence is given by Fried's coherence length, r_o,

$$r_o \simeq 0.114 \left(\frac{\lambda \cos z}{550} \right)^{0.6} \text{m} \qquad (1.1.73)$$

where λ is the operating wavelength in nm and z is the zenith angle. Fried's coherence length, r_o, is the distance over which the phase difference is one radian. Thus, for visual work telescopes of more than about 11.5 cm (4.5 in) diameter will always have their images degraded by atmospheric turbulence.

In an effort to reduce the effects of this turbulence (the 'seeing', 'scintillation' or 'twinkling'), many large telescopes are sited at high altitudes, or placed on board high-flying aircraft or balloons. The ultimate, though very expensive solution of course, is to orbit the telescope beyond the Earth's atmosphere out in space.

An alternative approach, however, to obtaining diffraction-limited performance for large telescopes, which is relatively inexpensive, and widely applicable, is to correct the distortions in the incoming light beam produced by the atmosphere. This atmospheric compensation is achieved through the use of adaptive optics. In such systems, one or more of the optical components can be changed rapidly and in such a manner that the undesired distortions in the light beam are reduced or eliminated. Although a relatively recent development in its application to large telescopes, adaptive optics is actually a very ancient technique familiar to us all. That is because the eye operates via an adaptive optic system in order to keep objects in focus, with the lens being stretched or compressed by the ciliary muscle (figure 1.1.1). The efficiency of an adaptive optics system is measured by the Strehl ratio which is the ratio of the intensity at the centre of the corrected image to that at the centre of a perfect diffraction-limited image of the same source. Strehl ratios up to 0.6 are currently being achieved, and may reach 0.8 in the near future.

Adaptive optics is not a complete substitute for spacecraft-based telescopes, however, because in the visual and near infrared, the correction only extends over a very small area (the isoplanatic patch, see below). Thus, for example, if applied to producing improved images of Jupiter, only a small percentage of the planet's surface would be seen sharply. Also, of course, ground-based telescopes are still limited in their wavelength coverage by atmospheric absorption.

There is often some confusion in the literature between adaptive optics and active optics. However, the most widespread usage of the two terms is that an adaptive optics system is a fast closed-loop system, and an active optics system a more slowly operating open- or closed-loop system. The division is made at a response time of a few seconds. Thus the tracking of a star by the telescope drive system can be considered as an active optics system which is open-loop if no guiding is used, and closed-loop if guiding is used. Large thin mirror optical telescopes and radio telescopes suffer distortion due to buffeting by the wind at a frequency of a tenth of a hertz or so. Correction of this would also be classified under active optics. There is also the term active support which refers to the mountings used for the optical components in either an adaptive or an active optics system.

An atmospheric compensation system contains three main components; a sampling system, a wavefront sensor and a correcting system. We will look at each of these in turn.

Sampling system. The sampling system provides the sensor with the distorted wavefront or an accurate simulacrum thereof. For astronomical adaptive optics systems, a beam splitter is commonly used. This is just a partially reflecting mirror which typically diverts about 10% of the radiation to the sensor, while allowing the other 90% to continue on to form the image. A dichroic mirror can also be used which allows all the light at the desired wavelength to pass into the image while diverting light of a different wavelength to the sensor. However, atmospheric effects change with wavelength, and so this latter approach may not give an accurate reproduction of the distortions unless the operating and sampling wavelengths are close together.

Since astronomers go to great lengths to gather photons as efficiently as possible, the loss of even 10% to the sensor is to be regretted. Many adaptive optics systems therefore use a guide star rather than the object of interest to determine the wavefront distortions. This becomes essential when the object of interest is a large extended object, since most sensors need to operate on point or near-point images. The guide star must be very close in the sky to the object of interest, or its wavefront will have undergone different atmospheric distortion (figure 1.1.53). For solar work small sunspots or pores can be used as the guide object. The region of the sky over which images have been similarly affected by the atmosphere is called the isoplanatic area or patch, and it can be as small as a few seconds of arc.

The small size of the isoplanatic area means that few objects have suitable guide stars. Recently therefore, artificial guide stars have been produced. This

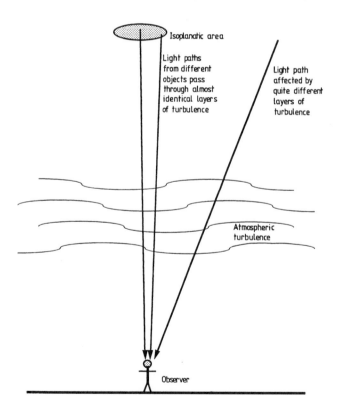

Figure 1.1.53. The isoplanatic area.

is accomplished using a powerful laser pointed skywards. The laser is tuned to one of the sodium D-line frequencies and excites the free sodium atoms in the atmosphere at a height of about 90 km. The glowing atoms appear as a star-like patch which can be placed as near to the object of interest as required. Guide stars at lower altitudes and at other wavelengths can be produced through back-scattering by air molecules of a laser beam. The laser light can be sent out through the main telescope or more usually using a nearby auxiliary telescope. Laser-produced guide stars have two problems, however, which limit their usefulness. Firstly, for larger telescopes, the relatively low height of the guide star means that the light path from it to the telescope differs significantly from that for the object being observed (figure 1.1.54). The use of multiple artificial guide stars may in the future reduce this problem and also extend the area capable of being corrected. Secondly, the outgoing laser beam is affected by atmospheric turbulence, and therefore the guide star moves with respect to the object, resulting in a blurred image on longer exposures.

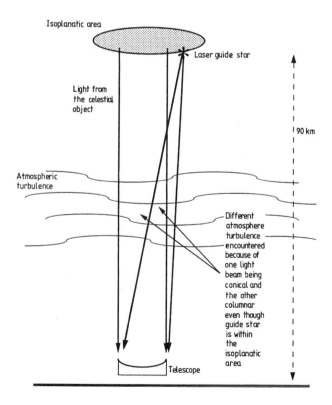

Figure 1.1.54. Light paths from a celestial object and a laser guide star to the telescope.

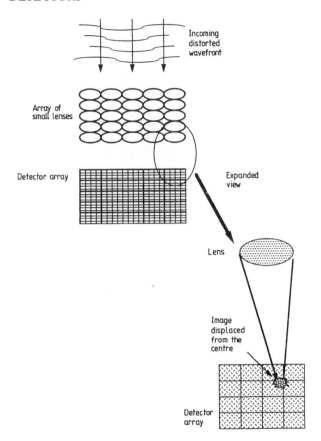

Figure 1.1.55. The Hartmann sensor.

Wavefront sensing.   The wavefront sensor detects the distortions in the incoming wavefront provided by the sampler.   The Hartmann (also known as the Hartmann–Shack or the Shack–Hartmann) sensor is in widespread use in astronomical adaptive optics systems.   This uses a two-dimensional array of small lenses (figure 1.1.55). Each lens produces an image which is sensed by an array detector. In the absence of wavefront distortion, each image will be centred on each detector. Distortion will displace the images from the centres of the detectors, and the degree of displacement and its direction is used to generate the error signal. An alternative sensor is based upon the shearing interferometer (figure 1.1.56). This is a standard interferometer but with the mirrors marginally turned so that the two beams are slightly displaced with respect to each other when they are recombined. The deformations of the fringes in the overlap region then provide the slopes of the distortions in the incoming wavefront.

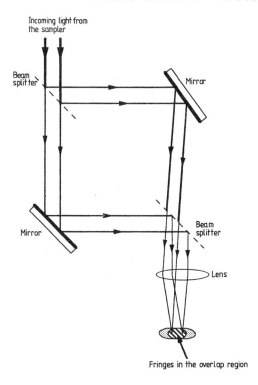

Figure 1.1.56. The shearing interferometer.

Wavefront correction. In most astronomical adaptive optics systems, the correction of the wavefront is achieved by distorting a subsidiary mirror. Since the atmosphere changes on a timescale of ten milliseconds or so, the sampling, sensing and correction have to occur in a millisecond or less. In the simplest systems only the overall tilt of the wavefront introduced by the atmosphere is corrected. That is accomplished by suitably tilting a plane or segmented mirror placed in the light beam from the telescope in the opposite direction (figure 1.1.57). An equivalent procedure, since the overall tilt of the wavefront causes the image to move, is 'shift and add'. Multiple short exposure images are shifted until their brightest points are aligned, and then added together. Even this simple correction, however, can result in a considerable improvement of the images.

More sophisticated approaches provide better corrections—either just of the relative displacements within the distorted wavefront, or of both displacement and fine scale tilt. In some systems the overall tilt of the wavefront is corrected by a separate system using a flat mirror whose angle can be changed. Displacement correction would typically use a thin mirror capable of being distorted by piezo-electric or other actuators placed underneath it. The error signal from the sensor is used to distort the mirror in the opposite manner to

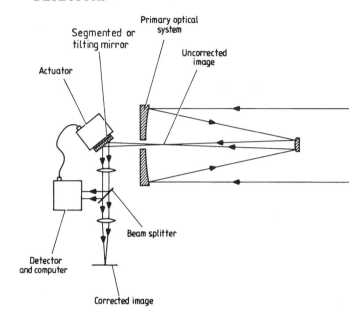

Figure 1.1.57. Schematic optical system for real-time atmospheric compensation.

the distortions in the incoming wavefront. The reflected wavefront is therefore almost flat. The non-correction of the fine scale tilt, however, does leave some small imperfections in the reflected wavefront. Nonetheless, currently operating systems using this approach can achieve diffraction-limited performance in the near infrared for telescopes of 3 or 4 m diameter (i.e. about 0.2″ at 2 μm wavelength). At visual wavelengths, reductions of the uncorrected image size by about a factor of ten are currently being reached.

Correction of fine scale tilt within the distorted wavefront as well as displacement is now being investigated in the laboratory. It requires an array of small mirrors rather than a single 'bendy' mirror. Each mirror in the array is mounted on four actuators so that it can be tilted in any direction as well as being moved linearly. At the time of writing it is not clear whether such systems will be applied to large astronomical telescopes, because at visual wavelengths few of them have good enough optical surfaces to take advantage of the improvements such a system would bring.

Future developments
As a general guide to the future we may look to the past. Figure 1.1.58 shows the way in which the collecting area of optical telescopes has increased with time. Simple extrapolation of the trends shown there suggests collecting areas of 800 m² by the year 2050 and of 4500 m² by the year 2100 (diameters of 32 m and 75 m respectively for filled circular apertures).

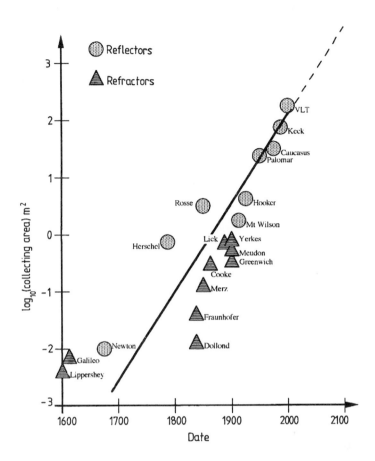

Figure 1.1.58. Optical telescope collecting area as a function of time.

Several possible paths along which telescopes may develop have already been mentioned: multimirror or segmented mirror systems, active surface control, aperture synthesis, space telescopes and so on. In this subsection some of these techniques and other more speculative ones are looked at in more detail.

The next stage after the Hubble space telescope might be an orbiting mirror twenty to thirty metres across. This could either be lifted in sections or possibly constructed in space. One possible manufacturing technique for large space mirrors might be to blow a large bubble in a suitable viscous liquid, attach this equatorially to a firm support, and then set the liquid. Solar heat or ultraviolet radiation could easily be used as the setting agent. Aluminising and then splitting the bubble along its equatorial support would then give two mirrors of almost any required size. Another similar proposal is to use a large spherical bubble filled with a low pressure gas as a weak lens. An aperture of 100 m with a focal length of 10^8 m might result from such a system. A somewhat different approach might be to use a thin membrane or pellicle which is held in shape by radiation pressure. Two mutually interfering lasers would produce a required interference pattern, and the pellicle would be positioned on its nodes. Pointing and guiding such large light structures would probably be quite difficult. The perturbations of the Echo I and II satellites by solar radiation pressure demonstrated how significant an influence this could be.

The multimirror concept may be extended or made easier by the development of fibre optics. Fused silica fibres can now be produced whose light losses and focal ratio degradation are acceptable for lengths of tens of metres. Thus mirrors on independent mounts with a total area of hundreds of square metres which all feed a common focus will become technically feasible within a decade.

Not all developments of the telescope, however, will be in terms of size. The use of adaptive optics and optical aperture synthesis mean that telescopes above a few centimetres in size can operate at or near to their diffraction limit. Current large telescopes, however, generally do not have optics of sufficient quality to take full advantage of adaptive optics. We are therefore likely to see telescopes of intermediate sizes and perhaps of new designs being developed which can operate at their diffraction limit.

A technique which is in occasional use today but which may well gain in popularity is daytime observation of stars. Suitable filters and large focal ratios (which diminish the sky background brightness, but not that of the star, see the earlier discussion) enable useful observations to be made of stars as faint as the seventh magnitude. The technique is particularly applicable in the infrared where the scattered solar radiation is minimal.

Lenses seem unlikely to make a comeback as primary light-gathering components, except perhaps for the previously mentioned space bubble. However, they are still used extensively in eyepieces and in ancillary equipment. Three major developments seem likely to affect these applications, principally by making lens systems simpler and/or cheaper. The first is the use of plastic to form the lens. High quality lenses then can be cheaply produced in quantity by moulding. The second advance is related to the first and it is the use of aspherical surfaces. Such surfaces have been used when they were essential in the past, but their production was expensive and time-consuming. The investment in an aspherical mould, however, would be small if that mould were to produce thousands of lenses. Hence cheap aspherical lenses could become available. The final possibility is for lenses whose refractive index varies across their diameter. Several techniques exist for introducing such variability to the refractive index of a material, of which those which appear most favourable at the time of writing are the diffusion of silver into glass, and the growing of crystals from a melt whose composition varies with time. Thus very highly corrected and relatively cheap lens systems seem likely to become available within a few years through the use of some or all of these techniques.

Observing sites

The selection of a site for a major telescope is at least as important for the future usefulness of that telescope as the quality of its optics. The main aim of site selection is to minimise the effects of the Earth's atmosphere, of which the most important are usually scattering, absorption and scintillation. It is a well known phenomenon to astronomers that the best observing conditions at any site occur immediately prior to the final decision to place a telescope there! Thus not all sites have fulfilled all their expectations.

Scattering by dust and molecules causes the sky background to have a certain intrinsic brightness, and it is this which imposes a limit upon the faintest detectable object through the telescope. For extended sources, it occurs when the source becomes comparable with the background in its surface brightness, while for point sources it occurs when the background density reaches the linear portion of a photographic plate's characteristic curve (section 2.2) so that no further improvement of the signal-to-noise ratio is possible. Other types of detectors also have limiting magnitudes due to the background which generally differ little from that of the photographic emulsion. The main source of the scattered light is artificial light, and most especially street lighting. Thus a first requirement of a site is that it be as far as possible from built-up areas. If an existing site is deteriorating due to encroaching suburbs etc then some improvement may be possible for some types of observation by the use of a light pollution rejection (LPR) filter which absorbs in the regions of the most intense sodium and mercury emission lines. Scattering can be worsened by the presence of industrial areas upwind of the site or by proximity to deserts, both of which inject dust into the atmosphere.

Absorption is due mostly to the molecular absorption bands of the gases forming the atmosphere. The two well known windows in the spectrum, wherein radiation passes through the atmosphere relatively unabsorbed, extend from about 360 nm to 100 μm and from 10 mm to 100 m. But even in these regions there is some absorption, so that visible light is decreased in its intensity by 10 to 20% for vertical incidence. The infrared region is badly affected by water vapour and other molecules to the extent that portions of it are completely obscured. Thus the second requirement of a site is that it be as high an altitude as possible to reduce the air paths to a minimum, and that the water content be as low as possible. We have already seen how balloons and satellites are used as a logical extension of this requirement, and a few high-flying aircraft are also used for this purpose.

Scintillation is the change in the wavefront due to the varying densities of the atmospheric layers. It is sometimes divided into seeing, which is due to low level turbulence, and scintillation proper, which arises at higher levels. It causes the image to shimmer when viewed through a telescope, changing its size, shape, position and brightness at frequencies from one to a thousand hertz. It is the primary cause of the low resolution of large telescopes, since the image of a point source is rarely less than one second of arc across due to this blurring.

Thus the third requirement for a good site is a steady atmosphere. Scintillation may be worsened by the ground-layer effects of the structures and landscape in the telescope's vicinity. A rough texture to the ground around the dome, such as low-growing bushes, seems to reduce scintillation when compared with that found for smooth (e.g. paved) and very rough (e.g. tree-covered) surfaces. Great care also needs to be taken to match the dome and telescope temperatures with the ambient temperature or the resulting convection currents can worsen the scintillation by an order of magnitude or more.

These requirements for an observing site restrict the choice very considerably and lead to a clustering of telescopes on the comparatively few optimum choices. Most are now found at high altitudes and on oceanic islands, or with the prevailing wind from the ocean. Built-up areas tend to be small in such places and the water vapour is usually trapped near sea level by an inversion layer. The long passage over the ocean by the winds tends to minimise dust and industrial pollution. The remaining problem is that there is quite a correlation of good observing sites and noted tourist spots so that it is somewhat difficult to persuade non-astronomical colleagues that one is going to, say, Tenerife or Hawaii, in order to work!

The recent trend in the reduction of the real cost of data transmission lines, whether these are via cables or satellites, is likely to lead to the greatly increased use of remote control of telescopes. In a decade or two most major observatories are likely to have only a few permanent staff physically present at the telescope wherever that might be in the world, and the astronomer would only need to travel to a relatively nearby control centre for his observing shifts. Since the travel costs for UK astronomers alone would suffice to buy two 1 m telescopes per year, such a development is likely to be greeted with relief by the funding agencies, if not by the more jet-setting astronomers! There are also a few observatories with completely robotic telescopes, in which all the operations are computer-controlled, with no staff on site at all during their use. Many of these robotic instruments are currently used for long-term photometric monitoring programs or for teaching purposes. However, their extension to more exacting observations such as spectroscopy is now starting to occur. The logical end point of these developments is for the telescope to become just another computer peripheral.

Exercises

1.1.1 Calculate the effective theoretical resolutions of the normal and the dark-adapted eye, taking their diameters to be 2 and 6 millimetres, respectively.

1.1.2 Calculate the focal lengths of the lenses of a cemented achromatic doublet, whose overall focal length is 4 m. Assume the contact surfaces are of equal radius, the second surface of the diverging lens is flat and that the components are made from crown and dense flint glasses. Correct it for wavelengths of 486 and 589 nm.

1.1.3   By ray tracing calculate the physical separation in the focal plane of two rays which are incident onto a parabolic mirror of focal length 4 m. The rays are parallel prior to reflection and at an angle of 1° to the optical axis. The first ray intersects the mirror at its centre, while the second intersects it 0.5 m from the optical axis, and in the plane containing the rays and the optical axis (figure 1.1.59).

Figure 1.1.59. Optical arrangement for exercise 1.1.3.

1.1.4   Calculate the maximum usable eyepiece focal length for the 5 m $f3.3/f16$ Mount Palomar telescope at its Cassegrain focus, and hence its lowest magnification.

1.1.5   If the original version Mount Hopkins multimirror telescope were to be used as a Michelson interferometer (section 2.5), what would be its resolution when used visually?

1.2   RADIO AND MICROWAVE DETECTION

Introduction

Radio astronomy is the oldest of the 'new' astronomies since it is now well over half a century since Jansky first observed radio waves from the galaxy. It has passed beyond the developmental stage wherein the other 'new' astronomies— infrared, ultraviolet, x-ray, neutrino astronomy etc—still remain. It thus has quite well established instruments and techniques which are not likely to change overmuch.

The reason why radio astronomy developed relatively early is that radio radiation penetrates to ground level. For wavelengths from about 10 mm to 10 m, the atmosphere is almost completely transparent. The absorption becomes almost total at about 0.5 mm wavelength, and between 0.5 and 10 mm there are a number of absorption bands which are mainly due to oxygen and water vapour, with more or less transparent windows between the bands. Radiation with wavelengths longer than about 50 m again fails to penetrate to ground level, but this time the cause is reflection by the ionosphere. Thus this section is concerned with the detection of radiation with wavelengths longer than 0.1

mm. That is frequencies less than 3×10^{12} Hz, or photon energies less than 2×10^{-21} J (0.01 eV).

The unit of intensity which is commonly used at radio wavelengths is the jansky (Jy)

$$1 \, \text{Jy} = 10^{-26} \, \text{W m}^{-2} \, \text{Hz}^{-1} \tag{1.2.1}$$

and detectable radio sources vary from about 10^{-3} to 10^6 Jy. Most radio sources of interest to astronomers generate their radio flux as thermal radiation, when their spectrum is given by the Rayleigh–Jeans law

$$\mathcal{F}_\nu = \frac{2\pi k}{c^2} T \nu^2 \tag{1.2.2}$$

or as synchrotron radiation from energetic electrons spiralling around magnetic fields, when the spectrum is of the form

$$\mathcal{F}_\nu \propto \nu^{-\alpha} \tag{1.2.3}$$

where $\mathcal{F}_\nu$ is the flux per unit frequency interval at frequency ν and α is called the spectral index of the source, and is related to the energy distribution of the electrons. For many sources $0.2 \leq \alpha \leq 1.2$. Radio detectors (see below) normally only accept a very narrow band of frequencies and only one plane of polarisation. Most radio telescopes are therefore automatically also radio spectroscopes (or monochromators) and radio polarimeters, the former by tuning the receiver to different frequencies, and the latter by rotating the aerial or its feed. When an image of an area of the sky is to be built up (see Chapter 2) the data are generally stored in a computer and displayed as contour maps viewed either from above or from the side. But many other display techniques are used by radio astronomers as may be convenient or which highlight interesting aspects of the data.

Detectors and receivers

The detection of radio signals is a two-stage process in which the detector or aerial or antenna produces an electrical signal which then has to be processed until it is in a form which is directly usable. Coherent detectors, which preserve the phase information of the signal are available for use over the whole of the radio spectrum, in contrast to the optical and infrared detectors discussed in section 1.1 which in general respond only to the total power of the signal. In the radio region, the antenna will be a dipole, such as the half-wave dipole shown in figure 1.2.1. The two halves of such a dipole are each a quarter of a wavelength long. Connection to the remainder of the system is by coaxial cable. In the microwave region a rectangular or circular horn antenna is used, with waveguides for the connection to the rest of the system. Both of these antennae can only be optimised at one wavelength and have very restricted bandwidths. Wider bandwidths and circular or elliptical polarisation detection are possible

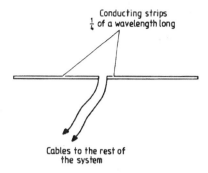

Figure 1.2.1. A half-wave dipole.

with logarithmic periodic antennas. These are helically wound antennas with the shape of a cone, which are ideally formed from a sheet conductor whose width and spacing decreases logarithmically towards the cone's apex. Elliptical polarisation can also be detected with helical antennas of constant width and spacing. Other wideband (but non-coherent) detectors include bolometers similar to those used in the infrared (section 1.1), Schottky diodes and superconducting tunnel junctions.

The signal from the antenna is carried to the receiver whose purpose is to convert the high frequency electrical currents into a convenient form. The behaviour of the receiver is governed by five parameters: sensitivity, amplification, bandwidth, receiver noise level and integration time.

The sensitivity and the other parameters are very closely linked, for the minimum detectable brightness, B_{min}, is given by

$$B_{min} = \frac{2k\nu^2 K T_s}{c^2 \sqrt{t \Delta \nu}} \qquad (1.2.4)$$

where T_s is the noise temperature of the system, t is the integration time, $\Delta \nu$ is the frequency bandwidth and K is a constant close to unity, which is a function of the type of receiver. The bandwidth is usually measured between the output frequencies whose signal strength is half the maximum when the input signal power is constant with frequency. The amplification and integration time are self-explanatory, so that only the receiver noise level remains to be explained. This noise originates as thermal noise within the electrical components of the receiver, and may also be called Johnson or Nyquist noise (see also section 1.1). The noise is random in nature and is related to the temperature of the component. For a resistor, the RMS voltage of the noise per unit frequency interval, $\bar{V}$, is given by

$$\bar{V} = 2\sqrt{kTR} \qquad (1.2.5)$$

where R is the resistance and T is the temperature. The noise of the system is then characterised by the temperature T_s which would produce the same noise

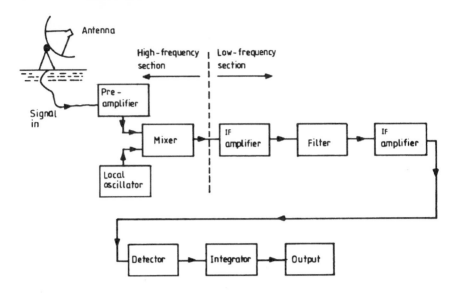

Figure 1.2.2. Block diagram of a basic superheterodyne receiver.

level for the impedance of the system. It is given by

$$T_s = T_1 + \frac{T_2}{G_1} + \frac{T_3}{G_1 G_2} + \cdots + \frac{T_n}{G_1 G_2 \ldots G_{n-1}} \ldots \qquad (1.2.6)$$

where T_n is the noise temperature of the nth component of the system and G_n is the gain (or amplification) of the nth component of the system. It is usually necessary to cool the initial stages of the receiver with liquid helium in order to reduce T_s to an acceptable level. Other noise sources which may be significant include shot noise which results from random electron emission in vacuum tubes, g–r noise due to a similar effect in semiconductors (section 1.1), noise from the parabolic reflector or other collector which may be used to concentrate the signal onto the antenna, radiation from the atmosphere, and last but by no means least, spill-over from radio taxis, microwave ovens and other artificial sources.

Many types of receiver exist, the simplest is a development of the superheterodyne system employed in the ubiquitous transistor radio. The basic layout of a superheterodyne receiver is shown in block form in figure 1.2.2. The pre-amplifier operates at the signal frequency and will typically have a gain of 10 to 1000. It is often mounted close to the feed and cooled to near absolute zero to minimise its contribution to the noise of the system (see above). Below 1 GHz, transistor amplifiers may be used for the pre-amplifier. From 1 to 40 GHz, FET, parametric and maser amplifiers are needed, while above 40 GHz the mixer, using Schottky diodes, must precede the pre-amplifier in order to reduce the frequency before it can be amplified.

The local oscillator produces a signal which is close to but different from the signal in its frequency. Thus when the signal and the local oscillator signal are combined by the mixer, the beat frequency between them (intermediate frequency or IF) is at a much lower frequency than that of the original signal. The relationship is given by

$$\nu_{SIGNAL} = \nu_{LO} \pm \nu_{IF} \qquad (1.2.7)$$

where ν_{SIGNAL} is the frequency of the original signal (i.e. the operating frequency of the radio telescope), ν_{LO} is the local oscillator frequency, and ν_{IF} is the intermediate frequency.

Normally, at lower frequencies, only one of the two possible signal frequencies given by equation (1.2.7) will be picked up by the feed antenna or passed by the pre-amplifier. At high frequencies, both components may contribute to the output.

The power of the intermediate frequency emerging from the mixer is directly proportional to the power of the original signal. The IF amplifiers and filter determine the pre-detector bandwidth of the signal and further amplify it by a factor of 10^6 to 10^9.

The detector is normally a square-law device, that is to say, the output *voltage* from the detector is proportional to the square of the input voltage. Thus the output voltage from the detector is proportional to the input *power*.

In the final stages of the receiver, the signal from the detector is integrated, usually for a few seconds, to reduce the noise level. Then it is fed to an output device, such as a voltmeter, chart recorder, etc from whence the signal strength may be determined by the observer, or it may be fed into an analogue-to-digital input to a computer for further processing.

The basic superheterodyne receiver has a high system temperature, and its gain is unstable. The temperature may be lowered by applying an equal and opposite voltage in the later stages of the receiver, and the stability of the gain may be greatly improved by switching rapidly from the antenna to a calibration noise source and back again, with a phase-sensitive detector (section 3.1) to correlate the changes. Such a system is then sometimes called a Dicke radiometer. The radiometer works optimally if the calibration noise source level is the same as that of the signal, and so it may be further improved by continuously adjusting the noise source to maintain the balance, and it is then termed a null-balancing Dicke radiometer. Since the signal is only being detected half the time the system is less efficient than the basic receiver, but its efficiency may be restored by using two alternately switched receivers. The value of T_s for receivers varies from 10 K at metre wavelengths to 10 000 K at millimetre wavelengths. The noise sources must therefore have a comparable range, and at long wavelengths are usually diodes, while at the shorter wavelengths a gas discharge tube inside the waveguide and inclined to it by an angle of about 10° is used.

As we have seen (equation (1.2.6)) the early stages of the receiver contribute most to the receiver noise temperature. It is therefore important to make these early stages with as low an intrinsic noise as possible. The radio-frequency amplification therefore uses low-noise amplifiers such as masers and parametric amplifiers for narrow band work, and travelling wave tubes and tunnel diodes for broad band work. Furthermore these amplifiers are usually cooled in liquid helium, and a noise temperature for the radio-frequency amplification stage as low as 5 K is then possible. Receivers are generally sky background limited just like terrestrial optical telescopes. The Earth's atmosphere radiates at 100 K and higher temperatures below a wavelength of about 3 mm. Only between 30 and 100 mm does its temperature fall as low as 2 K. Then, at longer wavelengths, the galactic emission becomes important, rising to temperatures of 10^5 K at wavelengths of 30 m.

Spectrographs at radio frequencies can be obtained in several different ways. The local oscillator may be tuned, producing a frequency sweeping receiver, or the receiver may be a multichannel device so that it registers several different discrete frequencies simultaneously. The latter is the more stable system and is generally to be preferred. High resolution scanning across molecular emission lines can be achieved by autocorrelation. Successive delays are fed into the signal and then recombined in a computer. The spectrum is obtained from the Fourier transform of the result. Alternatively the radio signal may be converted into a different type of wave, and the variations of this secondary wave studied instead. This is the basis of the acousto-optical radio spectrometer. The radio signal is converted into an ultrasonic wave whose intensity varies with that of the radio signal and whose frequency is also a function of that of the radio signal. Typically water is used for the medium in which the ultrasound propagates, and the wave is generated by a piezo-electric crystal driven either directly from the radio signal or by a frequency-reduced version of the signal. The cell containing the water is illuminated by a laser and a part of the light beam is diffracted by the sound wave. The angle of diffraction depends upon the sound wave's frequency, while the intensity of the diffracted light depends on the sound wave's intensity. Thus the output from the device is a fan beam of light, position within which ultimately depends upon the observed radio frequency, and whose intensity at that position ultimately depends upon the radio intensity. The fan beam may then simply be detected by photography, by a linear array of detectors, or by scanning, and the radio spectrum inferred from the result.

A major problem at all frequencies in radio astronomy is interference from artificial noise sources. In theory, certain regions of the spectrum (see table 1.2.1) are reserved partially or exclusively for use by radio astronomers. But leakage from devices such as microwave ovens, incorrectly tuned receivers, and illegal transmissions often overlap into these bands. The Russian *Glonass* satellite navigation system for example overlapped into the band reserved for the interstellar OH lines at 1.61 GHz. The use of highly directional aerials (see the next subsection) reduces the problem to some extent. But it is likely that radio astronomers will have to follow their optical colleagues to remote parts of the globe or place their aerials in space if their work is to continue in the future. Even the latter means of escape may be threatened by solar power satellites with their potentially enormous microwave transmission intensities, and by experiments such as the one which placed millions of tiny metal needles in orbit around the Earth.

Table 1.2.1. Radio astronomy reserved frequencies. The wave bands listed in this table are those current in Britain at the time of writing for the major radio observatories. Not all frequencies are kept free over all the country, and the allocations are continually changing. The reader should consult an up-to-date local source for specific information relative to his or her requirements.

Lower frequency	Upper frequency
37.75 MHz	38.25 MHz
80.5 MHz	82.5 MHz
150.05 MHz	153.0 MHz
326.5 MHz	328.5 MHz
406.1 MHz	410.0 MHz
606 MHz	614 MHz
1.380 GHz	1.400 GHz
1.400 GHz	1.427 GHz
1.6115 GHz	1.6125 GHz
1.6644 GHz	1.6684 GHz
1.720 GHz	1.721 GHz
2.670 GHz	2.690 GHz
2.690 GHz	2.700 GHz
4.825 GHz	4.835 GHz
4.950 GHz	4.990 GHz
4.990 GHz	5.000 GHz
10.6 GHz	10.68 GHz
10.68 GHz	10.7 GHz
15.35 GHz	15.4 GHz
22.21 GHz	22.5 GHz
23.07 GHz	23.12 GHz
23.6 GHz	24.0 GHz
31.3 GHz	31.5 GHz
42.5 GHz	43.5 GHz

Radio telescopes

The antenna and receiver, whilst they are the main active portions of a radio detecting system, are physically far less impressive than the large structures which serve to gather and concentrate the radiation and to shield the antenna from unwanted sources. Before going on, however, to the consideration of

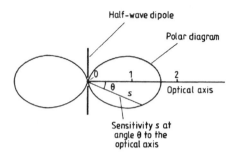

Figure 1.2.3. Polar diagram of a half-wave dipole.

these large structures which form most people's ideas of what comprises a radio telescope, we must look in a little more detail at the physical background of antennae.

The theoretical optics of light and radio radiation are identical, but different traditions in the two disciplines have led to differences in the mathematical and physical formulations of their behaviours. Thus the image in an optical telescope is discussed in terms of its diffraction structure (figure 1.1.26 for example), while that of a radio telescope is discussed in terms of its polar diagram. However, these are just two different approaches to the presentation of the same information. The polar diagram is a plot, in polar coordinates, of the sensitivity or voltage output of the telescope, with the angle of the source from the optical axis. (Note, the polar diagrams discussed herein are all far-field patterns, i.e. the response for a distant source, near-field patterns for the aerials may differ from those shown here.) The polar diagram may be physically realised by sweeping the telescope past a point source, or by using the telescope as a transmitter and measuring the signal strength around it.

The simplest antenna, the half-wave dipole (figure 1.2.1), accepts radiation from most directions, and its polar diagram is shown in figure 1.2.3. This is only a cross section through the beam pattern, the full three-dimensional polar diagram may be obtained by rotating the pattern shown in figure 1.2.3 about the dipole's long axis, and thus has the appearance of a toroid which is filled-in to the centre. The polar diagram, and hence the performance of the antenna, may be described by four parameters: the beam width at half-power points (BWHP), the beam width at first nulls (BWFN), the gain, and the effective area. The first nulls are the positions either side of the optical axis where the sensitivity of the antenna first decreases to zero, and the BWFN is just the angle between them. Thus the value of the BWFN for the half-wave dipole is 180°. The first nulls are the direct equivalent of the first fringe minima in the diffraction pattern of an optical image, and for a dish aerial type of radio telescope, their position is given by equation (1.1.27) thus

$$\text{BWFN} = 2 \times 1.22\lambda/D. \tag{1.2.8}$$

The Rayleigh criterion of optical resolution may thus be similarly applied to radio telescopes; two point sources are resolvable when one is on the optical axis and the other is in the direction of a first null. The half-power points may be best understood by regarding the radio telescope as a transmitter, they are then the directions in which the broadcast power has fallen to one half of its peak value. The BWHP is just the angular separation of these points. For a receiver they are the points at which the output voltage has fallen by a factor of the square root of two, and hence the output power has fallen by half. The maximum gain or directivity of the antenna is also best understood in terms of a transmitter. It is the ratio of the peak value of the output power to the average power. In a receiver it is a measure of the output from the system compared with that from a comparable (and hypothetical) isotropic receiver. The effective area of an antenna is the ratio of its output power to the strength of the incoming flux of the radiation which is correctly polarised to be detected by the antenna, i.e.

$$A_e = P_\nu/\mathcal{F}_\nu \tag{1.2.9}$$

where A_e is the effective area, P_ν is the power output by the antenna at frequency ν and $\mathcal{F}_\nu$ is the correctly polarised flux from the source at the antenna at frequency ν. The effective area and the maximum gain, g, are related by

$$g = \frac{4\pi}{c^2} \nu^2 A_e. \tag{1.2.10}$$

For the half-wave dipole, the maximum gain is about 1.6, and so there is very little advantage over an isotropic receiver.

The performance of a simple dipole may be improved by combining the outputs from several dipoles which are arranged in an array. In a collinear array, the dipoles are lined up along their axes and spaced at intervals of half a wavelength (figure 1.2.4). The arrangement is equivalent to a diffraction grating and so the sensitivity at an angle θ to the long axis of the array, $s(\theta)$, is given by

$$s(\theta) = s_0 \left(\frac{\sin(n\pi \sin \theta)}{\sin(\pi \sin \theta)} \right) \tag{1.2.11}$$

where n is the number of half-wave dipoles and s_0 is the maximum sensitivity (cf equation (4.1.54)). Figure 1.2.5 shows the polar diagrams for 1, 2 and 4 dipole arrays, their three-dimensional structure can be obtained by rotating these diagrams around a vertical axis so that they become lenticular toroids. The resolution along the axis of the array, measured to the first null is given by

$$\alpha = \sin^{-1}(1/n). \tag{1.2.12}$$

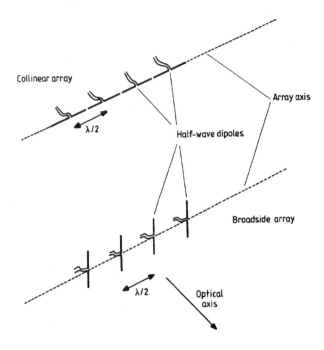

Figure 1.2.4. Dipole arrays.

The structure of the polar diagrams in figure 1.2.5 shows a new development. Apart from the main lobe whose gain and resolution increase with n as might be expected, a number of smaller side lobes have appeared. Thus the array has a sensitivity to sources which are at high angles of inclination to the optical axis. These side lobes correspond precisely to the fringes surrounding the Airy disc of an optical image (figure 1.1.26 etc). Although the resolution of an array is improved over that of a simple dipole along its optical axis, it will still accept radiation from any point perpendicular to the array axis. The use of a broadside array in which the dipoles are perpendicular to the array axis and spaced at half wavelength intervals (figure 1.2.4) can limit this 360° acceptance angle somewhat. For a four-dipole broadside array, the polar diagram in the plane containing the optical and array axes is given by polar diagram number II in figure 1.2.5, while in the plane containing the optical axis and the long axis of an individual dipole, the shape of the polar diagram is that of a single dipole (number I in figure 1.2.5), but with a maximum gain to match that in the other plane. The three-dimensional shape of the polar diagram of a broadside array thus resembles a pair of squashed balloons placed end to end. The resolution of a broadside array is given by

$$\alpha = \sin^{-1}(2/n) \qquad (1.2.13)$$

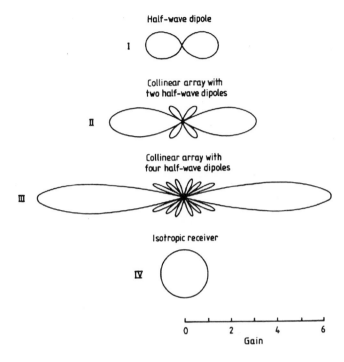

Figure 1.2.5. Polar diagrams for collinear arrays.

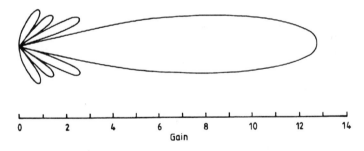

Figure 1.2.6. Polar diagram for a four-element collinear array with a mesh reflector.

along the array axis, and is that of a single dipole, i.e. 90°, perpendicular to this. Combinations of broadside and collinear arrays can be used to limit the beam width further if necessary.

With the arrays as shown, there is still a two-fold ambiguity in the direction of a source which has been detected, however narrow the main lobe may have been made, due to the forward and backward components of the main lobe. The backward component may easily be eliminated, however, by placing a reflector

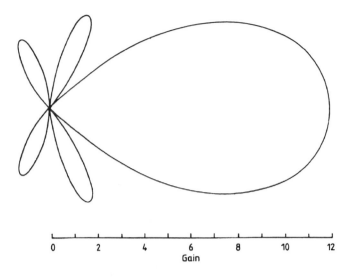

Figure 1.2.7. Polar diagram of a parasitic aerial.

behind the dipole. This is simply a conducting rod about 5% longer than the dipole and unconnected electrically with it. It is placed parallel to the dipole and about one eighth of a wavelength behind it. For an array, the reflector may be a similarly placed electrically conducting screen. The polar diagram of a four-element collinear array with a reflector is shown in figure 1.2.6 to the same scale as the diagrams in figure 1.2.5. It will be apparent that not only has the reflector screened out the backward lobe, but it has also doubled the gain of the main lobe. Such a reflector is termed a parasitic element since it is not a part of the electrical circuit of the antenna. Similar parasitic elements may be added in front of the dipole to act as directors. These are about 5% shorter than the dipole. The precise lengths and spacings for the parasitic elements can only be found empirically, since the theory of the whole system is not completely understood. With a reflector and several directors we obtain the parasitic or Yagi antenna, familiar from its appearance on so many roof-tops as a television aerial. A gain of up to 12 is possible with such an arrangement and the polar diagram is shown in figure 1.2.7. The Rayleigh resolution is 45° and it has a bandwidth of about 3% of its operating frequency. The main use of parasitic antennae in radio astronomy is as the receiving element (sometimes called the feed) of a larger reflector such as a parabolic dish.

The use of a single dipole, or even several dipoles in an array is the radio astronomy equivalent of a naked-eye observation. Just as in optical astronomy, the observations may be greatly facilitated by the use of some means of concentrating the signal from over a wide area onto the antenna. The most familiar of these devices are the large parabolic dishes which are the popular conception of a radio telescope. These are directly equivalent to an optical reflecting telescope. They are usually used at the prime focus or at the Cassegrain focus (section 1.1). The gain may be found roughly by substituting the dishes' area for the effective area in equation (1.2.10). The Rayleigh resolution is given by equation (1.1.28). The size of the dishes is so large because of the length of the wavelengths being observed; for example, to obtain a resolution of $1°$ at a wavelength of 0.1 m requires a dish 7 m across, which is larger than most optical reflectors for a resolution over 10^4 times poorer. The requirement on surface accuracy is the same as that for an optical telescope—deviations from the paraboloid to be less than $\lambda/8$ if the Rayleigh resolution is not to be degraded. Now, however, the longer wavelength helps since it means that the surface of a telescope working at 0.1 m, say, can deviate from perfection by over 10 mm without seriously affecting the performance. In practice a limit of $\lambda/20$ is often used, for as we have seen the Rayleigh limit does not represent the ultimate limit of resolution. These less stringent physical constraints on the surface of a radio telescope ease the construction problems greatly, more importantly, however, it also means that the surface need not be solid, as a wire mesh with spacings less than $\lambda/20$ will function equally well as a reflector. The weight and wind resistance of the reflector are thus reduced by very large factors. At the shorter radio wavelengths a solid reflecting surface may be more convenient, however, and at very short wavelengths (< 1 mm) active surface control to retain the accuracy is used along somewhat similar lines to the methods discussed in section 1.1 for optical telescopes. The shape of the mirror is monitored holographically. The dishes are usually of very small focal ratio, $f0.5$ is not uncommon, and the reason for this is so that the dish acts as a screen against unwanted radiation as well as concentrating the desired radiation. Fully steerable dishes up to 100 m across have been built, while fixed dishes up to 300 m across exist. These latter instruments act as transit telescopes and have some limited ability to track sources and look at a range of declinations by moving the feed antenna around the image plane. They may have a spherical surface in order to extend this facility and use a secondary reflector to correct the resulting spherical aberration. The feed antenna for such dishes may be made of such a size that it intercepts only the centre lobe of the telescope response (i.e. the Airy disc in optical terms). The effects of the side lobes are then reduced or eliminated. This technique is known as tapering the antenna and is the same as the optical technique of apodisation (section 4.1). The tapering function may take several forms and may be used to reduce background noise as well as eliminating the side lobes. Tapering reduces the efficiency and resolution of the dish, but this is usually more than compensated for by the improvement in the shape of the response function.

With a single feed, the radio telescope is a point-source detector only. Images have to be built up by scanning (section 2.4) or by interferometry (section 2.5). True imaging, however, can be achieved through the use of cluster or array feeds. These are simply multiple individual feeds arranged in a suitable array at the telescope's focus. Each feed is then the equivalent of a pixel in a CCD or other type of detector. At the time of writing the number of elements in such cluster feeds remains small compared with their optical equivalents.

Very many other systems have been designed to fulfil the same function as a steerable paraboloid but which are easier to construct. The best known of these are the multiple arrays of mixed collinear and broadside type, or similar constructions based upon other aerial types. They are mounted onto a flat plane which is oriented east–west and which is tiltable in altitude to form a transit telescope. Another system such as the 600 m RATAN telescope uses an off-axis paraboloid which is fixed and which is illuminated by a tiltable flat reflector. Alternatively the paraboloid may be cylindrical and tiltable itself around its long axis. For all such reflectors some form of feed antenna is required. A parasitic antenna is a common choice at longer wavelengths, while a horn antenna, which is essentially a flared end to a waveguide, may be used at higher frequencies.

A quite different approach is used in the Mills cross type of telescope. This uses two collinear arrays oriented north–south and east–west. The first provides a narrow fan beam along the north–south meridian, while the second provides a similar beam in an east–west direction. Their intersection is a narrow vertical pencil beam, typically 1° across. The pencil beam may be isolated from the contributions of the remainders of the fan beams by comparing the outputs when the beams are added in phase with when they are added out of phase. The in-phase addition is simply accomplished by connecting the outputs of the two arrays directly together. The out-of-phase addition delays one of the outputs by half a wavelength before the addition, and this is most simply done by switching in an extra length of cable to one of the arrays. In the first case radiation from objects within the pencil beam will interfere constructively, while in the second case there will be destructive interference. The signals from objects not within the pencil beam will be mutually incoherent and so will simply add together in both cases. Thus, looking vertically down onto the Mills cross, the beam pattern will alternate between the two cases shown in figure 1.2.8. Subtraction of the one from the other will then just leave the pencil beam. The pencil beam may be displaced by an angle θ from the vertical by introducing a phaseshift between each dipole. Again the simplest method is to switch extra cable into the connections between each dipole, the lengths of the extra portions, L, being given by

$$L = d \sin \theta \qquad (1.2.14)$$

where d is the dipole separation. The pencil beam may thus be directed around the sky as wished. In practice, the beam is only moved along the north–south plane, and the telescope is used as a transit telescope, since the alteration of the cable lengths between the dipoles is a lengthy procedure. The resolution of a Mills cross is the same as that of a parabolic dish whose diameter is equal to the array lengths. The sensitivity, however, is obviously much reduced from that of a dish, since only a tiny fraction of the aperture is filled by the dipoles. As well

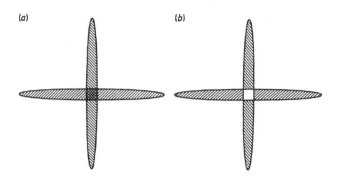

Figure 1.2.8. Beam patterns of a Mills cross radio telescope (*a*) with the beams added together (i.e. in phase); (*b*) with the beams subtracted from each other (i.e. 180° out of phase).

as being cumbersome to operate, the Mills cross suffers from the disadvantage that it can only operate at a single wavelength unless its dipoles are all changed. Furthermore, confusion of sources may arise if a strong source is in one of the pencil beams formed by the side lobes, since it will have the same appearance as a weak source in the main beam. An alternative and related system uses a single fan beam and observes the object at many position angles. The structure of the source is then retrievable from the observations in a relatively unambiguous manner.

The phasing of two dipoles, as we have seen, can be used to alter the angle of the beam to their axis. By continuously varying the delay the lobe can be swept across the sky, or accurately positioned on an object and then moved to track that object. This is an important technique for use with interferometers (section 2.5), and for solar work. It also forms the basis of a phased array. This latter instrument is basically a Mills cross in which the number of dipoles has been increased until the aperture is filled. It provides great sensitivity since a large area may be covered at a relatively low cost.

Great improvements in resolution and sensitivity of radio telescopes may be obtained through the use of interferometers and aperture synthesis, and these devices and their associated receivers and detectors are discussed in detail in section 2.5.

Construction

The large dishes that we have been discussing pose major problems in their construction. Both the gravitational and wind loads on the structure can be very large, and shadowing of parts of the structure can lead to inhomogenous heating and hence expansion and contraction-induced stresses.

The worst problem is due to wind, since its effect is highly variable. The force can be very large—1.5×10^6 N (150 tonnes) for a 50 m dish facing directly into a gale-force wind for example. A rough rule of thumb to allow scaling the degree of wind distortion between dishes of different sizes, is that the deflection, Δe, is given by

$$\Delta e \propto D^3 / A \qquad (1.2.15)$$

where D is the diameter of the dish and A is the cross-sectional area of its supporting struts. Thus, doubling the diameter of the design of a dish would require the supporting members' sizes to be tripled if it were to still work at the same wavelength. There are only two solutions to the wind problem; to enclose the dish, or to cease using it when the wind load becomes too great. Some smaller dishes, especially those working at short wavelengths where the physical constraints on the surface accuracy are most stringent, are enclosed in radomes, or space-enclosing structures built from non-conducting materials. But this is not usually practicable for the larger dishes. These must generally cease operating and be parked in their least wind-resistant mode once the windspeed rises above ten to fifteen metres per second.

The effects of gravity are easier to counteract. The problem arises from the varying directions and magnitudes of the loads placed upon the reflecting surface as the orientation of the telescope changes. Three approaches have been tried successfully for combating the effects of gravity. The first is the brute force approach, whereby the dish is made so rigid that its deformations remain within the surface accuracy limits. Except for very small dishes, this is an impossibly expensive option. The second approach is to compensate for the changing gravity loads as the telescope moves. The surface may be kept in adjustment by systems of weights, levers, guy ropes, springs etc or more recently by computer-controlled hydraulic jacks. The final approach is much more subtle and is termed the homological transformation system. The dish is allowed to deform, but its supports are designed so that its new shape is still a paraboloid, although in general an altered one, when the telescope is in its new position. The only active intervention which may be required is to move the feed antenna to keep it at the foci of the changing paraboloids.

There is little that can be done about inhomogeneous heating, other than painting all the surfaces white so that the absorption of the heat is minimised. Fortunately it is not usually a serious problem in comparison with the first two.

The supporting framework of the dish is generally a complex, cross-braced skeletal structure, whose optimum design requires a computer for its calculation. This framework is then usually placed onto an alt–az mounting (section 1.1), since this is generally the cheapest and simplest structure to build, and also because it restricts the gravitational load variations to a single plane, and so makes their compensation much easier. A few, usually quite small, radio telescopes do have equatorial mountings, however. There is now no difficulty in driving an alt–az mounting so that the telescope tracks an object across the sky, since a small microprocessor can calculate the required motor speeds for the drives in both axes in real time without any difficulty. The first large dishes, however, were built before the advent of these small, cheap computers, and they used analogue computers, which were essentially small equatorially mounted and driven telescopes, to provide the data for setting their axes and for tracking.

Exercises

1.2.1 Show that the HPBW of a collinear array with n dipoles is given by

$$\text{HPBW} \simeq 2\sin^{-1}\left(\frac{6n^2 - 3}{2n^4\pi^2 - \pi^2}\right)^{1/2}$$

when n is large.

1.2.2 Calculate the dimensions of a Mills cross to observe at a wavelength of 0.3 m with a Rayleigh resolution of $0.25°$.

1.2.3 Show that the maximum number of separable sources using the Rayleigh criterion, for a 60 m dish working at a wavelength of 0.1 m, is about 3.8×10^5 over a complete hemisphere.

1.3 X-RAY AND GAMMA-RAY DETECTION

Introduction

The electromagnetic spectrum comprises radiation of an infinite range of wavelengths, but the intrinsic nature of the radiation is unvarying. There is, however, a tendency to regard the differing wavelength regions from somewhat parochial points of view, and this tends to obscure the underlying unity of the processes that may be involved. The reasons for these attitudes are many; some are historical hang-overs from the ways in which the detectors for the various regions were developed, others are more fundamental in that different physical mechanisms predominate in the radiative interactions at different wavelengths. Thus high-energy gamma rays may interact directly with nuclei, at longer wavelengths we have resonant interactions with atoms and molecules producing electronic, vibrational and rotational transitions, while in the radio region currents are induced directly into conductors. But primarily the reason may be traced to the academic backgrounds of the workers involved in each of the regions. Because of the earlier reasons, workers involved in investigations in one spectral region will tend to have different bias in their backgrounds compared with workers in a different spectral region. Thus there will be different traditions, approaches, systems of notation, etc with a consequent tendency to isolationism and unnecessary failures in communication. The discussion of the point source image in terms of the Airy disc and its fringes by optical astronomers and in terms of the polar diagram by radio astronomers as already mentioned is one good example of this process, and many more exist. It is impossible to break out from the straitjacket of tradition completely, but an attempt to move towards a more unified approach has been made in this work by dividing the spectrum much more broadly than is the normal case. We only consider three separate spectral regions, within each of which the detection techniques bear at least a familial resemblance to each other. The overlap regions are fairly diffuse with some of the techniques from each of the major regions being applicable. We have already discussed two of the major regions, radio and microwaves, and optical and infrared. The third region, x-rays and gamma rays, is the high-energy end of the spectrum and is the most recent area to be explored. This region also overlaps to a very considerable extent, in the nature of its detection techniques, with the cosmic rays, which are discussed in the next section. None of the radiation discussed in this section penetrates down to ground level, so its study had to await the availability of observing platforms in space, or near the top of the Earth's atmosphere. Thus significant work on these photons has only been possible during the last three decades or so, and many of the detectors and observing techniques are still under development. This review will therefore especially need supplementing by more up-to-date texts for serious students, to cover any significant developments between the writing and publishing of this book.

The high-energy spectrum is fairly arbitrarily divided into:

- The extreme ultraviolet (EUV or XUV region): 10 to 100 nm wavelengths (12 to 120 eV photon energies)
- Soft x-rays: 1 to 10 nm (120 to 1200 eV)
- x-rays: 0.01 to 1 nm (1.2 to 120 keV)
- Soft gamma rays: 0.001 to 0.01 nm (120 to 1200 keV)
- Gamma rays: less than 0.001 nm (greater than 1.2 MeV).

We shall be primarily concerned with the last four regions in this section, i.e. with wavelengths less than 10 nm and photon energies greater than 120 eV. (Note that the electron volt, eV, is 1.6×10^{-19} J and is a convenient unit for use in this spectral region and also when discussing cosmic rays in the next section.) The main production mechanisms for high-energy radiation include electron synchrotron radiation, the inverse Compton effect, free–free radiation, and pion decay, while the sources include the Sun, supernova remnants, pulsars, bursters, binary systems, cosmic rays, the intergalactic medium, galaxies, Seyfert galaxies, and quasars. Absorption of the radiation can be by ionisation with a fluorescence photon or an Auger electron produced in addition to the ion and electron, by Compton scattering, or in the presence of matter, by pair production. This latter process is the production of a particle and its anti-particle, and not the pair production process discussed in section 1.1 which was simply the excitation of an electron from the valence band. The interstellar absorption in this spectral region varies roughly with the cube of the wavelength, so that the higher energy radiation can easily pass through the whole galaxy with little chance of being intercepted. At energies under about 2 keV, direct absorption by the heavier atoms and ions can be an important process. The flux of the radiation varies enormously with wavelength. The solar emission alone at the lower energies is sufficient to produce the ionosphere and thermosphere on the Earth. At 1 nm wavelength for example, the solar flux is 5×10^9 photons m^{-2} s^{-1}, while the total flux from all sources for energies above 10^9 eV is only a few photons per square metre per day.

Detectors

Geiger counters

The earliest detection of high energy radiation from a source other than the Sun took place in 1962 when soft x-rays from a source which later became known as Sco X-1 were detected by large area Geiger counters flown on a sounding rocket. Geiger counters and the related proportional counters, which are collectively known as gas-filled ionisation detectors, are still among the most frequently used detectors for high-energy radiation.

The principle of the Geiger counter is well known. Two electrodes inside an enclosure are held at such a potential difference that a discharge in the medium filling the enclosure is on the point of occurring. The entry of ionising radiation triggers this discharge, resulting in a pulse of current between the electrodes which may then be amplified and detected. The electrodes are usually arranged as the outer wall of the enclosure containing the gas and as a central coaxial wire (figure 1.3.1). The medium inside the tube is typically argon at a low pressure with a small amount of an organic gas, such as alcohol vapour, added. The electrons produced in the initial ionisation are accelerated towards the central electrode by the applied potential; as these electrons gain energy they cause further ionisation, producing more electrons, which in turn are accelerated towards the central electrode, and so on. The amplification factor can be as high as 10^8 electrons arriving at the central electrode for every one in the initial ionisation trail. The avalanche of electrons rapidly saturates, so that the detected

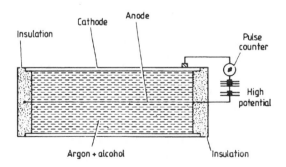

Figure 1.3.1. A typical arrangement for a Geiger counter.

pulse is independent of the original energy of the photon. This is a serious disadvantage if the device is used as the primary detector and not as a trigger for a more sophisticated detector such as a cloud chamber (section 1.4), and so Geiger counters have been replaced by proportional counters (see below) for this purpose. Another disadvantage of Geiger counters which also applies to many of the other detectors discussed in this and the following section, is that a response to one event leaves the detector inoperative for a short interval afterwards, which is known as the dead time. In the Geiger counter the cause of the dead time is that a discharge lowers the potential between the electrodes, so that it is momentarily insufficient to cause a second avalanche of electrons should another x-ray enter the device. The length of the dead time is typically 200 μs.

Proportional counters

These devices are very closely related to Geiger counters and are in effect Geiger counters operated at less than the trigger voltage. By using a lower voltage, saturation of the pulse is avoided, and its strength is then proportional to the energy of the original interaction. The gain of the system operated in this way is reduced to about 10^4 or 10^5, but it is still sufficient for further detection by conventional amplifiers etc. Provided that all the energy of the ionising radiation is absorbed within the detector, its original total energy may be obtained from the strength of the pulse, and we have a proportional counter. At low photon energies, a window must be provided for the radiation. These are typically made from thin sheets of plastic, mica, or beryllium, and absorption in the windows limits the detectors to energies above a few hundred electron volts. When a window is used, the gas in the detector has to be continuously replenished because of losses by diffusion through the window. At high photon energies the detector is limited by the requirement that all the energy of the radiation be absorbed within the detector's enclosure. To this end, proportional counters for high energy detection may have to be made very large. About 30 electron volts on average are required to produce one ion–electron pair, so that a 1 keV photon produces about 36 electrons, and a 10 keV photon about 360 electrons. This spectral energy resolution to two and a half standard deviations from the resulting statistical fluctuations of the electron numbers is thus about 40% at 1 keV and 12% at 10 keV. The quantum efficiencies of proportional counters approach 100% for energies up to 50 keV.

The position of the interaction of the x-ray along the axis of the counter may be obtained through the use of a resistive anode. The pulse is abstracted from both ends of the anode and a comparison of its strength and shape from the two ends then leads to a position of the discharge along the anode. The concept may easily be extended to a two-dimensional grid of anodes to allow genuine imaging. Spatial resolutions of about a tenth of a millimetre are possible. In this form the detector is called a position-sensitive proportional counter.

Many gases can be used to fill the detector; argon, methane, xenon, carbon dioxide, and mixtures thereof at pressures near that of the atmosphere, are among the commonest ones. The inert gases are to be preferred since there is then no possibility of the loss of energy into the rotation or vibration of the molecules.

Scintillation detectors

The ionising photons do not necessarily knock out only the outermost electrons from the atom or molecule with which they interact. Electrons in lower energy levels may also be removed. When this happens, a 'hole' is left behind into which one of the higher electrons may drop, with a consequent emission of radiation. Should the medium be transparent to this radiation, the photons may be observed, and the medium may be used as an x-ray detector. Each interaction produces a flash or scintilla of light, from which the name of the device is obtained. There are many materials which are suitable for this application. Commonly used ones include sodium iodide doped with an impurity such as thallium, and caesium iodide doped with sodium or thallium. For these materials the light flashes are detected by a photomultiplier (figure 1.3.2). There is no dead time for a scintillation detector, and the strength of the flash depends somewhat upon the original photon energy, so that some spectral resolution is possible. The noise level, however, is quite high since only about 3% of the x-ray's energy is converted into detectable radiation, with a consequent increase in the

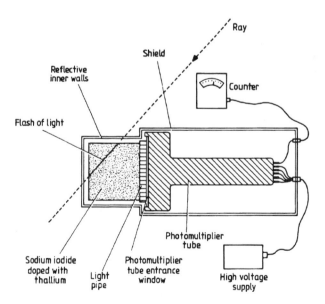

Figure 1.3.2. Schematic experimental arrangement of a scintillation counter.

statistical fluctuations in their numbers. The spectral resolution is thus about 6% at 1 MeV. Sodium iodide or caesium iodide are useful for x-ray energies up to several hundred keV, organic scintillators such as stilbene ($C_{14}H_{14}N_2$) and BGO ($Bi_4Ge_3O_{12}$) can be used for energies up to ten or more MeV.

Discrimination of the x-ray's arrival direction can be obtained by using sodium iodide and caesium iodide in two superimposed layers. The decay time of the pulses differs between the two compounds so that they may be separately identified, and the direction of travel of the photon inferred. This arrangement for a scintillation detector is frequently called a 'phoswich' detector. Several gases such as argon, xenon, nitrogen and their mixtures can also be used as scintillators, and combined with an optical system to produce another imaging device.

X-rays and cosmic rays (section 1.4) may be distinguished by the differences in their resulting pulses shapes. X-rays are rapidly absorbed, and their pulses are sharp and brief. Cosmic ray particles will generally have a much longer path and so their pulses will be comparatively broader and smoother than the x-ray pulse.

Gas scintillation proportional counters

A combination of the above two types of detector leads to a significant improvement in the low-energy spectral resolution. Resolutions as high as 8% at 6 keV have been achieved in practice with these devices. The x-radiation produces ion–electron pairs in an argon or xenon filled chamber. The electrons are then gently accelerated until they cause scintillations of their own in the gas. These scintillations can then be observed by a conventional scintillation counter system. These have been favoured detectors for launch on several of the more recent x-ray satellites because of their good spectral and positional discrimination, though in some cases their lifetimes in orbit have been rather short.

Charge coupled devices

The details of the operating principles for CCDs for optical detection are to be found in section 1.1. They are also, however, becoming increasingly widely used as primary detectors at EUV and x-ray wavelengths. CCDs become insensitive to radiation in the blue and ultraviolet parts of the spectrum because of absorption in the electrode structure on their surfaces. They regain sensitivity, however, at shorter wavelengths as the radiation is again able to penetrate that structure ($\lambda < 10$ nm or so). As with optical CCDs, the efficiency of the devices may be improved by using a very thin electrode structure, an electrode structure that only partially covers the surface (virtual phase CCDs), or by illuminating the device from the back. The reader is referred to section 1.1 for further details of CCDs.

Compton interaction detectors

Very-high-energy photons produce electrons in a scintillator through the Compton effect, and these electrons can have sufficiently high energies to produce scintillations of their own. Two such detectors separated by a metre or so can provide directional discrimination when used in conjunction with pulse analysers and time-of-flight measurements to eliminate other unwanted interactions.

Spark detectors

These are discussed in more detail in section 1.4 (figure 1.4.3). They are used for the detection of gamma rays of energies above 20 MeV, when the sparks follow the ionisation trail left by the photon. Alternatively a layer of tungsten or some similar material may be placed in the spark chamber. The gamma ray then interacts in the tungsten to produce an electron–positron pair, and these in turn can be detected by their trails in the spark chamber. Angular resolutions of 2°, energy resolutions of 50%, and detection limits of 10^{-3} photons m^{-2} s^{-1} are currently achievable.

Čerenkov detectors

Čerenkov radiation and the resulting detectors are referred to in detail in section 1.4. For x- and gamma radiation their interest lies in the detection of particles produced by the Compton interactions of the very-high-energy photons. If those particles have a velocity higher than the velocity of light in the local medium then Čerenkov radiation is produced. Čerenkov radiation produced in this manner high in the Earth's atmosphere may be detectable, and may form a minor noise source for the intensity interferometer (section 2.5). Such very-high-energy primary gamma rays (10^{15} eV and more) may also produce cosmic ray showers and so be detectable by the methods discussed in section 1.4.

Solid state detectors

There are a number of different varieties of these detectors including CCDs (see above and section 1.1), so that suitable ones may be found for use throughout most of the x- and gamma-ray region.

Germanium is used as a sort of solid proportional counter. A cylinder of germanium cooled by liquid nitrogen is surrounded by a cylindrical cathode and has a central anode (figure 1.3.3). A gamma ray scatters off electrons in the atoms until its energy has been consumed in ion—electron pair production. The number of released electrons is proportional to the energy of the gamma ray, and these are attracted to the anode where they may be detected. The spectral resolution is very high—0.2% at 1 MeV—so that detectors of this type are especially suitable for gamma-ray line spectroscopy. Other materials which

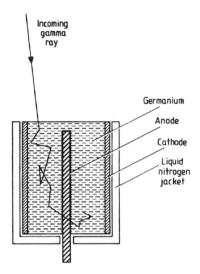

Figure 1.3.3. Germanium gamma-ray detector.

may replace the germanium include germanium doped with lithium, cadmium telluride and mercury-iodine.

At lower energies (0.4 to 4 keV) silicon based solid state detectors of the type discussed in section 1.4 may be used. Their energy resolution ranges from 4 to 30%.

Another type of detector altogether is a possibility for the future. This is a solid analogue of the cloud or bubble chamber (section 1.4). A superconducting solid would be held at a temperature fractionally higher than its critical temperature, or in a non-superconducting state and at a temperature slightly below the critical value. Passage of an ionising ray or particle would then precipitate the changeover to the opposing state, and this in turn could be detected by the change in a magnetic field.

Microchannel plates

For EUV and low-energy x-ray amplification and imaging, there is an ingenious variant of the photomultiplier (section 1.1). A thin plate is pierced by numerous tiny holes, each perhaps only about 25 μm across. Its top surface is an electrode with a negative potential of some few thousand volts with respect to its base. The top is also coated with a photoelectron emitter for the x-ray energies of interest. An impinging photon releases one or more electrons which are then accelerated down the tubes. There are inevitably collisions with the walls of the tube during which further electrons are released, and these in turn are accelerated down the tube and so on (figure 1.3.4). As many as 10^4 electrons can be produced for a

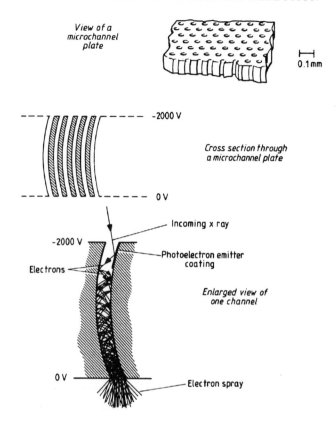

View of a
microchannel
plate

0.1mm

−2000 V

*Cross section through
a microchannel plate*

Incoming x ray

−2000 V

Photoelectron emitter
coating

Electrons

*Enlarged view of
one channel*

0 V

Electron spray

Figure 1.3.4. Schematic view of the operation of a microchannel plate.

single photon, and this may be increased to 10^6 electrons in future devices when the problems caused by ion feedback are reduced. The quantum efficiency can be up to 20%. The electrons spray out of the bottom of each tube, where they may be detected by a variety of the more conventional imaging systems, or they may be fed into a second microchannel plate for further amplification.

Nuclear emulsions

Although not normally used for high energy photon detection, the photographic emulsions used to detect the tracks of nuclear particles (sections 1.4 and 2.2) will respond to them. More correctly they pick up the tracks of the ejected electrons (or δ rays) produced by the passage of the photon. Direct interactions between the photon and a nucleus, or pair production will also show up if they occur in the emulsion.

Shielding

Very few of the detectors which we have just reviewed are used in isolation. Usually several will be used in modes which allow the rejection of information on unwanted interactions. This is known as active shielding of the detector. The range of possible configurations is very wide, and a few examples will suffice to give an indication of those in current use.

(1) The germanium solid state detectors must intercept all the energy of an incoming photon if they are to provide a reliable estimate of its magnitude. If the photon escapes, then the measured energy will be too low. The germanium is therefore surrounded by a thick layer of a scintillation crystal. A detection in the germanium simultaneously with two in the scintillation crystal is rejected as an escapee.

(2) The solid angle viewed by a germanium solid state detector can be limited by placing the germanium at the bottom of a hole in a scintillation crystal. Only those detections *not* occurring simultaneously in the germanium and the crystal are used, and these are the photons which have entered down the hole. Any required degree of angular resolution can be obtained by sufficiently reducing the size of the entrance aperture.

(3) Spark counters are surrounded by scintillation counters and Čerenkov detectors to discriminate between gamma-ray events and those due to cosmic rays.

(4) A sodium iodide scintillation counter may be surrounded, except for an entrance aperture, by one formed from caesium iodide to eliminate photons from the unwanted directions.

Recently, nuclear power sources on board other spacecraft have caused interference with γ-ray observations. These pulses have to be separated from the naturally occurring ones by their spectral signatures.

Passive shields are also used, and these are just layers of an absorbing material which screen out unwanted rays. They are especially necessary for the higher energy photon detectors in order to reduce the much higher flux of lower energy photons. The mass involved in an adequate passive shield, however, is often a major problem for satellite-based experiments with their very tight mass-budgets due to their launcher capabilities.

Imaging

Imaging of high-energy photons is a difficult task because of their extremely penetrating nature. Normal designs of telescope are impossible since in a reflector the photon would just pass straight through the mirror, while in a refractor it would be scattered or unaffected rather than refracted by the lenses. At energies under a few keV, forms of reflecting telescope can be made which work adequately, but at higher energies the only options are occultation, collimation and coincidence detection.

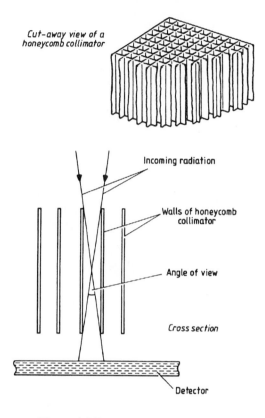

Figure 1.3.5. Honeycomb collimator.

Collimation

A collimator is simply any device which physically restricts the field of view of the detector without contributing any further to the formation of an image. The image is obtained by scanning the system across the object. Almost all detectors are collimated to some degree, if only by their shielding. The simplest arrangement is a series of baffles (figure 1.3.5) which may be formed into a variety of configurations, but all of which follow the same basic principle. Their general appearance leads to their name, honeycomb collimator, even though the cells are usually square rather than hexagonal. They can restrict angles of view down to a few minutes of arc, but they are limited at low energies by reflection off their walls and at high energies by penetration of the radiation through the walls. At high energies, the baffles may be formed from a crystal scintillator and pulses therefrom used to reject detections of radiation from high inclinations (cf active shielding). At the low energies the glancing reflection of the radiation can be used to advantage, and a more truly imaging collimator

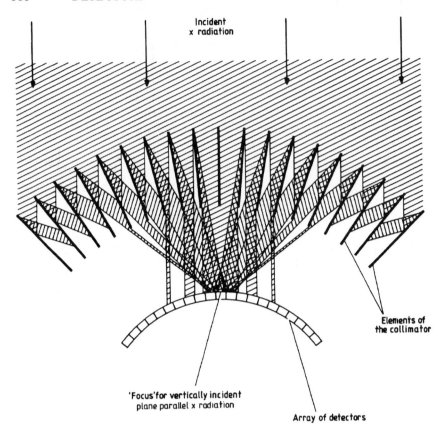

Figure 1.3.6. Cross section through a 'lobster eye' focusing wide-angle x-ray collimator.

produced. This is called a 'lobster eye' focusing collimator, and is essentially a honeycomb collimator curved into a portion of a sphere (figure 1.3.6), with a position-sensitive detector at its focal surface. The imaging is not of very high quality, and there is a high background from the unreflected rays and the doubly reflected rays (the latter not shown in figure 1.3.6). But it is a cheap system to construct compared with the others which are discussed later in this section, and it has the potential for covering very wide fields of view (tens of degrees) at very high resolutions (seconds of arc). In practice the device is constructed by fusing together thousands of tiny glass rods in a matrix of a different glass. The composite is then heated and shaped and the glass rods etched away to leave the required hollow tubes in the matrix glass.

Another system which is known as a modulation collimator or Fourier transform telescope, uses two or more parallel gratings which are separated by a short distance (figure 1.3.7). Since the bars of the gratings alternately obscure

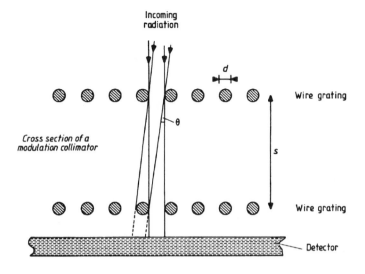

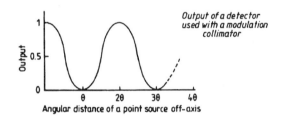

Figure 1.3.7. A modulation collimator.

the radiation and allow it to pass through, the output as the system scans a point source is a sine wave (figure 1.3.7). The resolution is given by the angle α

$$\alpha = d/s. \qquad (1.3.1)$$

To obtain unambiguous positions for the sources, or for the study of multiple or extended sources, several such gratings of different resolutions are combined. The image may then be retrieved from the Fourier components of the output (cf aperture synthesis, section 2.5). Two such grating systems at right angles can give a two-dimensional image. With these systems resolutions of a few tens of seconds of arc can be realised even for the high-energy photons.

A third type of collimating imaging system is a simple pinhole camera. A position-sensitive detector such as a resistive anode proportional counter, or a microchannel plate (see p. 114) is placed behind a small aperture. The quality of the image is then just a function of the size of the hole and its distance in front of the detector. Unfortunately, the small size of the hole also gives a low flux of radiation so that such systems are limited to the brightest objects. A better system replaces the pinhole with a mask formed from clear and opaque regions. The pattern of the mask is known, so that when sources cast shadows of it onto the detector, their position and structure can be reconstituted in a similar manner to that used for the modulation collimator. The technique is known as coded mask imaging, and resolutions of ten minutes of arc or better can be achieved. Since only half the aperture is obscured, the technique obviously uses the incoming radiation much more efficiently than the pinhole camera. A somewhat related technique known as Hadamard mask imaging is discussed in section 2.4. The disadvantage of coded mask imaging is that the detector must be nearly the same size as the mask. Only proportional counters are suitable as detectors therefore, but these are difficult to make in large sizes and can be unreliable. The image is extracted from the data by a cross correlation between the detected pattern and the mask pattern.

Coincidence detectors

A telescope, in the sense of a device which has directional sensitivity, may be constructed for use at any energy, and with any resolution, by using two or more detectors in a line, and by rejecting all detections except those which occur in both detectors and separated by the correct flight time (see figure 1.4.2 for example). Two separated arrays of detectors can similarly provide a two-dimensional imaging system.

Occultation

Although it is not a technique which can be used at will on any source, the occultation of a source by the Moon or other object can be used to give very precise positional and structural information. The use of occultations in this manner is discussed in detail in section 2.7. The technique is less important now than in the past since other imaging systems are available, but it was used in 1964, for example, to provide the first indication that the x-ray source associated with the Crab nebula was an extended source.

Reflecting telescopes

At energies below one or two keV, photons may be reflected with about 50% efficiency off metal surfaces, when their angle of incidence approaches 90°. Mirrors in which the radiation just grazes the surface (grazing incidence optics) may therefore be built and configured to form reflecting telescopes. Several systems have been devised, but the one which has achieved most practical use is formed from a combination of annular sections of very deep paraboloidal and

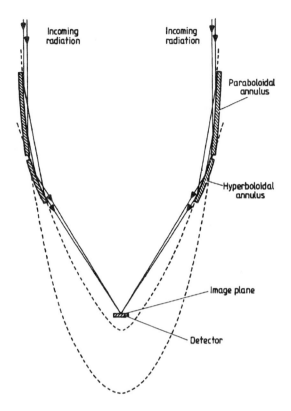

Figure 1.3.8. Cross section through a grazing incidence x-ray telescope.

hyperboloidal surfaces (figure 1.3.8), and known as a Wolter type I telescope after its inventor. Other designs which are also due to Wolter are shown in figure 1.3.9, and have also been used in practice, although less often than the first design. A simple paraboloid can also be used, though the resulting telescope then tends to be excessively long. The aperture of such telescopes is a thin ring, since only the radiation incident onto the paraboloidal annulus is brought to the focus. To increase the effective aperture, and hence the sensitivity of the system, several such confocal systems of differing radii may be nested inside each other (figure 1.3.10). A position-sensitive detector at the focal plane can then produce images with resolutions as high as a second of arc. The limit of resolution is due to surface irregularities in the mirrors, rather than the diffraction limit of the system. The irregularities are about 0.3 nm in size for the very best of the current production techniques, i.e. comparable with the wavelength of photons of about 1 keV. The mirrors are produced by electro-deposition of nickel onto a mandrel of the inverse shape to that required for the mirror, the mirror shell then being separated from the mandrel by cooling. At energies of a few keV,

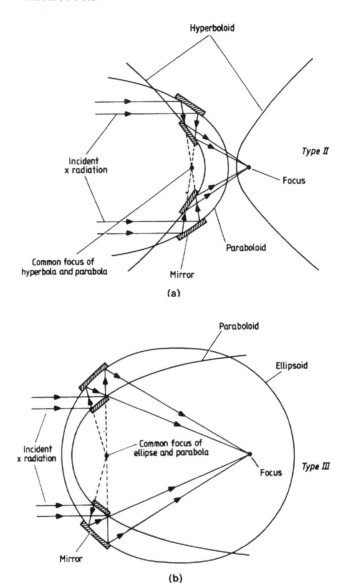

Figure 1.3.9. Cross sections through alternative designs for grazing incidence x-ray telescopes.

glancing incidence telescopes with low but usable resolutions can be made using foil mirrors. The incident angle is less than 1° so a hundred or more mirrors are needed. They may be formed from thin aluminium foil with a lacquer coating to

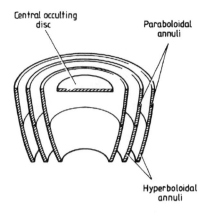

Figure 1.3.10. Section through a nested grazing incidence x-ray telescope.

provide the reflecting surface. The angular resolution of around 1′ means that the mirror shapes can be simple cones, rather than the paraboloids and hyperboloids of a true Wolter telescope, and so fabrication costs are much reduced.

An alternative glancing incidence system which has fewer production difficulties is based upon cylindrical mirrors. Its collecting efficiency is higher but its angular resolution poorer than the three-dimensional systems discussed above. Two mirrors are used which are orthogonal to each other (figure 1.3.11). The surfaces can have a variety of shapes, but two paraboloids are probably the commonest arrangement. The mirrors can again be stacked in multiples of the

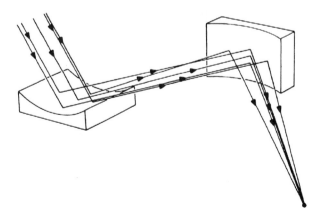

Figure 1.3.11. X-ray imaging by a pair of orthogonal cylindrical mirrors.

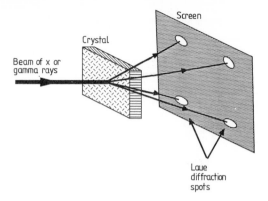

Figure 1.3.12. Laue diffraction of x- and gamma rays.

basic configuration to increase the collecting area.

At lower energies, in the EUV and soft x-ray regions, near normal incidence reflection with efficiencies up to 20% is possible using multilayer coatings. These are formed from tens, hundreds or even thousands of alternate layers

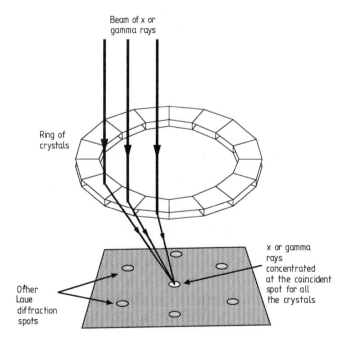

Figure 1.3.13. A Laue diffraction lens for x- or gamma rays (only three crystals shown illuminated for clarity).

of (for example) tungsten and carbon, aluminium and gold or magnesium and gold, each about 1 nm thick. The reflection is essentially monochromatic, with the wavelength depending upon the orientation of the crystalline structure of the layers, and upon their thickness. Recently direct images of the Sun at wavelengths down to 4 nm have been obtained using such mirrors.

At energies of tens or hundreds of keV, Bragg reflection (see p. 127) and Laue diffraction can be used to concentrate the radiation and to provide some limited imaging. For example, the Laue diffraction pattern of a crystal comprises a number of spots into which the incoming radiation has been concentrated (figure 1.3.12). With some crystals, such as germanium, only a few spots are produced. A number of crystals can be mutually aligned so that one of the spots from each crystal is directed towards the same point (figure 1.3.13), resulting in a crude type of lens. With careful design, efficiencies of 25% to 30% can be achieved at some wavelengths. As with the glancing incidence telescopes, several such ring lenses can be nested to increase the effective area of the telescope. A position-sensitive detector such as an array of germanium crystals (figure 1.3.3) can then provide direct imaging at wavelengths down to 0.001 nm (1 MeV).

A completely different approach to imaging gamma ray sources is based upon cloud or bubble chambers (section 1.4). In these devices the tracks of the gamma rays are seen directly and can be followed back to their point of origin in the sky. Liquid xenon seems likely to be particularly useful at MeV energies since it has up to 50% detection efficiency and can provide information upon the spectral distribution and the polarization properties of the photons.

Resolution and image identification

At the low-energy end of the x-ray spectrum, resolutions and positional accuracies comparable with those of Earth-based optical systems are possible, as we have seen. There is therefore usually little difficulty in identifying the optical counterpart of an x-ray source, should it be bright enough to be visible. The resolution rapidly worsens, however, as higher energies are approached. The position of a source which has been detected can therefore only be given to within quite broad limits. It is customary to present this uncertainty in the position of a source by specifying an error box. This is an area of the sky, usually rectangular in shape, within which there is a certain specified probability of finding the source. For there to be a reasonable certainty of being able to find the optical counterpart of a source unambiguously, the error box must be smaller than about one square minute of arc. Since high energy photon imaging systems have resolutions measured in tens of minutes of arc or larger, such identification is not usually possible. The exception to this occurs when an 'unusual' optical object lies within the error box. This object may then, rather riskily, be presumed also to be the x-ray source. Examples of this are the Crab nebula and Vela supernova remnant which lie in the same direction as two strong

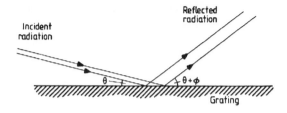

Figure 1.3.14. Optical paths in a grazing incidence reflection grating.

sources at 100 MeV and are therefore assumed to be those sources, even though the uncertainties are measured in many degrees.

Spectroscopy

Many of the detectors which were discussed earlier are intrinsically capable of separating photons of differing energies. At photon energies above about 10 keV, it is only this inherent spectral resolution which can provide information on the energy spectrum, although some additional resolution may be gained by using several similar detectors with varying degrees of passive shielding—the lower energy photons will not penetrate through to the most highly shielded detectors. At low and medium energies, the gaps between the absorption edges of the materials used for the windows of the detectors can form wideband filters. Devices akin to the more conventional idea of a spectroscope, however, can only be used at the low end of the energy spectrum and these are discussed below.

Grating spectrometers

Gratings may be either transmission or grazing incidence reflection. The former may be ruled gratings in a thin metal film which has been deposited onto a substrate transparent to the x-rays, or they may be formed on a pre-ruled substrate by vacuum deposition of a metal from a low angle. The shadow regions where no metal is deposited then form the transmission slits of the grating. The theoretical background for x-ray gratings is identical with that for optical gratings and is discussed in detail in section 4.1. Typical transmission gratings have around one thousand lines per millimetre. The theoretical resolution (section 4.1) is between 10^3 and 10^4, but is generally limited in practice to 50 to 100 by other aberrations.

Reflection gratings are also similar in design to their optical counterparts. Their dispersion differs, however, because of the grazing incidence of the radiation. If the separation of the rulings is d then, from figure 1.3.14, we may easily see that the path difference, ΔP, of two rays which are incident onto adjacent rulings is

$$\Delta P = d[\cos\theta - \cos(\theta + \varphi)]. \qquad (1.3.2)$$

We may expand this via the Taylor series and neglect powers of θ and φ higher than two since they are small angles, to obtain

$$\Delta P = \tfrac{1}{2}d(\varphi^2 - 2\theta\varphi). \tag{1.3.3}$$

In the mth order spectrum, constructive interference occurs for radiation of wavelength λ if

$$m\lambda = \Delta P \tag{1.3.4}$$

so that

$$\varphi = \left(\frac{2m\lambda}{d} + \theta^2\right)^{1/2} - \theta \tag{1.3.5}$$

and

$$\frac{d\varphi}{d\lambda} = \left(\frac{m}{2d\lambda}\right)^{1/2} \tag{1.3.6}$$

where we have neglected θ^2, since θ is small. The dispersion for a glancing incidence reflection grating is therefore inversely proportional to the square root of the wavelength, unlike the case for near normal incidence, when the dispersion is independent of wavelength. The gratings may be plane or curved, and may be incorporated into many different designs of spectrometer (section 4.2). Resolutions of up to 10^3 are possible for soft x-rays, but again this tends to be reduced by the effects of other aberrations. Detection may be by scanning the spectrum over a detector (or vice versa), or by placing a position-sensitive detector in the image plane so that the whole spectrum is detected in one go.

Bragg spectrometers

Background
The planes of atoms in a crystal are separated by distances ranging from 0.1 to 10 nanometres. This is comparable with the wavelengths of x-rays, and so a beam of x-rays interacts with a crystal in a complex manner. The details of the interaction were first explained by the Braggs (father and son). Typical radiation paths are shown in figure 1.3.15. The path differences for rays such as a, b and c, or d and e, are given by multiples of the path difference, ΔP, for two adjacent layers

$$\Delta P = 2d \sin\theta. \tag{1.3.7}$$

There will be constructive interference for path differences which are whole numbers of wavelengths. So that the reflected beam will consist of just those wavelengths, λ, for which this is true

$$M\lambda = 2d \sin\theta. \tag{1.3.8}$$

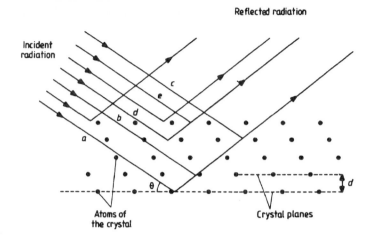

Figure 1.3.15. Bragg reflection.

The spectrometer

The Bragg spectrometer uses a crystal to produce monochromatic radiation of known wavelength. If a crystal is illuminated by a beam of x-rays of mixed wavelengths, at an approach angle, θ, then only those x-rays whose wavelength is given by equation (1.3.8) will be reflected. The first-order reflection ($m = 1$) is by far the strongest and so the reflected beam will be effectively monochromatic with a wavelength of

$$\lambda_\theta = 2d \sin \theta. \qquad (1.3.9)$$

The intensity of the radiation at that wavelength may then be detected with, say, a proportional counter, and the whole spectrum scanned by tilting the crystal to alter θ (figure 1.3.16). An improved version of the instrument uses a bent crystal and a collimated beam of x-rays so that the approach angle varies over the crystal. The reflected beam then consists of a spectrum of all the wavelengths (figure 1.3.17), and this may then be detected by a single observation using a position-sensitive detector. The latter system has the major advantage for satellite-borne instrumentation of having no moving parts, and so significantly higher reliability, and it also has good time resolution. High spectral resolutions are possible—up to 10^3 at 1 keV—but large crystal areas are necessary for good sensitivity, and this may present practical difficulties on a satellite. Many crystals may be used. Among the commonest are lithium fluoride, lithium hydride, tungsten disulphide, graphite, and potassium acid phthalate (KAP).

Many variants upon the basic spectrometer can be devised. It may be used as a monochromator and combined with a scanning telescope to produce spectroheliograms, for faint sources it may be adapted for use at the focus of a telescope and so on. Designs are rapidly changing and the reader desiring completely up-to-date information must consult the current literature.

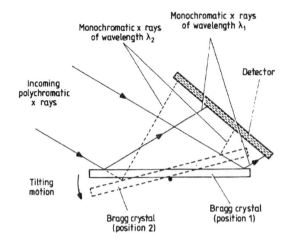

Figure 1.3.16. Scanning Bragg crystal x-ray spectrometer.

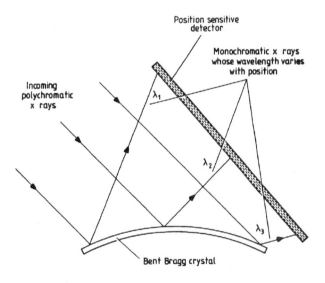

Figure 1.3.17. Bent Bragg crystal x-ray spectrometer.

Polarimetry

Bragg reflection of x-rays is polarisation dependent. For an angle of incidence of 45°, photons which are polarised perpendicularly to the plane containing the incident and reflected rays will be reflected, while those polarised in this plane will not. Thus a crystal and detector at 45° to the incoming radiation and which may be rotated around the optical axis will function as a polarimeter.

The efficiency of such a system would be very low, however, due to the narrow energy bandwidth of Bragg reflection. This is overcome by using many randomly orientated small crystals. The crystal size is too small to absorb radiation significantly. If it should be aligned at the Bragg angle for the particular wavelength concerned, however, that radiation will be reflected. The random orientation of the crystals ensures that overall, many wavelengths are reflected, and we have a polarimeter with a broad bandwidth.

A second type of polarimeter looks at the scattered radiation due to Thomson scattering in blocks of lithium or beryllium. If the beam is polarised, then the scattered radiation is asymmetrical, and this may be measured by surrounding the block with several pairs of detectors.

Observing platforms

The Earth's atmosphere completely absorbs x- and gamma-ray radiation, so that all the detectors and other apparatus discussed above, with the possible exception of the Čerenkov detectors, have to be lifted above at least 99% of the atmosphere. The three systems for so lifting equipment are balloons, rockets, and satellites. Satellites give the best results in terms of their height, stability, and the duration of the mission. Their cost is very high, however, and there are weight and space restrictions to be complied with; nonetheless there have been quite a number of satellites launched for exclusive observation of the x- and gamma-ray region, and appropriate detectors have been included on a great many other missions as secondary instrumentation. Balloons can carry much heavier equipment far more cheaply than satellites, but only to a height of about 40 km, and this is too low for many of the observational needs of this spectral region. Their mission duration is also comparatively short being a few days to a week for even the most sophisticated of the self-balancing versions. Complex arrangements have to be made for communication and for retrieval of the payload because of the unpredictable drift of the balloon during its mission. The platform upon which the instruments are mounted has to be actively stabilised to provide sufficient pointing accuracy for the telescopes etc. Thus there are many drawbacks to set against the lower cost of a balloon. Sounding rockets are even cheaper still, and they can reach heights of several hundred kilometres without difficulty. But their flight duration is measured only in minutes and the weight and size restrictions are far tighter even than for most satellites. Rockets still, however, find uses as rapid response systems for events such as solar flares. A balloon- or satellite-borne detector may not be available when the event occurs, or may take some time to bear onto the target, whereas a rocket may be held on stand-by on Earth, during which time the cost commitment is minimal, and then launched when required at only a few minutes notice.

1.4 COSMIC RAY DETECTORS

Background

Cosmic rays comprise two quite separate populations of particles, primary cosmic rays, and secondary cosmic rays. The former are the true cosmic rays, and consist mainly of atomic nuclei in the proportions 84% hydrogen, 14% helium, 1% other nuclei, and 1% electrons and positrons, moving at velocities from a few per cent of the speed of light to $0.999\,999\,999\,999\,999\,999\,9999c$. That is, the energies per particle range from 10^{-13} to 10 J. As previously mentioned (section 1.3), a convenient unit for measuring the energy of sub-nuclear particles is the electron volt. This is the energy gained by an electron in falling through a potential difference of one volt, and is about 1.602×10^{-19} J. Thus the primary cosmic rays have energies in the range 10^6 to 10^{20} eV, with the bulk of the particles having an energy near 10^9 eV, or a velocity of $0.9c$. These are obviously relativistic velocities for most of the particles, for which the relationship between kinetic energy and velocity is

$$v = \frac{(E^2 + 2Emc^2)^{1/2}c}{E + mc^2} \tag{1.4.1}$$

where E is the kinetic energy of the particle and m is the rest mass of the particle. The flux in space of the primary cosmic rays of all energies is about 10^4 particles $m^{-2}\,s^{-1}$.

Secondary cosmic rays are produced by the interaction of the primary cosmic rays with nuclei in the Earth's atmosphere. A primary cosmic ray has to travel through about $800\,kg\,m^{-2}$ of matter on average before it collides with a nucleus. A column of the Earth's atmosphere one square metre in cross section contains about 10^4 kg, so that the primary cosmic ray usually collides with an atmospheric nucleus at a height of 30 to 60 km. The interaction results in numerous fragments, nucleons, pions, muons, electrons, positrons, neutrinos, gamma rays etc and these in turn may still have sufficient energy to cause further interactions, producing more fragments, and so on. For high energy primaries ($> 10^{11}$ eV), a few such secondary particles will survive down to sea level and be observed as secondary cosmic rays. At higher energies ($> 10^{13}$ eV), large numbers of the secondary particles ($> 10^9$) survive to sea level, and a cosmic ray shower or extensive air shower is produced. At altitudes higher than sea level, the secondaries from the lower energy primary particles may also be found.

Detectors

The methods of detecting cosmic rays may be divided into:

(*a*) Real-time methods. These observe the particles instantaneously and produce information on their direction as well as their energy and composition. They are the standard detectors of nuclear physicists, and there is considerable overlap between these detectors and those discussed in section 1.3. They are generally used in conjunction with passive and/or active shielding so that directional and spectral information can also be obtained.

(b) Residual track methods. The path of a particle through a material may be found some time (hours to millions of years) after its passage.

(c) Indirect methods. Information on the flux of cosmic rays at large distances from the Earth, or at considerable times in the past may be obtained by studying the consequent effects of their presence.

Real-time methods

Cloud chamber

This is the standard cloud or bubble chamber of the nuclear physicist. It has numerous variants, but they are all based upon the same principle, i.e. initiation of droplet or bubble formation onto ions produced along the track of a rapidly moving charged particle (or an x- or gamma ray). In a cloud chamber there is a supersaturated vapour which condenses out onto the ions because of their electrostatic attraction. Although the vapour is supersaturated before the particles pass through it, condensation does not occur until the ions or other nucleation centres are present because very small droplets are not in equilibrium with the vapour pressure if there is only a slight degree of supersaturation. Thus they evaporate or fail to form in the first place. As an example, a vapour pressure of twice the saturation pressure is required for the continued existence of water droplets of radius 10^{-9} m. In a bubble chamber a liquid, such as liquid hydrogen, is maintained a little above its boiling point, and in a similar manner to the cloud chamber, bubble formation does not start until nucleation centres are present. This time the reason why nucleation centres are required arises from the very high pressures inside small bubbles, the presence of the centres increases the size and reduces the pressure required for bubble initiation. If the chamber is illuminated, the tracks of the particles are visible as lines of tiny droplets or bubbles and are usually recorded by being photographed. The droplets or bubbles may be evaporated or recondensed by lowering or raising the pressure of the medium. If the pressure is then restored to its previous value, the material will again be supersaturated or superheated, and the chamber is ready for the next detection. The pressure may be conveniently changed using a piston, and cycle times as short as half a second are attainable.

In cosmic ray work the chamber is activated by the passage of the cosmic ray particle so that photographic film is not wasted on featureless images. The particle is detected by one or more pairs of Geiger counters (section 1.3) above and below the chamber (figure 1.4.1), and the piston is actuated when a near simultaneous detection occurs. Information on the nature of the particle may be obtained by applying magnetic or electrostatic fields and measuring the resulting curvature of the tracks.

Techniques such as this one and others in this section might seem to fit

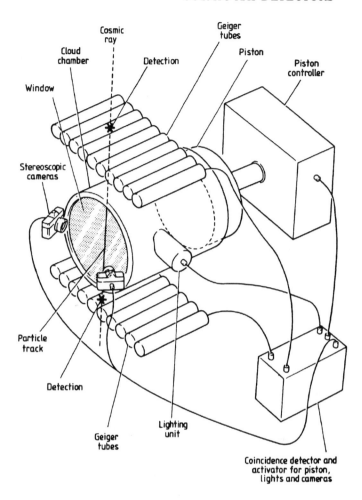

Figure 1.4.1. Schematic experimental arrangement for the observation of cosmic rays using a cloud chamber.

more appropriately into imaging in Chapter 2. However, the imaging of the tracks in a bubble chamber, is the imaging of the information carrier and not of the source, and so counts simply as detection. In any case the sources of cosmic rays cannot currently be imaged because of the extremely tortuous paths of the particles through the galactic and solar magnetic fields.

Geiger counters
We have already reviewed the principles and operational techniques of Geiger counters in section 1.3. Little change in them is required for their use in the detection of cosmic rays. They are now primarily used to trigger the operation of

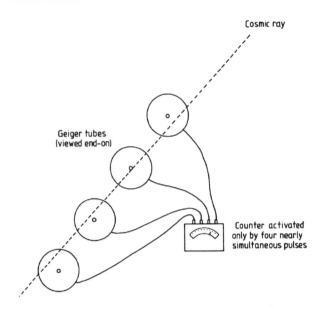

Figure 1.4.2. A cosmic ray telescope utilising Geiger counters.

more informative detectors, but have in the past been used as primary detectors.

The azimuthal variation in the cosmic ray intensity was first found by using several counters in a line (figure 1.4.2) and by registering only those particles which triggered all the detectors, so that some directivity was conferred to the system. They have also found application, along with other detectors, within muon energy spectrometers. Arrays of the detectors are stacked above and below a powerful magnet to detect the angular deflection of the muons.

Spark detector

A closely related device to the Geiger counter in terms of its operating principle is the spark chamber. A series of plates separated by small gaps are charged to successively higher potential differences. The passage of an ionising particle through the stack results in a series of sparks which follow the path of the particle (figure 1.4.3). The system is also used for the detection of gamma rays (section 1.3). Then a plate of a material of a high atomic weight such as tantalum may be interposed, and the ray detected by the resulting electron–positron pairs rather then by its own ionisations.

The operation of a spark chamber is similar to that of a cloud chamber; the track is photographed when Geiger counters above and below the detector give simultaneous pulses.

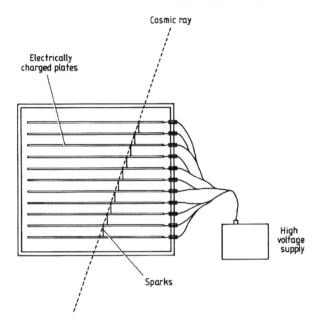

Figure 1.4.3. A spark chamber.

Flash tubes
If a high voltage pulse is applied to a tube filled with a low-pressure gas, shortly after a particle has passed through it, then the electrons along the charged particle's path will initiate a glow discharge in the tube. This can then easily be detected by a photomultiplier or other optical detector. Arrays of such tubes can provide some limited pointing accuracy to the detection, as in the case of Geiger tubes (figure 1.4.2).

Scintillation detectors
See section 1.3 for a discussion of these devices. There is little change required in order for them to detect cosmic ray particles.

Čerenkov detectors
Background. When a charged particle is moving through a medium with a speed greater than the local speed of light in that medium, it causes the atoms of the medium to radiate. This radiation is known as Čerenkov radiation, and it arises from the abrupt change in the electric field near the atom as the particle passes by. At subphotic speeds, the change in the field is smoother and little or no radiation results. The radiation is concentrated into a cone spreading outward from the direction of motion of the particle (figure 1.4.4), whose half angle, θ,

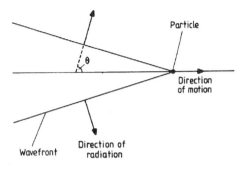

Figure 1.4.4. Čerenkov radiation.

is

$$\theta = \tan^{-1}\left[\left(\mu_\nu^2 \frac{v^2}{c^2} - 1\right)^{1/2}\right] \qquad (1.4.2)$$

where μ_ν is the refractive index of the material at frequency ν and v is the particle's velocity ($v > c/\mu_\nu$). Its spectrum is given by

$$I_\nu = \frac{e^2\nu}{2\epsilon_0 c^2}\left(1 - \frac{c^2}{\mu_\nu^2 v^2}\right) \qquad (1.4.3)$$

where I_ν is the energy radiated at frequency ν per unit frequency interval, per unit distance travelled by the particle through the medium. The peak emission depends upon the form of the variation of refractive index with frequency. For a proton in air, peak emission occurs in the visible when its energy is about 2×10^{14} eV.

Detectors. Čerenkov detectors are very similar to scintillation detectors except in the manner in which their flashes of visible radiation are produced (see above). A commonly used system employs a tank of very pure water surrounded by photomultipliers for the detection of the heavier cosmic ray particles, while high pressure carbon dioxide is used for the electrons and positrons. With adequate observation and the use of two or more different media, the direction, energy, and possibly the type of the particle may be deduced.

As mentioned previously (section 1.3), the flashes produced in the atmosphere by the primary cosmic rays can be detected. A large light 'bucket' and a photomultiplier are needed, preferably with two or more similar systems observing the same part of the sky so that non-Čerenkov events can be eliminated by anti-coincidence discrimination. The stellar intensity interferometer (section 2.4) is an example of such a system, although there the Čerenkov events are a component of its noise spectrum, and not usually detected for their own interest.

An intriguing aside to Čerenkov detectors arises from the occasional flashes seen by astronauts when in space. These are thought to be Čerenkov radiation from primary cosmic rays passing through the material of the eyeball. Thus cosmic ray physicists could observe their subjects directly. It is also just possible that they could listen to them as well! A large secondary cosmic ray shower hitting a water surface will produce a 'click' sound which is probably detectable by the best of the current hydrophones. Unfortunately a great many other events produce similar sounds so that the cosmic ray clicks are likely to be well buried in the noise. Nonetheless the very highest energy cosmic ray showers might even be audible directly to a skin diver.

Solid state detectors

These are essentially solid ionisation chambers. They consist of a thick layer (up to 5 mm) of a semiconductor such as silicon. It is formed into an n–p junction by lithium doping, and has a high resistivity by virtue of being back biased to a hundred volts or so. When a charged particle passes through it, a pulse of current is detected as the electron–hole pairs which it produces are collected at the electrodes on the top and bottom of the silicon slice. Detection of the pulses can then proceed by any of the usual methods. Solid state detectors have several advantages which suit them particularly for use in satellite-borne instrumentation—simplicity, reliability, low power consumption, high stopping power for charged particles, no entrance window needed, high counting rates possible, etc. Their main disadvantages are that their size is small compared with many other detectors, so that their collecting area is also small, and that unless the particle is stopped within the detector's volume, only the rate of energy loss may be found and not the total energy of the particle. This latter disadvantage, however, also applies to most other detectors.

Nucleon detector

Generally only the products of the high-energy primaries survive to be observed at the surface of the Earth. A proton will lose about 2×10^9 eV by ionisation of atoms in the atmosphere even if it travels all the way to the surface without interacting with a nucleus, and there is only about one chance in ten thousand of it doing this. Thus the lower energy primaries will not produce secondaries which can be detected at ground level, except if a neutron is one of its products. There is then a similar chance of non-interaction with atmospheric nuclei, but since the neutron will not ionise the atoms, even the low energy ones will still be observable at sea level if they do miss the nuclei. Thus neutrons and their reaction products from cosmic ray primaries of any energy can be detected using nucleon detectors.

These instruments detect low energy neutrons by, for example, their interaction with boron in boron fluoride (BF_3)

$$^{10}_{5}B + n \rightarrow {}^{7}_{3}Li + {}^{4}_{2}He + \gamma. \tag{1.4.4}$$

The reaction products are detected by their ionisation of the boron fluoride

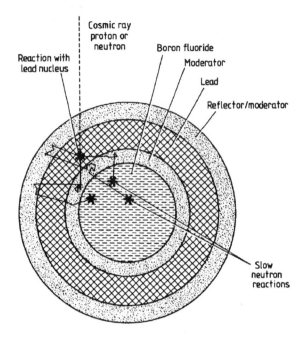

Figure 1.4.5. Cross section through a nucleon detector.

when this is used as a proportional counter (section 1.3). The slow neutrons are produced by the interaction of the cosmic ray neutrons and the occasional proton with lead nuclei, followed by deceleration in a hydrogen-rich moderator such as polyethylene (figure 1.4.5).

Residual track detectors

Photographic emulsions
In a sense the use of photographic emulsion for the detection of ionising radiation is the second oldest technique available to the cosmic ray physicist after the electroscope (see p. 143), since the existence of ionising particles was discovered by Becquerel in 1896 by their effect upon a photographic plate. However, the early emulsions only hinted at the existence of the individual tracks of the particles, and were far too crude to be of any real use. The emulsions now used had thus to await the optimisation of their properties for this purpose by C F Powell some forty years later, before they became of possible use as detectors.

The grain size of the silver halide in nuclear emulsions is small, but not unusually so—ranging from 0.03 to 0.5 μm compared with the range for emulsions used to detect electromagnetic waves of 0.05 to 1.7 μm—(an extensive discussion of the properties of the photographic emulsion will be found in section 2.2). Where the grains are unusual is in the very high degree of uniformity of

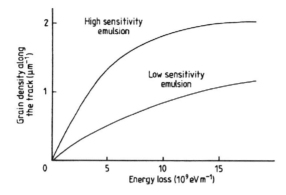

Figure 1.4.6. Response curves for nuclear emulsions (the tracks become opaque at a grain density of about 1.5 grains per micron).

their size which leads to a high degree of uniformity of response. Also very few of the grains are chemically fogged (section 2.2) so that long developing times can be used to show the fainter tracks. The emulsion has a very high silver halide content; up to 80% by weight, so that the emulsion has a very high spatial resolution. The high density of the emulsion also gives it a very high stopping power for the particles, their path lengths are only about 0.05% of the distances in air. Individual layers of the emulsion are quite thick, a millimetre or more not being uncommon. Many such layers are then stacked on top of each other to produce a block of emulsion whose volume may be several litres. Since 80% of such a block is silver, it is obviously a very costly item, however, since only a very tiny fraction of the silver is used to form the images of the particle tracks, the bulk of it may be recovered during processing. Emulsions may be produced to have differing sensitivities in the sense of their responses to differing rates of energy loss by the particles. A typical response curve is shown in figure 1.4.6. Using the least sensitive emulsions, energy loss values up to 1.5×10^{10} eV m^{-1} may be evaluated. For high-energy particles the rate of energy loss in a uniform medium, dE/dx, is given by a modification of the Bethe–Bloch formula

$$\frac{dE}{dx} = -\frac{e^2 Z^2 A N}{8\pi \epsilon_0^2 m_e v^2} \left[\ln\left(\frac{2m_e v^2 W_m}{\chi^2 [1 - (v^2/c^2)]} \right) - \frac{2v^2}{c^2} \right] \qquad (1.4.5)$$

where Z is the charge of the incident particle, v the velocity of the incident particle, A the atomic number of the stopping atoms, N the number density of the stopping atoms, W_m the maximum amount of energy transferable from the particle to an electron and χ the average ionisation potential of the stopping atoms. At low velocities the function is dominated by the v^{-2} term, while at high velocities v approaches asymptotically towards c and so is effectively constant and the $(1 - v^2/c^2)^{-1}$ term then predominates. Thus there is a minimum in

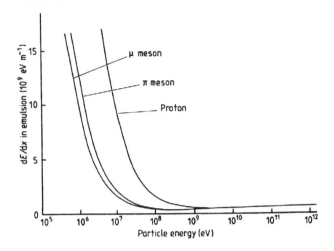

Figure 1.4.7. Rates of energy loss in nuclear emulsion.

the rate of energy loss (figure 1.4.7) and hence also of the track density. This minimum can be used to identify particles since their velocities are then similar to each other, enabling the mass of the particle to be found. The energies at minimum track density are listed in the table below for some of the commoner cosmic ray particles.

Particle	Energy (eV)
Electron	10^6
μ meson	2×10^8
π meson	3×10^8
Proton	2×10^9
Deuteron	4×10^9

Equation (1.4.5) also shows that the rate of energy deposition is proportional to Z^2. Hence the tracks of heavy nuclei are enormously thicker than those of protons and so the nature of the nucleus can be determined from the track width. If particles with a wide range of energy and composition are to be adequately separated from each other, it may be necessary to interleave emulsions with different stopping powers. A practical measure of the track width which is frequently used is the number of high energy electrons (often called δ rays) whose tracks extend more than 1.58 μm from the main track. Nuclei up to iron can then be unequivocally identified. The nuclei beyond iron become progressively harder to distinguish, but there is some evidence that elements up to and perhaps beyond uranium exist in the primary cosmic rays.

The precise mechanism of image formation in the emulsion is not fully understood (section 2.2). There seem to be two possibilities; uniform or discontinuous energy loss. A high-energy proton loses energy at a rate of about 5×10^8 eV m^{-1} in the emulsion. This corresponds to about 100 eV per 0.2 μm grain. The outermost electrons require some 6 eV for their ionisation, so that at most 16 electrons will be available within each grain for the formation of the latent image. This is close to the threshold for stable latent image formation (section 2.2). But for most of the time fewer than the maximum number of electrons will be produced, so that continuous energy loss within the emulsion is a dubious mechanism for the track formation. An alternative hypothesis is that the particle loses its energy discontinuously in the production of occasional high energy electrons. These electrons can then have an energy ranging up to many millions of electron volts depending upon the original interaction. Those with energies below a few keV will be halted within the grain by their own ionisation of the atoms in the grain, so that several hundred or even a thousand electrons may be released and a stable latent image formed. Other grains along the particle's track, however, will not contribute to the image unless they also have high energy electrons produced within them. Electrons with energies above a few keV will have path lengths longer than the size of a single grain, and so will leave the original grains and go on to produce their own short tracks. An electron with an energy of 10 keV for example, has a path length of about 2 μm. The existence of such short electron tracks has already been mentioned, and they provide some support for this second hypothesis of the mode of image formation in nuclear emulsions.

The use of nuclear emulsions to detect cosmic rays is very simple. A block of the emulsion is exposed to cosmic rays, while shielded from light of course, for a period of time from several hours to several weeks. Exposure is achieved by the emulsion being left on top of a mountain or by being carried to higher altitudes by a balloon. The emulsion is then developed (see below), and the particle tracks studied through a microscope.

The simplicity of the exposure of the nuclear emulsion stack is more than counteracted by the complexity of the processing technique. For consistent results a long and involved sequence of stages must be followed precisely and repeatably. The major problems are caused by the thickness of the individual emulsion layers, which can easily lead to non-uniformity of response and/or distortion of the tracks if the procedure is not adhered to. The principal steps in the processing of the emulsion are, in outline:

(*a*) Separation of each layer of the emulsion and its mounting onto a glass supporting plate, followed by exposure of the emulsion to a calibration grid. The high density of the emulsion means that the grid is printed into only a very thin surface layer and does not affect the bulk of the emulsion.

(*b*) Each layer of emulsion is placed into developer which has been chilled to below 5°C, and left for several hours. At this temperature the developer is inactive but permeates the whole layer until it is uniformly distributed throughout the emulsion. If the emulsion were placed straight into active developer, then its surface layers would be much more highly developed than its centre because of the time taken for the solution to diffuse through the emulsion. A developer with a low pH is used to minimise swelling of the gelatine.

(c) The developer is activated by raising its temperature to 20 to 25°C for about half an hour. The development is halted by cooling back to less than 5°C.

(d) The emulsion is immersed in an acid stop bath for several hours.

(e) The emulsion is fixed at 5°C or lower for several days with several changes of the fixing solution.

(f) The emulsion is then washed at a low temperature for several days. Initially the washing solution is fixer at a concentration slightly lower than that already in the emulsion. The concentration of the washing solution is progressively reduced as the fixer leaves the emulsion until finally pure water is being used. This procedure is necessary in order to reduce the swelling and distortion of the emulsion.

(g) The final stage is drying for several days under conditions in which the humidity is progressively reduced, in a similar manner and for similar reasons to the reduction in the strength of the washing solution.

The processed emulsion must be examined layer by layer and measured, and usually this is by hand while viewing through a microscope, since the process is not easily amenable to automation. It is obviously a long and tedious job to do this for the several square metres of emulsion in a typical stack but nonetheless, the nuclear emulsion remains one of the most useful tools of the cosmic ray astronomer, and has provided most of the direct observations to date of primary cosmic rays.

Ionisation damage detectors

These are a relatively recent development and they provide a selective detector for nuclei with masses above about 150 amu. Their principle of operation is allied to that of the nuclear emulsions in that it is based upon the ionisation produced by the particle along its track through the material. The material is usually a plastic with relatively complex molecules. As an ionising particle passes through it, the large complex molecules are disrupted, leaving behind short, chemically reactive segments, radicals etc. The higher chemical reactivity along the track of a particle may be revealed by etching the plastic. A conical pit then develops along the line of the track. By stacking many thin layers of plastic, the track may be followed to its conclusion. The degree of damage to the molecules, and hence the characteristics of the pit which is produced, is a function of the particle's mass, charge and velocity. A particular plastic may be calibrated in the laboratory so that these quantities may be inferred from the pattern of the sizes and shapes of the pits along the particle's track. Cellulose nitrate and polycarbonate plastics are the currently favoured materials. The low weight of the plastic and the ease with which large-area detectors can be formed make this a particularly suitable method for use in space when there is an opportunity for returning the plastic to Earth for processing, as for example when the flight is a manned one.

Similar tracks may be etched into polished crystals of minerals such as feldspar and rendered visible by infilling with silver. Meteorites and lunar samples can thus be studied and provide data on cosmic rays which extend back into the past for many millions of years. The majority of such tracks appear to be attributable to iron group nuclei, but the calibration is very uncertain. Because of the uncertainties involved the evidence has so far been of more use to the meteoriticist in dating the meteorite, than to the cosmic ray astronomer.

Indirect detectors

Electroscope
The earliest and simplest observation of cosmic ray effects can be repeated by the science department of any moderately well equipped secondary school. A simple gold leaf electroscope with an applied electric charge will show that charge slowly leaking away even if it is well isolated electrically. The leakage is due to the non-zero conductivity of air through the formation of ion–electron pairs by cosmic radiation. Near sea level this rate averages about 2×10^6 m^{-1} s^{-1}. The electroscope was used during the very early stages of the investigation of cosmic rays, and revealed, for example, the variation of their intensity with height, but now it is of historical interest only.

100 MeV gamma rays
Primary cosmic rays occasionally collide with nuclei in the interstellar medium. Even though the chance of this occurring is only about 0.1% if the particle were to cross the galaxy in a straight line, it happens sufficiently often to produce detectable results. In such collisions π^0 mesons will frequently be produced, and these will decay rapidly into two gamma rays, each with an energy of about 100 MeV. π^0 mesons may also be produced by the interaction of the cosmic ray particles and the 3 K microwave background radiation. This radiation when 'seen' by a 10^{20} eV proton is doppler shifted to a gamma ray of 100 MeV energy, and neutral pions result from the reaction

$$p^+ + \gamma \rightarrow p^+ + \pi^0. \tag{1.4.6}$$

Inverse Compton scattering of starlight or the microwave background by cosmic ray particles can produce an underlying continuum around the line emission produced by the pion decay.

Gamma rays with energies as high as these are little affected by the interstellar medium, so that they may be observed, wherever they may originate within the galaxy, by artificial satellites (section 1.3). The 100 MeV gamma ray flux thus gives an indication of the cosmic ray flux throughout the galaxy and beyond.

Diffuse radio emission

Cosmic ray electrons are only a small proportion of the total flux, and the reason for this is that they lose significant amounts of energy by synchrotron emission as they interact with the galactic magnetic field. This emission lies principally between 1 MHz and 1 GHz and is observable as diffuse radio emission from the galaxy. However, the interpretation of the observations into electron energy spectra etc is not straightforward, and is further complicated by the lack of a proper understanding of the galactic magnetic field.

Fluorescence

The very highest energy extensive air showers are detectable via weak fluorescent light from atmospheric nitrogen. This is produced through the excitation of the nitrogen by the electron–photon component of the cascade of secondary cosmic rays. The equipment required is a light bucket and a detector (cf Čerenkov detectors), and detection rates of a few tens of events per year for particles in the 10^{19} to 10^{20} eV range are achieved by several automatic arrays.

Solar cosmic rays

Very high fluxes of low energy cosmic rays can follow the eruption of a large solar flare. The fluxes can be high enough to lower the Earth's ionosphere and to increase its electron density. This in turn can be detected by direct radar observations, or through long-wave radio communication fade-outs, or through decreased cosmic radio background intensity as the absorption of the ionosphere increases.

Carbon 14

The radioactive isotope $^{14}_{6}C$ is produced from atmospheric $^{14}_{7}N$ by neutrons from cosmic ray showers

$$^{14}_{7}N + n \rightarrow {}^{14}_{6}C + p^{+}. \tag{1.4.7}$$

The isotope has a half-life of 5730 years and has been studied intensively as a means of dating archaeological remains. Its existence in ancient organic remains shows that cosmic rays have been present in the Earth's vicinity for at least 20 000 years. The flux seems, however, to have varied markedly at times from its present day value, particularly between about 4000 and 1000 years BC. But this is probably attributable to increased shielding of the Earth from the low energy cosmic rays at times of high solar activity, rather than to a true variation in the number of primary cosmic rays.

Arrays

Primary cosmic rays may be studied by single examples of the detectors that we have considered above. The majority of the work on cosmic rays, however, is on the secondary cosmic rays, and for these a single detector is not very informative. The reason is that the secondary particles from a single high energy primary particle have spread over an area of ten square kilometres or more by the time they have reached ground level from their point of production some fifty kilometres up in the atmosphere. Thus to deduce anything about the primary particle that is meaningful, the secondary shower must be sampled over a significant fraction of its area. Hence, arrays of detectors are used rather than single ones. These are typically spread out over an area of several square kilometres, with perhaps a hundred individual detectors all linked to a central computer. Any of the real-time detectors may be used, but scintillation counters and cloud chambers are probably the commonest.

The analysis of the data from such arrays is difficult. Only a very small sample, typically less than 0.01%, of the total number of particles in the shower is normally caught. The properties of the original particle have then to be inferred from this small sample. However, the nature of the shower varies with the height of the original interaction, and there are numerous corrections to be applied as discussed below. Thus normally the observations are fitted to a grid of computer-simulated showers. The original particle's energy can usually be obtained within fairly broad limits by this process, but its further development is limited by the availability of computer time, and by our lack of understanding of the precise nature of these extraordinarily high energy interactions.

Correction factors

Atmospheric effects

The secondary cosmic rays are produced within the Earth's atmosphere, and so its changes may affect the observations. The two most important variations are caused by air mass and temperature.

The air mass depends upon two factors—the zenith angle of the axis of the shower, and the barometric pressure. The various components of the shower are affected in different ways by changes in the air mass. The muons are more penetrating than the nucleons and so the effect upon them is comparatively small. The electrons and positrons come largely from the decay of muons, and so their variation tends to follow that of the muons. The corrections to an observed intensity are given by

$$I(P_0) = I(P) \exp\left[K(P - P_0)/P_0\right] \qquad (1.4.8)$$
$$I(0) = I(\theta) \exp[K(\sec\theta - 1)] \qquad (1.4.9)$$

where P_0 is the standard pressure, P is the instantaneous barometric pressure at the time of the shower, $I(P_0)$ and $I(P)$ are the shower intensities at pressures P_0 and P respectively, $I(0)$ and $I(\theta)$ are the shower intensities at zenith angles of zero and θ respectively and K is the correction constant for each shower component. K has a value of 2.7 for the muons, electrons and positrons and 7.6 for the nucleons. In practice a given detector will also have differing sensitivities for different components of the shower, and so more precise correction factors must be determined empirically.

The temperature changes primarily affect the muon and electron components. The scale height of the atmosphere, which is the height over which the pressure changes by a factor e^{-1}, is given by

$$H = \frac{kR^2T}{GMm} \qquad (1.4.10)$$

where R is the distance from the centre of the Earth, T the atmospheric temperature, M the mass of the Earth and m the mean particle mass for the atmosphere. Thus the scale height increases with temperature and so a given pressure will be reached at a greater altitude if the atmospheric temperature increases. The muons, however, are unstable particles and will have a longer time in which to decay if they are produced at greater heights. Thus the muon and hence the electron and positron intensity decreases as the atmospheric temperature increases. The relationship is given by

$$I(T_0) = I(T)\exp\left[0.8(T - T_0)/T\right] \qquad (1.4.11)$$

where T is the atmospheric temperature, T_0 the standard temperature and $I(T_0)$ and $I(T)$ are the muon (or electron) intensities at temperatures T_0 and T respectively. Since the temperature of the atmosphere varies with height, equation (1.4.11) must be integrated up to the height of the muon formation in order to provide a reliable correction. Since the temperature profile will only be poorly known, it is not an easy task to produce accurate results.

Solar effects

The Sun affects cosmic rays in two main ways. Firstly it is itself a source of low-energy cosmic rays whose intensity varies widely. Secondly the extended solar magnetic field tends to shield the Earth from the lower energy primaries. Both these effects vary with the sunspot cycle and also on other time scales, and are not easily predictable.

Terrestrial magnetic field

The Earth's magnetic field is essentially dipolar in nature. Lower energy charged particles may be reflected by it, and so never reach the Earth at all, or particles which are incident near the equator may be channelled towards the poles. There is thus a dependence of cosmic ray intensity on latitude (figure 1.4.8). Furthermore, the primary cosmic rays are almost all positively charged and they are deflected so that there is a slightly greater intensity from the west. The vertical concentration (equation (1.4.9) and figure 1.4.9) of cosmic rays near sea level makes this latter effect only of importance for high altitude balloon or satellite observations.

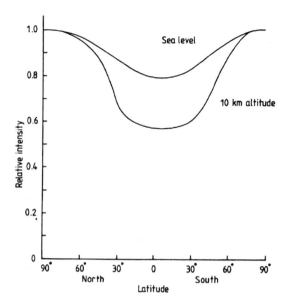

Figure 1.4.8. Latitude effect on cosmic rays.

Exercises

1.4.1 Show that the true counting rate, C_t, of a Geiger counter whose dead time is of length Δt, is related to its observed counting rate, C_0, by

$$C_t = \frac{C_0}{1 - \Delta t C_0}.$$

(section 1.3 is also relevant to this problem).

If the effective range of a Geiger counter is limited to $C_t \le 2C_0$, calculate the maximum useful volume of a Geiger counter used to detect secondary cosmic rays at sea level if its dead time is 250 μs.

1.4.2 The minimum particle energy required for a primary cosmic ray to produce a shower observable at ground level when it is incident vertically onto the atmosphere is about 10^{14} eV. Show that the minimum energy required to produce a shower when the primary particle is incident at a zenith angle θ, is given by

$$E_{\min}(\theta) = 6.7 \times 10^{12} \, e^{2.7 \, \sec \theta} \text{ (eV)}$$

(Hint: use equation (1.4.9) for muons and assume that the number of particles in the shower at its maximum is proportional to the energy of the primary particle.)

The total number of primary particles, $N(E)$, whose energy is greater than or equal to E, is given at high energies by

$$N(E) \simeq 10^{22} E^{-1.85} \ \left(\text{m}^{-2} \ \text{s}^{-1} \ \text{str}^{-1}\right)$$

for E in eV. Hence show that the number of showers, $N(\theta)$, observable from the ground at a zenith angle of θ is given by

$$N(\theta) \simeq 0.019 \ e^{-5.0 \ \sec \theta} \ \left(\text{m}^{-2} \ \text{s}^{-1} \ \text{str}^{-1}\right)$$

1.4.3   By numerical integration, or otherwise, of the formula derived in problem 1.4.2, calculate the total flux of showers of all magnitudes onto a detector array covering one square kilometre. (Assume that the primary particle flux is isotropic.)

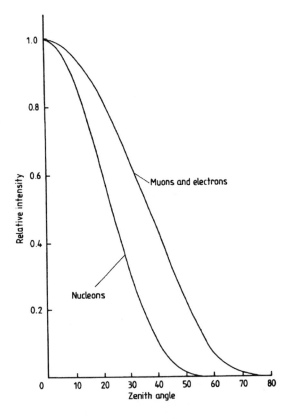

Figure 1.4.9. Zenithal concentration of secondary cosmic ray shower components as given by equation (1.4.9).

1.5 NEUTRINO DETECTORS

Background

The neutrino was postulated in 1930 by Pauli in order to retain the principle of conservation of mass and energy in nuclear reactions. It was necessary in order to provide a mechanism for the removal of residual energy in some β-decay reactions. From other conservation laws, its properties could be defined quite well; zero charge, zero or very small rest mass, zero electric moment, zero magnetic moment, and a spin of one half. A quarter of a century passed before the existence of this hypothetical particle was confirmed experimentally. The reason for the long delay in its confirmation lay in the very low probability of the interaction of neutrinos with other matter; a neutrino originating at the centre of the Sun would have only one chance in 10^{10} of interacting with any other particle during the whole of its 700 000 km journey to the surface of the Sun. The interaction probability for particles is measured by their cross sectional area for absorption, σ, given by

$$\sigma = 1/\lambda N \tag{1.5.1}$$

where N is the number density of target nuclei and λ is the mean free path of the particle. Even for high-energy neutrinos the cross section for the reaction

$$\nu + {}^{37}_{17}\text{Cl} \rightarrow {}^{37}_{18}\text{Ar} + \text{e}^- \tag{1.5.2}$$

which is commonly used for their detection (see later in this section) is only $10^{-46}\,\text{m}^2$ (figure 1.5.1), so that such a neutrino would have a mean free path of over one parsec even in pure liquid ${}^{37}_{17}\text{Cl}$!

Three varieties of neutrino are known, plus their anti-particles. The electron neutrino, ν_e, is the type originally postulated to save the β reactions, and which is therefore involved in the archetypal decay—the decay of a neutron

$$\text{n} \rightarrow \text{p}^+ + \text{e}^- + \tilde{\nu}_\text{e} \tag{1.5.3}$$

where $\tilde{\nu}_\text{e}$ is an anti-electron neutrino. The electron neutrino is also the type commonly produced in nuclear fusion reactions and hence to be expected from the Sun and stars. For example the first stage of the proton–proton cycle is

$$\text{p}^+ + \text{p}^+ \rightarrow {}^2_1\text{H} + \text{e}^+ + \nu_\text{e} \tag{1.5.4}$$

and the second stage of the carbon cycle

$$^{13}_7\text{N} \rightarrow {}^{13}_6\text{C} + \text{e}^+ + \nu_\text{e} \tag{1.5.5}$$

and so on.

The other two types of neutrino are the muon neutrino, ν_μ, and the tauon neutrino, ν_τ. These are associated with reactions involving the heavy electrons called muons and tauons. For example, the decay of a muon

$$\mu^+ \rightarrow \text{e}^+ + \nu_\text{e} + \tilde{\nu}_\mu \tag{1.5.6}$$

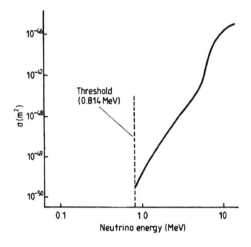

Figure 1.5.1. Neutrino absorption cross sections for the $^{37}_{17}$Cl $\rightarrow$ $^{37}_{18}$Ar reaction.

which involves an anti-muon neutrino amongst other particles.

While these other types were not until recently expected to be of much interest to the neutrino astronomer, they may now turn out to be very important. This change in attitude arises from some recent work which indicates that the neutrino's rest mass may not be zero, but may be between 4 and 22 electron volts. This is still very small when compared with the half million electron volts required for the rest mass of even an electron, but associated with a non-zero rest mass there may be an ability for the neutrino to change its type. Thus a neutrino may oscillate between the three types during its existence (figure 1.5.2). Now, the major work on the detection of neutrinos from astronomical sources has been that of Davis (see later in this section) with his detection of neutrinos from the Sun. But his detection rate has been consistently too low by a factor of about 3.3, when compared with the predictions of the astrophysicists. Confirmation of this observation has recently come from the Kamiokande and SAGE detectors (see below). However, the situation has also been further complicated by the possible detection of a correlation between neutrino flux and sunspot activity, which may imply that solar flares are a significant source of neutrinos, or that solar flares affect neutrino detectors in some other way.

The factor of 3 discrepancy is known as the solar neutrino problem and has caused a major reconsideration of our ideas of the processes occurring inside the Sun, to the extent that even the basic conversion of hydrogen into helium has been questioned. However, none of the many suggestions has satisfactorily explained the discrepancy between the theory and observation. But most experiments detect only electron neutrinos, so that if those produced in the centre of the Sun are oscillating between the three varieties, only one third of

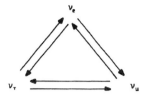

Figure 1.5.2. Neutrino oscillation.

the neutrinos passing through Davis' apparatus will be detectable at any one time. The discrepancy is then reduced to a factor of 1.1 which is well inside the experimental errors and so not significant. The non-zero rest mass of the neutrino remains to be confirmed at the time of writing as does the interchange of type even if the rest mass is non-zero, but perhaps the solution of the solar neutrino problem is in sight.

Chlorine 37 detectors

The problems involved in any attempt to detect neutrinos arise mainly from the extreme rarity of the interactions that are involved. This not only dictates the use of very large detectors and very long integration times, but also means that the neutrino-induced reactions may be swamped by other commoner radioactive processes. The first neutrino 'telescope' to be built is designed to detect electron neutrinos through the reaction given in equation (1.5.2). The threshold energy of the neutrino for this reaction is 0.814 MeV (see figure 1.5.1), so that 80% of the neutrinos from the Sun which are detectable by this experiment are expected to arise from the decay of boron in a low probability side chain of the proton–proton reaction

$$\\,^8_5\\mathrm{B} \\rightarrow \\,^8_4\\mathrm{Be} + \\mathrm{e}^+ + \\nu_e. \\qquad (1.5.7)$$

The full predicted neutrino spectrum of the Sun, based upon the conventional astrophysical models, and not taking account of the modifications suggested to try and solve the solar neutrino problem, is shown in figure 1.5.3.

The detector began operating in 1968 under the auspices of R Davis Jr and his group. It consists of a large tank of over 600 tonnes of tetrachloroethene (C_2Cl_4). About one chlorine atom in four is the $^{37}_{17}Cl$ isotope, so on average each of the molecules contains one of the required atoms, giving a total of about $2 \\times 10^{30}$ $^{37}_{17}Cl$ atoms in the tank. Tetrachloroethene is chosen rather than liquid chlorine for its comparative ease of handling, and because of its cheapness—it is a common industrial solvent. The interaction of a neutrino (equation (1.5.2)) produces a radioactive isotope of argon. The half-life of the argon is 35 days, and it decays back to $^{37}_{17}Cl$ by capturing one of its own inner orbital electrons, at the same time ejecting a 2.8 keV electron. The experimental procedure is to allow the argon atoms to accumulate for some time, then to bubble helium

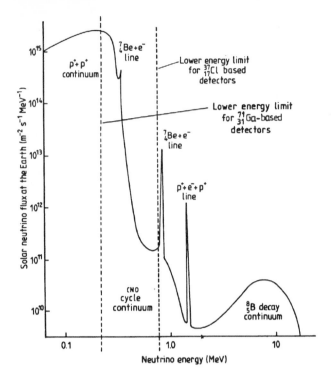

Figure 1.5.3. Postulated solar neutrino spectrum.

through the tank. The argon is swept up by the helium and carried out of the tank. It can be separated from the helium by passing the gas stream through a charcoal cold trap. The number of $^{37}_{18}$Ar atoms can then be counted by counting the 2.8 keV electrons as they are emitted.

The experimental procedure must be carried out with extraordinary care if the aim of finding a few tens of atoms in 10^{31} is to succeed. Pre-eminent amongst the precautions are:

(a) Siting the tank 1.5 km below the ground to shield it against cosmic rays and natural and artificial radiation sources. The neutrinos of course pass almost without effect through this depth of material, indeed they can pass right through the Earth without any measurable drop in their intensity.

(b) Surrounding the tank with a thick water jacket to shield against neutrons.

(c) Monitoring the efficiency of the argon extraction by introducing a known quantity of $^{36}_{18}$Ar into the tank before it is swept with helium, and determining the percentage of this which is recovered.

(d) Shielding the sample of $^{37}_{18}$Ar during counting of its radioactive decays and using anti-coincidence techniques to eliminate extraneous pulses.

(e) Use of long integration times (50 to 100 days) to accumulate the argon in the tank.

(f) Measurement of and correction for the remaining noise sources.

Currently about one neutrino is detected every other day. This corresponds to a detection rate of 2.2 ± 0.4 SNU. (A solar neutrino unit, SNU, is 10^{-36} captures per second per target atom.) The expected rate from the Sun is 7.3 ± 1.5 SNU, and the disparity between the observed and predicted rates is the previously mentioned solar neutrino problem.

Water-based detectors

The next generation of working neutrino telescopes did not appear until nearly two decades after the chlorine-based detector. Neither of these instruments, the Kamiokande detector in Japan nor the IMB (Irvine–Michigan–Brookhaven) detector in the USA, were built to be neutrino detectors. Both were originally intended to look for proton decay, and were only later converted for use with neutrinos. The design of both detectors is similar (figure 1.5.4), they differ primarily only in size (Kamiokande: 3000 tonnes, IMB: 8000 tonnes).

The principle of operation is the detection of Čerenkov radiation from the products of neutrino interactions. These may take two forms, electron scattering and inverse β-decay. In the former process, a collision between a high-energy neutrino and an electron sends the latter off at a speed in excess of the speed of light in water ($225\,000\,\mathrm{km\,s^{-1}}$) and in roughly the same direction as the neutrino. In inverse β-decay, an energetic positron is produced via the reaction

$$\tilde{\nu}_e + p \rightarrow n + e^+ \tag{1.5.8}$$

but the positron can be emitted in any direction and thus gives no clue to the original direction of the incoming neutrino. Both positron and electron, travelling at superphotic speeds in the water, emit Čerenkov radiation (section 1.4) in a cone around their direction of motion. That radiation is then picked up by arrays of photomultipliers (section 1.1) which surround the water tank on all sides. The pattern of the detected radiation can be used to infer the energy of the original particle, and in the case of a scattering event, also to give some indication of the arrival direction. The minimum detectable energy is several MeV.

Both detectors were fortunately in operation in February 1987 and detected the burst of neutrinos from the supernova in the Large Magellanic Cloud. This was the first (and so far only) detection of neutrinos from an astronomical source other than the Sun, and it went far towards confirming the theoretical models of supernova explosions. The Kamiokande detector has also recently confirmed the reality of the solar neutrino problem.

Large versions of these water-based detectors are currently under construction. Super Kamiokande for example will contain some 50 000 tonnes of water. Even larger versions may be produced by lowering strings of

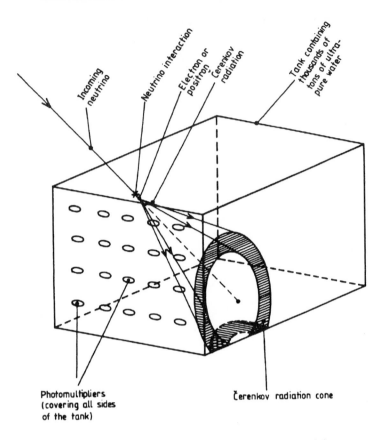

Figure 1.5.4. Principle of neutrino detection by water-based detectors.

photodetectors directly into the ocean to monitor up to a cubic kilometre of water (DUMAND2 in the Pacific, NESTOR in the Mediterranean and NT36 in Lake Baikal), or down holes drilled into the Antarctic ice cap (AMANDA).

Another and different water-based detector is currently under construction in Canada. This is the SNO detector (Sudbury Neutrino Observatory). This detector uses 1000 tonnes of ultra-pure heavy water (deuterium oxide), and will detect neutrinos mainly via the reaction

$$\nu_e + {}_1^2H \rightarrow p^+ + p^+ + e^- \tag{1.5.9}$$

the actual detection being of the Čerenkov radiation from the electron in the reaction. The SNO detector is predicted to be fifty times more sensitive than any existing neutrino detector.

Gallium-based detectors

At the time of writing (1997) there are two further operative detectors, both based upon gallium. The SAGE detector (Soviet-American Gallium Experiment) uses 60 tonnes of liquid metallic gallium buried beneath the Caucasus Mountains, and detects neutrinos via the reaction

$$^{71}_{31}\text{Ga} + \nu_e \rightarrow \, ^{71}_{32}\text{Ge} + e^-. \qquad (1.5.10)$$

The germanium product of the reaction is radioactive, with a half-life of 11.4 days. It is separated out by chemical processes on lines analogous to those of the chlorine-based detector. The detection threshold is only 0.236 MeV so that the p–p neutrinos (figure 1.5.3) are detected directly. Preliminary results support the deficit measured by other detectors in the number of neutrinos coming from the Sun. The other gallium-based detector, 'Gallex', is based upon the same detection mechanism as SAGE, but uses 30 tonnes of gallium chloride ($GaCl_3$). It is buried in the Gran Sasso Tunnel in Italy.

Other types of detectors

Detectors based upon $^{37}_{17}\text{Cl}$ etc can only detect the high-energy neutrinos, while the bulk of the solar neutrinos are of comparatively low energies (figure 1.5.3). Furthermore, no indication of the direction of the neutrinos is possible. Numerous alternative detectors which are designed to overcome such restrictions are therefore under construction, proposed or imagined at the time of writing. A brief survey of the range of possibilities is given below, but the reader should be a little wary of the viability of some of the proposals; some of them require the use of several times the entire current world stocks of an element! We may roughly classify the proposals into four types—direct interaction, indirect interaction, geological, and scattering—based upon the nature of their interactions.

Direct interaction type detectors

These are mostly variants of Davis' experiment with another element used in place of the chlorine. Indium may perhaps be usable as a more genuine neutrino telescope, with the direction and energy of the neutrino being determined rather than just a simple detection. The reaction in this case is

$$^{115}_{49}\text{In} + \nu_e \rightarrow \, ^{115}_{50}\text{Sn} + e^-. \qquad (1.5.11)$$

The tin nucleus is in an excited state after the interaction and emits two gamma rays with energies of 116 and 498 keV, 3 μs, on average, after its formation. A neutrino detection therefore has a characteristic signature of an electron emission followed after a brief delay by two distinctive gamma rays, which would allow it to be unambiguously separated from other interactions. The electron's energy and direction are related to those of the neutrino so allowing these to be found. The threshold is again low enough to capture some of the proton–proton neutrinos. About 10 tonnes of indium might be needed to detect one neutrino per day.

Another possibility uses lithium and is based upon the reaction

$$^{7}_{3}\text{Li} + \nu_e \rightarrow {}^{7}_{4}\text{Be} + e^-. \tag{1.5.12}$$

It has the disadvantage of a high threshold (0.862 MeV), but the advantage of comparative cheapness, only about 15 tonnes being needed for one detection per day. There is also the possibility of the detection of neutrinos via the reaction

$$^{37}_{19}\text{K} + \nu_e \rightarrow {}^{37}_{18}\text{Ar} + e^+ \tag{1.5.13}$$

and a detector using 6 tonnes of potassium hydroxide is currently operating alongside the chlorine detector in the Homestake mine.

Other direct interaction detectors which are based upon different principles have recently been proposed, for example, the detection of the change in the nuclear spin upon neutrino absorption by $^{115}_{49}\text{In}$ and its conversion to $^{115}_{50}\text{Sn}$. Up to 20 solar neutrinos per day might be found by a detector based upon 10 tonnes of superfluid helium held at a temperature below 0.1 K. Neutrinos would deposit their energy into the helium leading to the evaporation of helium atoms from the surface of the liquid. Those evaporated atoms would then be detected from the energy (heat) that they deposit into a thin layer of silicon placed above the helium.

A detector based upon 100 m^3 of a liquid scintillator, is currently operating under Mont Blanc and attempting to detect neutrinos via inverse β-decay.

Indirect interaction type detectors

These work upon the principle of detecting the products of neutrino interactions which have occurred outside the detector itself. They are most suitable for neutrinos with energies of 10^9 eV or above, such as might arise from the action of cosmic rays on nuclei in the Earth's atmosphere, or from presupernova collapses. Scintillation counters may be used to detect the products of the reaction. For example a recent proposal has been to monitor a cubic kilometre of the ocean depths with several thousand Čerenkov detectors to look for radiation from charged particles resulting from such high energy interactions; an angular resolution of a degree or better might be possible with such a system.

Geological detectors

This class of detectors offers the intriguing possibility of looking at the neutrino flux over the past few thousand years. They rely upon determining the isotope ratios for possible neutrino interactions in elements in natural deposits. This is a function of the half-life of the product and the neutrino flux in the recent past, assuming that a state of equilibrium has been reached. Possible candidate elements are tabulated below.

Element	Reaction	Product half-life (yrs)	Threshold energy (MeV)
Thallium	$^{205}_{81}\text{Tl} + \nu_e \rightarrow {}^{205}_{82}\text{Pb} + e^-$	3×10^7	0.046
Molybdenum	$^{98}_{42}\text{Mo} + \nu_e \rightarrow {}^{97}_{43}\text{Tc} + n + e^-$	2.6×10^6	8.96
Bromine	$^{81}_{35}\text{Br} + \nu_e \rightarrow {}^{81}_{36}\text{Kr} + e^-$	2.1×10^5	0.490
Potassium	$^{41}_{19}\text{K} + \nu_e \rightarrow {}^{41}_{20}\text{Ca} + e^-$	8×10^4	2.36

Scattering detectors

These are based upon neutrino–electron scattering events. The electrons are detected by scintillation counters, and some guide to the neutrino's energy and direction may be obtained from that of the electron. Alternatively electrons may be stored in a superconducting ring, and coherent scattering events detected by the change in the current. With this latter system detection rates for solar neutrinos of one per day may be achievable with volumes for the detector as small as a few millilitres.

Exercises

1.5.1 Show that if an element and its radioactive reaction product are in an equilibrium state with a steady flux of neutrinos, then the number of decays per second of the reaction product is given by

$$N_p T_{1/2}^{-1} \sum_{n=1}^{\infty} \frac{1}{n} \left(\frac{1}{2}\right)^n$$

when $T_{1/2} \gg$ one second. N_p is the number density of the reaction product, $T_{1/2}$ is its half-life.

Hence show that the equilibrium ratio of product to original element is given by

$$\frac{N_p}{N_e} = \sigma_\nu \mathcal{F}_\nu T_{1/2} \left\{ \left[\sum_{n=1}^{\infty} \frac{1}{n} \left(\frac{1}{2}\right)^n \right]^{-1} \right\}$$

where N_e is the number density of the original element, σ_ν is the neutrino capture cross section for the reaction and $\mathcal{F}_\nu$ is the neutrino flux.

1.5.2 Calculate the equilibrium ratio of $^{41}_{20}\text{Ca}$ to $^{41}_{19}\text{K}$ for ^8_5B solar neutrinos ($T_{1/2} = 80\,000$ years, $\sigma_\nu = 1.45 \times 10^{-46}\,\text{m}^{-2}$, $\mathcal{F}_\nu = 3 \times 10^{10}\,\text{m}^{-2}\,\text{s}^{-1}$).

1.5.3 If Davis' chlorine 37 neutrino detector were allowed to reach equilibrium, what would be the total number of $^{37}_{18}\text{Ar}$ atoms to be expected? (Hint: see problem 1.5.1, and note that $\sum_{n=1}^{\infty}(1/n)(1/2)^n = 0.693$).

1.6 GRAVITATIONAL RADIATION

Introduction

This section differs from the previous ones because none of the techniques that are discussed have yet indisputably detected gravitational radiation. Furthermore, while most researchers accept the probable existence of gravitational radiation, not all do, and even amongst the converted there is still a continuing debate over the theoretical background of its production and propagation.

The basic concept of a gravity wave† is simple; if an object with mass changes its position then, in general, its gravitational effect upon another object will change, and the information on that changing gravitational field propagates outward through the space–time continuum at the speed of light. The propagation of the changing field obeys equations analogous to those for electromagnetic radiation provided that a suitable adaption is made for the lack of anything in gravitation which is equivalent to positive and negative electric charges. Hence we speak of gravitational radiation and gravitational waves. Their frequencies for astronomical sources are anticipated to run from a few kilohertz for collapsing or exploding objects to a few microhertz for binary star systems. A pair of binary stars coalescing into a single object will emit a burst of waves whose frequency rises rapidly with time: a characteristic 'chirp' if we could hear it.

Theoretical difficulties with gravitational radiation arise from the multitudinous metric, curved space–time theories of gravity which are currently extant. The best known of these are due to Einstein (general relativity), Brans and Dicke (scalar-tensor theory) and Hoyle and Narlikar (C-field theory), but there are many others. Einstein's theory forbids dipole radiation, but this is allowed by most of the other theories. Quadrupole radiation is thus the first allowed mode of gravitational radiation in general relativity, and will be about two orders of magnitude weaker than the dipole radiation from a binary system which is predicted by the other theories. Furthermore, there are only two polarisation states for the radiation as predicted by general relativity, compared with six for most of the other theories. The possibility of decisive tests for general relativity is thus a strong motive for aspiring gravity wave observers, in addition to the information which may be contained in the waves on such matters as collapsing and colliding stars, supernovae, close binaries, pulsars, early stages of the big bang etc.

The detection problem for gravity waves is best appreciated with the aid of a few order-of-magnitude calculations on their expected intensities. The quadrupole gravitational radiation from a binary system of moderate eccentricity ($e \leq 0.5$), is given in general relativity by

$$L_G \simeq \frac{2 \times 10^{-63} M_1^2 M_2^2 (1 + 30e^3)}{(M_1 + M_2)^{2/3} P^{10/3}} \text{(W)} \qquad (1.6.1)$$

† The term 'gravity wave' is also used to describe oscillations in the Earth's atmosphere arising from quite different processes. There is not usually much risk of confusion.

where M_1 and M_2 are the masses of the components of the binary system, e is the orbital eccentricity and P is the orbital period. Thus for a typical dwarf nova system with

$$M_1 = M_2 = 1.5 \times 10^{30} \text{ kg} \tag{1.6.2}$$

$$e = 0 \tag{1.6.3}$$

$$P = 10^4 \text{ s} \tag{1.6.4}$$

we have

$$L_G = 2 \times 10^{24} \text{ W}. \tag{1.6.5}$$

But for an equally typical distance of 250 pc, the flux at the Earth is only

$$F_G = 3 \times 10^{-15} \text{ W m}^{-2}. \tag{1.6.6}$$

This energy will be radiated predominantly at twice the fundamental frequency of the binary with higher harmonics becoming important as the orbital eccentricity increases. Even for a nearby close binary such as ι Boo (distance 23 pc, $M_1 = 2.7 \times 10^{30}$ kg, $M_2 = 1.4 \times 10^{30}$ kg, $e = 0$, $P = 2.3 \times 10^4$ s), the flux only rises to

$$F_G = 5 \times 10^{-14} \text{ W m}^{-2}. \tag{1.6.7}$$

Rotating elliptical objects radiate quadrupole radiation with an intensity given approximately by

$$L_G \simeq \frac{GM^2\omega^6 r^4(A+1)^6(A-1)^2}{64c^5} \text{ (W)} \tag{1.6.8}$$

where M is the mass, ω the angular velocity, r the minor axis radius and A the ratio of the major and minor axis radii. So that for a pulsar with

$$\omega = 100 \text{ rad s}^{-1} \tag{1.6.9}$$

$$r = 15 \text{ km} \tag{1.6.10}$$

$$A = 0.999\,98 \tag{1.6.11}$$

we obtain

$$L_G = 1.5 \times 10^{26} \text{ W} \tag{1.6.12}$$

which for a distance of 1000 pc leads to an estimate of the flux at the Earth of

$$F_G = 10^{-14} \text{ W m}^{-2}. \tag{1.6.13}$$

Objects collapsing to form black holes within the galaxy or nearby globular clusters or coalescing binary systems may perhaps produce transient fluxes up to three orders of magnitude higher than these continuous fluxes.

Now these fluxes are relatively large compared with, say, those of interest to radio astronomers, whose faintest sources may have an intensity of

10^{-29} W m^{-2} Hz^{-1}. But the gravitational detectors which have been built to date and those planned for the future have all relied upon detecting the strain ($\partial x/x$) produced in a test object by the tides of the gravitational wave rather than the absolute gravitational wave flux, and this in practice means that measurements of changes in the length of the test object of only 5×10^{-21} m, or about 10^{-12} times the diameter of the hydrogen atom, must be obtained even for the detection of the radiation from ι Boo.

In spite of the difficulties which the detection of such small changes must obviously pose, a detector was built by J Weber which appeared to have detected gravity waves from the centre of the galaxy early in 1969. Unfortunately this was not confirmed by other workers, and the source of Weber's pulses remains a mystery, although they are not now generally attributed to gravity waves. The pressure to confirm Weber's results, however, has led to a wide variety of gravity wave detectors being proposed and built, and their rate of development is such that an unequivocal detection should occur within the next decade. The detectors so far built, under construction or proposed are of two types; direct detectors in which there is an attempt to detect the radiation experimentally, and indirect detectors in which the existence of the radiation is inferred from the changes that it incidentally produces in other observable properties of the object. The former group may be further subdivided into resonant, or narrow bandwidth detectors, and non-resonant or wideband detectors.

Detectors

Direct resonant detectors

The vast majority of the working gravity wave telescopes fall into this category. They are similar to, or improvements on, Weber's original system. This used a massive (> 1 tonne) aluminium cylinder which was isolated by all possible means from any external disturbance, and whose shape was monitored by piezo-electric crystals attached around its 'equator'. Two such cylinders separated by up to 1000 km were utilised, and only coincident events were regarded as significant in order to eliminate any remaining external interference. With this system, Weber could detect a strain of 10^{-16} (cf 5×10^{-21} for the detection of ι Boo). Since his cylinders had a natural vibration frequency of 1.6 kHz, this was also the frequency of the gravitational radiation which they would detect most efficiently. Weber detected about three pulses per day at this frequency with an apparent sidereal correlation and a direction corresponding to the galactic centre or a point 180° away from the centre. If originating in the galactic centre, some 500 solar masses would have to be totally converted into gravitational radiation each year in order to provide the energy for the pulses. Unfortunately (or perhaps fortunately from the point of view of the continuing existence of the galaxy!) the results were not confirmed even by workers using identical equipment to Weber's, and it is now generally agreed that there have been no definite detections of gravitational waves to date.

Adaptations of Weber's system have been made by Aplin, who used two cylinders with the piezo-electric crystals sandwiched in between them, and by Papini, who used large crystals with very low damping constants so that the vibrations would continue for days following a disturbance. Other systems cooled to liquid helium temperatures and/or based upon niobium bars have also been used. Ultimately bar detectors might achieve strain sensitivities of 10^{-21} or 10^{-22}.

Direct, non-resonant detectors

These are of two types and have the potential advantage of being capable of detecting gravity waves with a wide range of frequencies. The first uses a Michelson type interferometer (section 2.5) to detect the relative changes in the positions of two or more test masses (known as LIGOs—laser interferometer gravity wave observatories). A possible layout for the system is shown in figure 1.6.1, usually, however, the path length will be amplified by up to a factor of 300 by multiple reflections from the mirrors though these are not shown in the figure in order to preserve simplicity. Four such systems are under construction at the time of writing, two in the USA, and two in Europe. A fifth one is planned for Australia. Though there are differences in detail all the detectors have the same basic characteristics. The arm lengths are several kilometres, and are increased to several hundred kilometres by multiple reflection from ultra-high reflectivity mirrors. The mirror surfaces have to be flat to $\frac{1}{2}\%$ of the operating wavelength. The light sources are high-stability neodymium–yttrium–

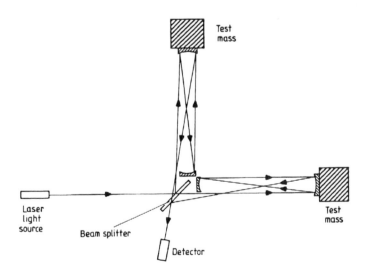

Figure 1.6.1. A possible layout for an interferometric gravity wave detector.

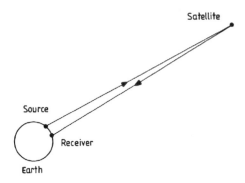

Figure 1.6.2. Arrangement for a satellite-based Doppler tracking gravity wave detector.

garnet lasers with outputs of 100 W or more. The mirrors are mounted on test masses on pendulum suspensions, and the whole system, including the light paths, operates in a vacuum. Path length changes are detected by looking at the interference pattern between the two orthogonal beams. Strain sensitivities of 10^{-22} over a bandwidth of 1 kHz are predicted for these systems, enabling detections of collapsing neutron stars, supernovae, and coalescing binaries to be made out to distances of 10^7 to 10^8 pc. The distribution of these four (or five) detectors over the Earth will not only provide confirmation of detections, but also enable the arrival directions of the waves to be pinpointed to a few minutes of arc.

There is also a proposal for a space-based interferometer system. This would employ three or more spacecraft at the corners of a triangle with sides 5 million kilometres long. Laser beams would circulate along the sides of the triangle. If other sources of noise can be eliminated, the system might be able to reach a strain sensitivity of 10^{-21} or so.

An ingenious idea underlies the second method in this class. An artificial satellite and the Earth will be independently influenced by gravity waves whose wavelength is less than the physical separation of the two objects. If an accurate frequency source on the Earth is broadcast to the satellite, and then returned to the Earth (figure 1.6.2), its frequency will be changed by the Doppler shift as the Earth or satellite is moved by a gravity wave. Each gravity wave pulse will be observed three times enabling the reliability of the detection to be improved. The three detections will correspond to the effect of the wave on the transmitter, the satellite, and the receiver, although not necessarily in that order. A drag-free satellite (figure 1.6.3) would be required in order to reduce external perturbations arising from the solar wind, radiation pressure, etc. A lengthy series of X-band observations using the Ulysses spacecraft aimed at detecting gravity waves by this method have been made recently, but so far without any successful detections.

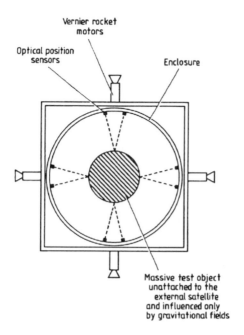

Figure 1.6.3. Schematic cross section through a drag-free satellite. The position of the massive test object is sensed optically, and the external satellite driven so that it keeps the test object centred.

Indirect detectors

Proposals for these so far only involve binary star systems. The principle of the method is to attempt to observe the effect on the period of the binary of the loss of angular momentum due to the gravitational radiation. A loss of angular momentum by some process is required to explain the present separations of dwarf novae and other close binary systems with evolved components. These are so small that the white dwarf would have completely absorbed its companion during its earlier giant stage, thus the separation at that time must have been greater than it is now. Gravitational radiation can provide an adequate orbital angular momentum loss to explain the observations, but it is not the only possible mechanism. Stellar winds, turbulent mass transfer and tides may also operate to reduce the separation, so that these systems do not provide unequivocal evidence for gravitational radiation.

The prime candidate for study for evidence of gravitational radiation is the binary pulsar (PSR 1913 + 16). Its orbital period is 2.8×10^4 seconds, and it was found to be decreasing at a rate of 10^{-4} seconds per year soon after the system's discovery. This was initially attributed to gravitational radiation, but it now appears that a helium star may be a part of the system, and so tides or other interactions could again explain the observations. On the other hand, the effect of a rapid pulsar on the local interstellar medium might mimic the appearance of a helium star when viewed from a distance of 5000 pc. Thus the detection of gravity waves from the binary pulsar, and indeed by all other methods remains 'not proven' at the time of writing.

Exercises

1.6.1   Show that the gravitational radiation of a planet orbiting the Sun is approximately given by

$$I_G \simeq 7 \times 10^{21} \ M^2 P^{-10/3} \ (\text{W})$$

where M is the planet's mass in units of the solar mass and P the planet's orbital period in days.

1.6.2   Calculate the gravitational radiation for each planet in the solar system and show that every planet radiates more energy than the combined total radiated by those planets whose individual gravitational luminosities are lower than its own.

1.6.3   Find the distance at which a typical dwarf nova system might be detectable by a Michelson interferometer type of direct detector, assuming that the hoped-for sensitivity of such detectors can be achieved.

Chapter 2

Imaging

2.1 THE INVERSE PROBLEM

A problem which occurs throughout much of astronomy and other remote sensing applications is how best to interpret noisy data so that the resulting deduced quantities are real and not artefacts of the noise. This problem is termed the inverse problem.

For example, stellar magnetic fields may be deduced from the polarisation of the wings of spectrum lines (section 5.2). The noise (errors, uncertainties) in the observations, however, will mean that a range of field strengths and orientations will fit the data equally well. The three-dimensional distribution of stars in a globular cluster must be found from observations of its two-dimensional projection onto the plane of the sky. Errors in the measurements will lead to a variety of distributions being equally good fits to the data. Similarly the images from radio and other interferometers (section 2.5) contain spurious features due to the side lobes of the beams. These spurious features may be removed from the image if the effects of the side lobes are known. But since there will be uncertainty in both the data and the measurements of the side lobes, there can remain the possibility that features in the final image are artefacts of the noise, or are incompletely removed side lobe effects etc.

The latter illustration is an instance of the general problem of instrumental degradation of data. Such degradation occurs from all measurements, since no instrument is perfect: even a perfectly constructed telescope will spread the image of a true point source into the Airy diffraction pattern (figure 1.1.29). If the effect of the instrument on a point source or its equivalent is known (the instrumental profile or spread function) then an attempt may be made to remove its effect from the data. The process of removing instrumental effects from data can be necessary for any type of measurement, but is perhaps best studied in relation to imaging, when the process is generally known as deconvolution.

165

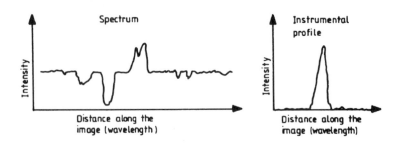

Figure 2.1.1. Representation of a one-dimensional image and the instrumental profile plotted as intensity versus distance along image.

Deconvolution

This form of the inverse problem is known as deconvolution, because the true image *convolves* with the instrumental profile to give the observed image. Inversion of its effect is thus deconvolution. Convolution is most easily illustrated in the one-dimensional case (such as the image of a spectrum or the output from a Mills Cross radio array), but applies equally well to two-dimensional images.

A one-dimensional image, such as a spectrum, may be completely represented by a graph of its intensity against the distance along the image (figure 2.1.1.). The instrumental profile may similarly be plotted and may be found, in the case of a spectrum, by observing the effect of the spectroscope on a monochromatic source.

If we regard the true spectrum as a collection of adjoining monochromatic intensities, then the effect of the spectroscope will be to broaden each monochromatic intensity into the instrumental profile. At a given point (wavelength) in the observed spectrum therefore, some of the original energy will have been displaced out to nearby wavelengths, while energy will have been added from the spreading out of nearby wavelengths (figure 2.1.2). The process may be expressed mathematically by the convolution integral:

$$O(\lambda_1) = \int_0^\infty T(\lambda_2) I(\lambda_1 - \lambda_2) \, d\lambda_2 \qquad (2.1.1)$$

where

$O(\lambda_1)$ is the intensity in the observed spectrum at wavelength λ_1
$T(\lambda_2)$ is the intensity in the true spectrum at wavelength λ_2
$I(\lambda_1 - \lambda_2)$ is the response of the instrument (spectroscope) at a distance $(\lambda_1 - \lambda_2)$ from its centre.

Equation (2.1.1) is normally abbreviated to

$$O = T * I \qquad (2.1.2)$$

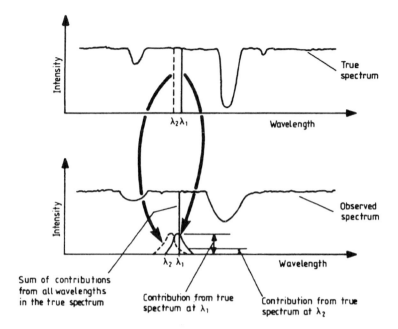

Figure 2.1.2. Convolution of the true spectrum with the instrumental profile to produce the observed spectrum.

where * is the convolution symbol.

The inversion of equation (2.1.1) to give the true spectrum cannot be accomplished directly but involves the use of Fourier transforms.

With the Fourier transform and its inverse in the form

$$F(s) = \mathcal{F}(f(x)) = \int_{-\infty}^{\infty} f(x)e^{-2\pi ixs}\, dx \qquad (2.1.3)$$

$$f(x) = \mathcal{F}^{-I}(F(s)) = \int_{-\infty}^{\infty} F(s)e^{2\pi ixs}\, ds \qquad (2.1.4)$$

then the convolution theorem states that:

Convolution of two functions corresponds to the multiplication of their Fourier transforms.

Thus taking Fourier transforms of equation (2.1.2) we have

$$\mathcal{F}(O) = \mathcal{F}(T * I) \qquad (2.1.5)$$

$$= \mathcal{F}(T) \times \mathcal{F}(I) \qquad (2.1.6)$$

and so the true spectrum (etc) may be found from inverting equation (2.1.6) and taking its inverse Fourier transform:

$$T = \mathcal{F}^{-1}\left[\frac{\mathcal{F}(O)}{\mathcal{F}(I)}\right]. \qquad (2.1.7)$$

In practice, obtaining the true data (or source function) via equation (2.1.7) is complicated by two factors. Firstly most data is sampled at discrete intervals and so is not the continuous function required by equations (2.1.3) and (2.1.4), and also it is not available over the complete range from $-\infty$ to $+\infty$. Secondly, the presence of noise will produce ambiguities in the calculated values of T.

The first problem may be overcome by using the discrete version of the Fourier transform

$$F_D(S_n) = \mathcal{F}_D(f(x))_n = \sum_{k=0}^{N-1} f(x_k)e^{-2\pi ikn/N}\Delta \qquad (2.1.8)$$

where

$F_D(S_n)$ is the nth value of the discrete Fourier transform of $f(x)$
N is the total number of measurements
$f(x_k)$ is the kth measurement
Δ is the step length between measurements.

and the inverse transform

$$f(x_k) = \mathcal{F}_D^{-1}(F_D(S_n)) = \sum_{n=0}^{N-1} F_D(S_n)e^{2\pi ikn/N}\frac{1}{N} \qquad (2.1.9)$$

and setting the functions to zero outside the measured range. Now a function which has a maximum frequency of f is completely determined by sampling at $2f$ (the sampling theorem). Thus, the use of the discrete Fourier transform involves no loss of information providing the sampling frequency $(1/\Delta)$ is twice the highest frequency in the source function. The highest frequency that can be determined for a given sampling interval $(1/2\Delta)$ is known as the Nyquist or critical frequency. If the source function contains frequencies higher than the Nyquist frequency, then these will not be determined by the measurements and the finer detail in the source function will be lost when it is reconstituted via equation (2.1.7). Rather more seriously, however, the higher frequency components may beat with the measuring frequency to produce spurious components at frequencies lower than the Nyquist frequency. This phenomenon is known as aliasing and can give rise to major problems in finding the true source function.

The actual evaluation of the transforms and inverse transforms may nowadays be relatively easily accomplished using the fast Fourier transform algorithm on even quite small computers. The details of this algorithm are outside the scope of this book, but may for example be found in *Numerical Recipes* (see the Bibliography).

The one-dimensional case just considered may be directly extended to two or more dimensions, though the number of calculations involved then increases dramatically. Thus, for example, the two-dimensional Fourier transform equations are

$$F(s_1 s_2) = \mathcal{F}(f(x_1, x_2)) = \int_{-\infty}^{\infty} \int_{-\infty}^{\infty} f(x_1 x_2) e^{-2\pi i x_1 s_1} e^{-2\pi i x_2 s_2} \, dx_1 \, dx_2$$

(2.1.10)

$$f(x_1 x_2) = \mathcal{F}^{-1}(F(s_1 s_2)) = \int_{-\infty}^{\infty} \int_{-\infty}^{\infty} F(s_1 s_2) e^{2\pi i x_1 s_1} e^{2\pi i x_2 s_2} \, ds_1 \, ds_2.$$

(2.1.11)

Some reduction in the noise in the data may be achieved by operating on its Fourier transform. In particular, cyclic noise such as 50 or 60 Hz mains hum or the stripes on scanned images may be reduced by cutting back on or removing the corresponding frequencies in the transform domain. Random noise may be reduced by using the optimal (or Wiener) filter defined by

$$W = \frac{[\mathcal{F}(O)]^2}{[\mathcal{F}(O)]^2 + [\mathcal{F}(N)]^2}$$

(2.1.12)

where $\mathcal{F}(O)$ is the Fourier transform of the observations, without the effect of the random noise, and $\mathcal{F}(N)$ is the Fourier transform of the random noise. (The noise and the noise-free signal are separated by assuming the high-frequency tail of the power spectrum to be just due to noise, and then extrapolating linearly back to the lower frequencies.) Equation (2.1.7) then becomes

$$T = \mathcal{F}^{-1} \left[\frac{\mathcal{F}(O) W}{\mathcal{F}(I)} \right].$$

(2.1.13)

While noise in the data may be reduced by processes such as those outlined above, it can never be totally eliminated. The effect of the residual noise, as previously mentioned, is to cause uncertainties in the deduced quantities. The problem is often ill-conditioned, that is to say, the uncertainties in the deduced quantities may, proportionally, be very much greater than those in the original data.

Recently, several methods have been developed to aid choosing the 'best' version of the deduced quantities from the range of possibilities. Termed 'non-classical' methods, they aim to stabilise the problem by introducing additional information not inherently present in the data as constraints, and thus to arrive at a unique solution. The best known of these methods is the Maximum Entropy Method (MEM). The MEM introduces the external constraint that the intensity cannot be negative, and finds the solution which has the least structure in it that is consistent with the data. The name derives from the concept of entropy as the inverse of the structure (or information content) of a system. The maximum entropy solution is thus the one with the least structure (the least information content or the smoothest solution) that is consistent with the data. The commonest measure of the entropy is

$$s = -\Sigma p_i \ln p_i \qquad (2.1.14)$$

where

$$p_i = \frac{d_i}{\Sigma_j d_j} \qquad (2.1.15)$$

and d_i is the ith datum value, but other measures can be used.

A solution obtained by an MEM has the advantage that any features in it must be real and not artefacts of the noise. However, it also has the disadvantages of perhaps throwing away information, and that the resolution in the solution is variable, being highest at the strongest values. Other approaches can be used either separately or in conjunction with MEM to try and improve the solution. The CLEAN method, much used on interferometer images, is discussed in section 2.5. The method of regularisation stabilises the solution by minimising the norm of the second derivative of the solution as a constraint on the smoothness of the solution.

2.2   PHOTOGRAPHY

Introduction

Photography is such an essential weapon in the armoury of an observational astronomer, that it is difficult to realise that it has only been around for a little over a century. Its pre-eminence as a recording device arises primarily from its ability to provide a permanent record of an observation which is (largely) independent of the observer. But it also has many other major advantages including its cheapness, its ability to accumulate light over a long period of time and so to detect sources fainter than those visible to the eye, its very high information density, and its ease of storage for future reference. These undoubted advantages coupled with our familiarity with the technique have, however, tended to blind us to its many disadvantages. The most important of these are probably its slow speed and very low quantum efficiency. The human eye has an exposure time of about a tenth of a second, but a photograph would need an exposure of ten to a hundred times longer in order to show the same detail. The impression that photography is far more sensitive than the eye only arises because most photographs of astronomical objects have exposures ranging from minutes to hours, and so show far more detail than is seen directly. The processing of a photograph is a complex procedure and there are many possibilities for the introduction of errors. The long delay between an exposure and the developed image can lead to waste of telescope time should there be any problems with the system; there can exist few observational astronomers who have never taken an exposure only to find after development that they had forgotten to remove the dark slide, or to take off one of the mirror covers so that several hours of work have produced only a clear piece of glass! Furthermore, and by no means finally in this list of problems of the photographic method, the final image is recorded as a density, and the conversion of this back to intensity is a tedious process. The whole business of photography in fact bears more resemblance to alchemy than to science, especially when it has to be used near the limit of its capabilities.

A number of alternative imaging techniques, mostly electronic in nature, which overcome some of the disadvantages of photography are therefore in use or under development (sections 1.2, 2.3, 2.4 and 4.1). The use of these on large telescopes has become the primary means of imaging within the last decade. The major disadvantage of the currently existing and proposed alternatives to photography, however, is cost. Though this is coming down rapidly: a cooled CCD camera can now be purchased for less than the price of an amateur's 0.2 m Schmidt–Cassegrain telescope, while when the first edition of this book was published (1984) the cost was that of a 1 m telescope. Nonetheless, the cost per pixel for electronic systems is still many orders of magnitude greater than that of a photograph. Also the hard copy of the image produced by the electronic imaging systems is quite likely to be a photograph taken from a television screen or produced by a computer. Photography will therefore remain a vital technique for at least the next generation of astronomers.

Structure of the photographic emulsion

Many materials exhibit a sensitivity to light which might be potentially usable to produce images. In practice, however, a compound of silver and one or more of the halogens is used almost exclusively. This arises from the relatively high sensitivity of such compounds allied to their ability to retain the latent image (see p. 174) for long periods of time. For most purposes the highest sensitivity is realised using silver bromide with a small percentage of iodide ions in solid solution. The silver halogen is in the form of tiny crystals, and these are supported in a solid, transparent medium, the whole structure is called the photographic emulsion. The supporting medium is usually gelatine, which is a complex organic material manufactured from animal collagen, whose molecules may have masses up to half a million amu. Gelatine has the advantage of adding to the stability of the latent image by absorbing the halogens which are released during its formation. It also allows easy penetration of the processing chemicals, and forms the mixing medium during the manufacture of the emulsion. Its organic nature may create problems of stability and swelling when it is used in thick layers, such as for nuclear emulsions (section 1.4) and it absorbs radiation from wavelengths of about 235 nm downwards creating problems for its use in the ultraviolet (section 1.3).

The size of the silver halide crystals, or grains as they are commonly known, is governed by two conflicting requirements. The first of these is resolution, or

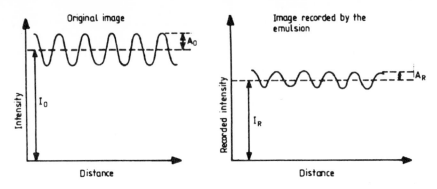

Figure 2.2.1. Schematic representation of the reproduction of an image varying spatially in a sinusoidal manner by an emulsion.

the ability of the film to reproduce fine detail. It is affected by both the grain size and by scattering of the light within the emulsion, and is measured by the maximum number of lines per millimetre that can be distinguished in images of gratings. It ranges from about 20 to 2000 lines per millimetre, with generally the smaller the grain size the higher the resolution. More precisely the resolution of an emulsion may be specified by the Modulation Transfer Function (MTF). The definition of the MTF is in terms of the recording of an image which varies sinusoidally in intensity with distance. In terms of the quantities shown in figure 2.2.1, the modulation transfer, T, is given by

$$T = \frac{(A_O/I_O)}{(A_R/I_R)}. \tag{2.2.1}$$

The MTF is then the manner in which T varies with spatial frequency. Schematic

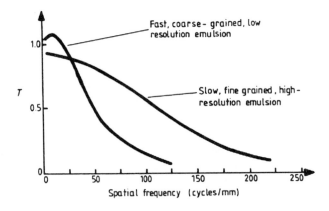

Figure 2.2.2. Schematic MTF curves for typical emulsions.

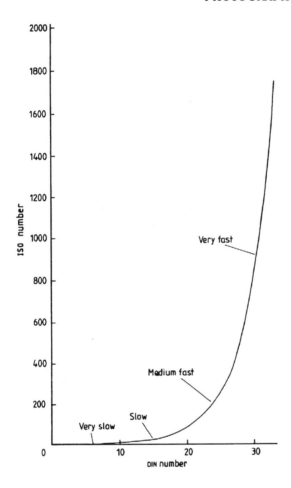

Figure 2.2.3. Emulsion speeds (approximate relationship).

examples of the MTF for some emulsions are shown in figure 2.2.2. Using an MTF curve, the resolution may be defined as the spatial frequency at which the recorded spatial variations become imperceptible. For visual observation, this usually occurs for values of T near 0.05.

The second requirement is that of the speed of the emulsion, and this is essentially the reciprocal of a measure of the time taken to reach a given image density under some standard illumination. The two systems in common use for measuring the speed of an emulsion are ISO, which is based upon the exposure times for image densities of 0.1 and 0.9, and DIN, which is based only on the exposure time to reach an image density of 0.1. ISO is an arithmetic scale, while DIN is logarithmic. Their relationship is shown in figure 2.2.3, together with an indication of the speeds of normal films. The two numbers are frequently combined into the exposure index number (EI) which is simply the ISO number followed by the DIN number, as in EI 100/21°. The speed of an emulsion is proportional to the volume of a grain for unsensitised emulsions, and to the surface area of a grain for dye-sensitised (see below) emulsions. Thus in either case, higher speed requires larger grains. The conflict between these two requirements means that high resolution films are slow, and that fast films have poor resolution. Grain sizes range from 50 nm for a very high resolution film, through 800 nm for a normal slow roll film to 1100 nm for a fast roll film. Not all the grains are the same size, except in nuclear emulsions (section 1.4), and the standard deviation of the size distribution curve ranges from about 1% of the particle diameter for very high resolution films to 50% or more for normal commercial films.

A normal silver halide emulsion is sensitive only to short wavelengths, i.e. to the blue, violet and near ultraviolet parts of the spectrum. The fundamental discovery which was to render lifelike photography possible and to extend the sensitivity into the red and infrared was made by Vogel in 1873. This discovery was that certain dyes when adsorbed onto the surfaces of the silver halide grains would absorb light at their own characteristic frequencies, and then transfer the absorbed energy to the silver halide. The latent image would then be produced in the same manner as normal. The effect of the dye can be very great, a few spectral response curves are shown in figure 2.2.4 for unsensitised and sensitised emulsions. Dyes may be used in combinations, normally they then tend to counteract each other's effects, but they will occasionally act synergistically, so that almost any desired response curve can be produced. In particular the panchromatic or isochromatic emulsions have an almost uniform sensitivity throughout the visible, while the orthochromatic emulsions mimic the response of the human eye. Colour film would of course be quite impossible without the use of dyes since it uses three emulsions with differing spectral responses. The advantages of dye sensitisation are very slightly counteracted by the reduction in the sensitivity of the emulsion in its original spectral region, due to the absorption by the dye, and more seriously by the introduction of some chemical fogging.

The photographic image

When an emulsion is exposed to light, the effect is to form the latent image, that is an image which requires the further step of development (p. 182) in order to be visible. Its production on silver halide grains is thought to arise in the following manner.

(a) Absorption of a photon by an electron occurs in the valence band of the silver halide. The electron thereby acquires sufficient energy to move into the conduction band (see section 1.1 for a discussion of solid state energy levels).

(b) The removal of the electron leaves a positive hole in the valence band.

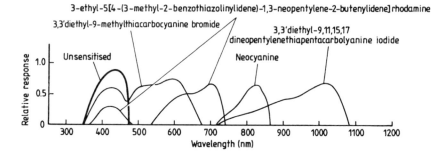

Figure 2.2.4. Effects of dyes upon the spectral sensitivity of photographic emulsion.

(c) Both the electron and the hole are mobile. If they meet again, they will recombine and emit a photon. There will then have been no permanent effect. To avoid this recombination, the electron and the hole must be separated by their involvement in alternative reactions. The electrons are immobilised at chemical or physical traps formed from impurities or crystal defects. Thus very pure silver halides do not form a permanent latent image, their electrons and holes recombine in under a millionth of a second. The positive hole's motion may eventually carry it to the surface of the grain. There it can be consumed directly by reaction with the gelatine, or two holes can combine releasing a halogen atom which will then in turn be absorbed by the gelatine.

(d) The electron, now immobilised, may neutralise a mobile silver ion leaving a silver atom in the crystal structure.

(e) The effect of the original electron trap is now enhanced by the presence of the silver atom, and it may more easily capture further conduction electrons. In this way, a speck of pure silver containing from a few atoms to a few hundred atoms may be formed within the silver halide crystal.

(f) Developers (see p. 182) are reducing solutions which convert silver halide into silver. They act only very slowly, however, on pure silver halide, but the specks of silver act as catalysts, so that those grains containing silver specks are reduced rapidly to pure silver. Since a 1 μm grain will contain 10^{10} to 10^{11} silver atoms, this represents an amplification of the latent image by a factor of 10^9. Adjacent grains which do not contain an initial silver speck will be unaffected and will continue to react at the normal slow rate.

(g) Thus the latent image consists of those grains which have developable specks of silver on them. For the most sensitive emulsions, three to six silver atoms per grain will suffice, but higher numbers are required for less sensitive emulsions, and specks comprising ten to twelve silver atoms normally represent the minimum necessary for a stable latent image in normal emulsions. The latent image is turned into a visible one therefore by developing for a time which is long enough to reduce the grains containing the specks of silver, but which is not long

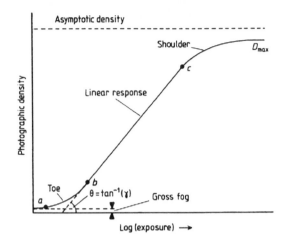

Figure 2.2.5. Schematic characteristic curve of an emulsion.

enough to affect those grains without such specks. The occasional reduction of an unexposed grain occurs through other processes and is a component of chemical fog (see p. 184). After the development has stopped, the remaining silver halide grains are removed by a second chemical solution known as the fixer.

The final visible image therefore consists of a number of silver grains dispersed in the emulsion. These absorb light, and so the image is darkest at those points where it was most brightly illuminated, i.e. the image is the negative of the original. Astronomers customarily work directly with negatives since the production of a positive image normally requires the formation of a photograph of the negative with all its attendant opportunities for the introduction of distortions and errors. This practice may seem a little strange at first, but the student will rapidly become so familiar with negatives that the occasional positive, when encountered, requires considerable thought for its interpretation.

It is possible to produce positive images in a single step. This is known confusingly as reversal, and it is used to produce black and white or colour slides. Several properties of the emulsion may be invoked as a basis for reversal. The most important of these is called solarisation. If the characteristic curve (figure 2.2.5 for example) is continued to very high levels of illumination, then it starts to decrease. A photograph taken in this region will be a positive image. Other reversal techniques rely upon multiple exposures at very high intensity, or upon exposure during development. Reversal films have a small dynamic range, and hence little room for latitude in their exposure. In addition uniform results are difficult to obtain, and so reversal films are never used for direct photography at the telescope except for the production of publicity material for posters etc.

One of the problems encountered in using photographic images to determine the original intensities of sources is that the response of the emulsion is non-linear. The main features of the response curve, or characteristic curve as it is more commonly known, are shown in figure 2.2.5. Photographic density is plotted along the vertical axis. This is more properly called the optical density since it is definable for any partially transparent medium, and it is defined by

$$D = \log(F_i/F_t) \tag{2.2.2}$$

where F_i is the incident flux and F_t is the transmitted flux (figure 2.2.6). The transmittance, T, of a medium is defined similarly by

$$T = F_t/F_i = 10^{-D} \tag{2.2.3}$$

and the opacity, A, by

$$A = F_i/F_t = T^{-1} = 10^D \tag{2.2.4}$$

(where D is the optical density defined in equation (2.2.2)). The intensity of the exposure is plotted on the horizontal axis as a logarithm. Since density is also a logarithmic function (equation (2.2.2)), the linear portion of the characteristic curve (b to c in figure 2.2.5) does represent a genuine linear response of the emulsion to illumination. The response is non-linear above and below the linear portion. The gross fog level is the background fog of the unexposed but developed emulsion and of its supporting material. The point marked a on figure 2.2.5 is called the threshold and is the minimum exposure required for a detectable image. The asymptotic density is the ultimate density of which the emulsion is capable, the maximum density achieved for a particular developer, D_{max}, will generally be below this value. In practice the image is ideally formed in the linear portion of the response curve, although this may not always be possible for objects with a very wide range of intensities, or for very faint objects.

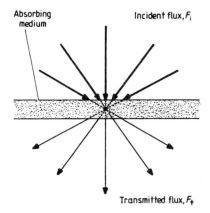

Figure 2.2.6. Quantities used to define photographic density.

The contrast of a film is a measure of its ability to separate regions of differing intensity. It is determined by the steepness of the slope of the linear portion of the characteristic curve. A film with a high contrast has a steep linear section, and a low contrast film a shallow one. Two actual measures of the contrast are used: gamma, and the contrast index. Gamma is simply the slope of the linear portion of the curve

$$\gamma = \tan\theta. \tag{2.2.5}$$

The contrast index is the slope of that portion of the characteristic curve which is actually used to form most images. It usually includes some of the toe, but does not extend above a density of 2.0. Contrast and response of emulsions both vary markedly with the type of developer and with the conditions during development. Thus for most astronomical applications a calibration exposure (see p. 191) must be obtained which is developed along with the required photograph, and which enables the photographic density to be converted back into intensity.

We have already seen how the use of dyes can change the spectral sensitivity of an emulsion (figure 2.2.4). In many applications the response of the film can be used to highlight an item of interest, for example comparison of photographs of the same area of sky with blue- and red-sensitive emulsions is a simple way of finding very hot and very cool stars. Often the wavelength range of the emulsion has to be curtailed by the use of filters. This is especially the case for infrared films whose blue sensitivity may be many times their infrared sensitivity, thus a short wave rejection filter must be used so that the infrared images are not swamped.

The resolution of an emulsion is affected by the emulsion structure (see the earlier discussion and table 2.2.1) and by the processing stages. In addition to the grain size and light scattering in the emulsion, there may also be scattering or reflection from the supporting material (figure 2.2.7). This is known as halation and it may be reduced by the addition of an absorbing coating on the back of the plate. This coating is logically called the anti-halation backing, and it is automatically removed during processing. When plates are to be hypersensitised, such a backing may cause problems by being only partially removed during this process and/or deposited onto the emulsion side of the plate. Thus plates required for hypersensitising, at least by the liquid-based methods, should be specified as unbacked when they are ordered from the supplier. Processing affects resolution through the change in the concentration of the processing solutions within the emulsion since these are more rapidly consumed in well exposed regions and almost unused elsewhere. The basic effect of this on the image may best be demonstrated at a sharp boundary between a uniformly exposed region and an unexposed region. The availability of the developer from the unexposed region at the edge of the exposed region leads to its greater development and so greater density (figure 2.2.8). In astronomy this has important consequences for photometry of small images. If two small regions are given exposures of equal length and equal intensity per unit area, then the smaller image will be darker

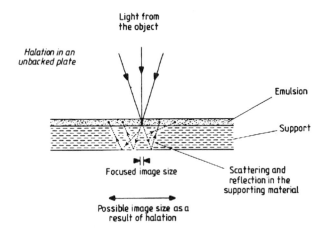

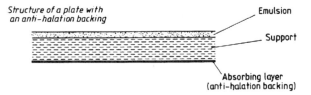

Figure 2.2.7. Halation.

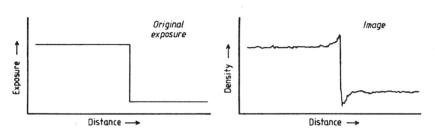

Figure 2.2.8. Edge effect in photographic emulsions.

than the larger one by up to 0.3 in density units. This corresponds to an error in the finally estimated intensities by a factor of two, unless account is taken of the effect. A similar phenomenon occurs in the region between two small close images. The developer becomes exhausted by its work within the two images, leaving the region between them underdeveloped. Images of double stars, close spectrum emission lines, etc may therefore appear more widely separated because of this than is the true case.

The quantum efficiency of an emulsion and its speed are two related concepts. We may measure the quantum efficiency by the number of photons required to produce a detectable image compared with an ideal detector (see also section 1.1). Now all detectors have a certain level of background noise, and we may regard a signal as ideally detectable if it exceeds the noise level by one standard deviation of the noise. In a photographic image the noise is due to the granularity of the image and the background fog. For a Poisson variation in granularity, the standard deviation of the noise is given by

$$\sigma = (N\omega)^{1/2} \qquad (2.2.6)$$

where σ is the noise equivalent power (section 1.1), N is the number of grains in the exposed region and ω is the equivalent background exposure in terms of photons per grain. (All sources of fogging may be regarded as due to a uniform illumination of a fog-free emulsion, the equivalent background exposure is therefore the exposure required to produce the gross fog level in such an emulsion.) Now for our practical, and non-ideal, detector let the actual number of photons per grain required for a detectable image be Ω. then the detective quantum efficiency (section 1.1) is given by

$$\text{DQE} = (\sigma/N\Omega)^2 \qquad (2.2.7)$$

or

$$\text{DQE} = \omega/N\Omega^2. \qquad (2.2.8)$$

A given emulsion may further reduce its quantum efficiency by failing to intercept all the photons which fall onto it. Also between two and twenty photons are required to be absorbed before any given grain reaches developability, as we have already seen. So the quantum efficiency rarely rises above 1% in practice (figure 2.2.9). The speed of an emulsion and its quantum efficiency are related since the speed is a measure of the exposure required to reach a given density in the image. Aspects of the speed have already been mentioned (figure 2.2.3 etc), and it is also inherent in the position of the characteristic curve along the horizontal axis (figure 2.2.5). It can also be affected by processing; the use of more active developers and/or longer developing times usually increases the speed. Hypersensitisation can also improve the speed, for some emulsions the gain can be by factors measured in the hundreds, so that its use is essential and not merely desirable. Other factors which affect the speed include temperature, humidity, the spectral region being observed, the age of the emulsion etc.

One very important factor for astronomical photography which influences the speed very markedly is the length of the exposure. The change in the speed for a given level of total illumination as the exposure length changes is called reciprocity failure. If an emulsion always gave the same density of image for the same number of photons, then the exposure time would be proportional to the reciprocal of the intensity of the illumination. However, for very long and very short exposures this reciprocal relationship breaks down; hence the name reciprocity failure. A typical example of the variation caused by exposure time

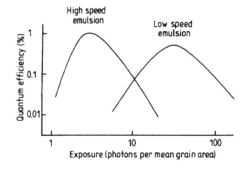

Figure 2.2.9. Quantum efficiency curves for photographic emulsions.

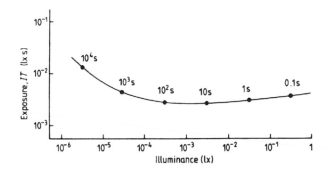

Figure 2.2.10. A typical reciprocity curve for a photographic emulsion.

is shown in figure 2.2.10. Here the total exposure, IT, which is required to produce a given density in the final image is plotted against the illumination. Exposure times are also indicated. If the reciprocity law held, then the graph would be a horizontal straight line. The effect of reciprocity failure is drastic; some commonly available roll films for example may have a speed of a few per cent of their normal ratings for exposures of an hour or more. Most of the emulsions commonly used in astronomy have reduced reciprocity failure but still suffer from the effect to some extent. The temperature of an emulsion during exposure has a major effect upon the reciprocity curves (figure 2.2.11). Frequently therefore the emulsion may be cooled with solid carbon dioxide while it is being exposed in order to try and improve its response. Colour films contain three separate emulsions which in general will have different reciprocity curves, so that the colour balance becomes incorrect at long exposures. Colour reversal film retains its colour balance for exposures up to about fifteen minutes long, but colour negative films are much less tolerant. For both types of colour film cooling may be advantageous, this time because it reduces the differences between the reciprocity curves.

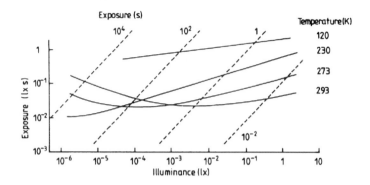

Figure 2.2.11. Schematic effect of the temperature during the exposure of the emulsion upon its reciprocity curve.

A somewhat similar effect may arise if the exposure is intermittent. This is usually the case in spectroscopy (Chapter 4) since the spectrum is widened by trailing the star's image up and down the length of the entrance slit to the spectrograph. If more than about one hundred instalments go into an exposure then it appears that the effect may usually be disregarded, many observatories, however, 'play safe' by interrupting the calibration exposure in a similar manner to that of the main exposure so that the effects are the same for both.

The photographic emulsion is a weak and easily damaged material and it must therefore have a support. This is usually either a glass plate or a plastic film. Both are used in astronomy, but plates are preferred for applications requiring high dimensional stability and/or retention for comparative purposes for long periods of time. In both cases great care must be taken in the storage of the material to avoid scratches, high humidity, mould, breakages, distortion etc which can all too easily destroy the usefulness of the information on the photograph.

Processing

Processing is the means whereby the latent image is converted into a visible and permanent image. For monochrome work there are two main stages; development and fixing. Colour processing is much more complex and is rarely required for astronomical work, the interested reader is therefore referred to specialist books on the subject and to the manufacturers' literature for further information.

The action of a developer is to reduce to a grain of silver those silver halide grains which contain a latent image. Four factors principally affect the ability of a developer to carry out this function: the chemical activity of the developer, the length of the developing time, the temperature of the solution, and the degree of agitation (mixing) during development. The activity of the developer is affected

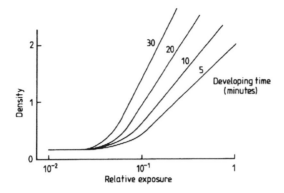

Figure 2.2.12. Effect of changing the developing time upon the characteristic curve of an emulsion.

by its consumption in the reduction of silver halide and also by oxidation by the atmosphere. Its effect may be further inhibited by the presence of reaction products from the development itself. Some developers may be kept active by the addition of replenishers to restore their consumed constituents. However, for most astronomical work, and certainly when a density to intensity conversion is required, fresh developer should always be used at the start of each observing session. In general, the developer should always be replaced long before it has processed the amount of material claimed possible for it by the manufacturer; the small amount of waste involved in this is trivial when compared to the possible waste of telescope and observer's time if the processing should go amiss. Development time and the temperature of development are interrelated. For high quality and reproducible work a standard value for each should be adopted and always used. Normally this is 20°C and the appropriate developing time specified by the manufacturer for that temperature. For non-critical work, there is a useful rule-of-thumb which allows development at other temperatures

$$L_T = L_{20} \times (1.1)^{(20-T)} \qquad (2.2.9)$$

where L_T is the developing time that is required at a temperature T and L_{20} is the developing time at 20°C. The characteristic curve may be altered to some extent by varying the developing time, in particular the contrast may be changed if so desired (figure 2.2.12). However, both the fog and the granularity of the image are increased by increasing the developing time, so the technique is not without its drawbacks. Agitation is the movement of the developer and/or the emulsion to ensure an even mix of the developer. A commonly used technique is to bubble nitrogen from a high pressure cylinder (after passage through a pressure-reducing valve) through the solution. Bursts of bubbles two or three seconds long repeated at ten or twenty second intervals are usually adequate. For critical work however the agitation should be continuous. For large plates great care is required to ensure that the agitation is uniform over the whole plate. Since streams of bubbles can tend to stick to particular 'tramlines', other techniques may be necessary for such plates. Movement of the tank or dish containing the developer, or movement of a soft brush through the liquid are two alternative possibilities. Whichever technique is used, it should be standardised for all work of any particular type.

Some developers produce smaller grain sizes in a given emulsion than others, and can be selected when high resolution is required. However, the finer grains are usually obtained at the cost of a reduced effective speed for the emulsion, so that an emulsion with an intrinsically high resolution may well give a better overall performance. Other developers may also be chosen to enhance the normal speed of the emulsion. But these tend to increase the level of fog, so again an intrinsically faster emulsion may be a better choice. Fog is caused partially by the development of grains which do not have a latent image. Most of the manufacturers' recommendations on developing times, solution strengths etc are intended to minimise the production of fog, and so should only be varied if absolutely essential. In the limit, of course, all developers will develop all the grains, so that longer or stronger developing almost inevitably worsens the fog level.

The second main stage of processing is fixation. The purpose of this is to remove the remaining silver halide which is left in the emulsion after development. Commonly, sodium thiosulphate (known as 'hypo'), or ammonium thiosulphate are used for this purpose. Their reaction products are water soluble and they do not attack the photographic materials. If fixation is over-extended however, the fixer will start to attack the silver image, and so the manufacturers' recommended times should again be used. It is good practice to use two fixing baths with the fresher one used second. As the first bath approaches exhaustion, it is replaced by the second bath, and a new second bath of fresh fixer started.

In addition to these two active stages of processing, the chemicals must be thoroughly washed from the emulsion at each stage. It is also good practice to halt the development sharply by immersing the plate in a bath of acetic acid, or one of the commercially available stop baths. The main stages of processing, in summary, are therefore: development; stop bath; washing; fixation; washing; drying in a dust-free atmosphere.

Most of the processing stages must be carried out in total darkness. Some emulsions do have safelights; that is, they may be illuminated in a wavelength region to which they are insensitive, but this is not possible with all emulsions, and in any case it is recommended to dispense with safelights if possible, since the emulsion usually has some residual sensitivity at all visible wavelengths. Small quantities of developing may be undertaken in the commercially available developing tanks, and with makeshift solution preparation and water supplies. For larger quantities of emulsion and regular usage however, a darkroom is almost essential. This should be a light-tight, dust-free room which is temperature controlled to 20°C. It also needs adequate water and electricity supplies, drainage and timing facilities.   The processing materials and the processing baths may then be arranged conveniently and permanently. Even in such an environment, however, the temperature of the solutions should always be checked before use since evaporation from open tanks can lower the temperature of the solutions quite markedly.

Hypersensitisation

Hypersensitisation includes any process which when applied to an emulsion causes it to have a faster response than normal.   Its use in astronomical applications is very widespread since even a 10% saving on an exposure of an hour is worthwhile, and the savings in exposure time are frequently very much higher than this. Most of the time the mechanism which produces the hypersensitisation is not well understood. Some or all of the following processes may be occurring in particular cases, however.

(a) Chemicals can be removed from the emulsion which have the effect of slowing down its response. These chemicals are usually preservatives which are added to the emulsion to give it a reasonable shelf life. Thus emulsions which may have been hypersensitised by this process must be used very rapidly.

(b) The fog level may be reduced.

(c) The absorption of the radiation may be improved.

(d) The stability and/or developability of the latent image may be improved.

(e) The threshold for the formation of the latent image may be lowered.

The hypersensitisation is usually applied before exposure, but it may also be resorted to after an exposure. In the latter case it is often called latensification, since it is the latent image which is being intensified. The number of different methods of hypersensitisation is very large, to the extent that one gets the feeling that almost anything which is done to an emulsion will improve it! Some of the methods act synergistically, while others tend to cancel out each other's defects. Many of them worsen the fog level, or more seriously, can have non-uniform effects. The use of hypersensitisation has therefore to be a cautious one, but with care it is usually, on balance, advantageous.

A brief review of the main methods used by astronomers is given below, further details of these and other methods may be found fairly easily in the technical literature or from the emulsion manufacturer.   Processes such as dye sensitisation and extended development fall within the definition of hypersensitisation as given earlier.   However, by common consent they are regarded as a part of the normal production of an image and so are not included here.

Temperature effects

The effect of cooling the emulsion during the exposure has already been discussed. The commonest method of hypersensitisation in use in astronomy consists simply of baking the emulsion prior to exposure. For blue-sensitive plates, 24 to 72 hours at a temperature of 50 to 75°C can increase their speed by a factor of two or three. The optimum baking conditions vary between emulsions and even from one batch of a particular emulsion to the next, and so must be determined empirically in each case. The shelf life of baked plates is severely reduced, and there is some increase in the fog level. Red-sensitive emulsions do not seem to benefit by the process. The improvement in the emulsion's response is enhanced if the baking occurs in an inert atmosphere or in a vacuum. The underlying mechanism behind this method of hypersensitisation is thought to be the elimination of moisture and oxygen from the emulsion. During long exposures it may be necessary to continue to bathe the plate in a dry inert gas in order to prevent reabsorption of water and oxygen.

Liquid bathing effects

These methods may be applied most usefully to red- and infrared-sensitive emulsions. Their mode of operation may be that of increasing the concentration of silver ions by removal of halide ions. All the methods increase the fog level, and the optimum gain is usually obtained when the fog level is about double that of the unhypersensitised emulsion. A major problem with this class of methods is that non-uniform drying may cause irregular fogging, although the chances of this happening may be minimised by vigorous agitation during immersion in the liquids and by rapid drying. The shelf life of the emulsion is reduced to a few tens of hours, and the effect of the hypersensitisation on the response is temporary, thus plates hypersensitised in this way must be used within a few hours. It is essential to hypersensitise far-infrared-sensitive emulsions, since the gain in their speed can be as high as a factor of four hundred. Lower, but still substantial, gains are obtained on near infrared materials. The details of the process for an individual emulsion may be found from the manufacturer's literature, commonly, however, the liquids employed are water, silver nitrate solution, or ammonia.

Gas immersion effects

These are frequently combined with baking in order to enhance the effects of both methods. Their action is probably mainly that of the removal of the moisture and oxygen content of the emulsion, but some additional enhancement by chemical reaction may also occur. For optimum results, it may be necessary to retain the gas around the plate during its exposure. The main gases which are used are as follows.

(a) Nitrogen. Immersion in an atmosphere of dry nitrogen for several days, or for shorter periods if baking is simultaneously applied, produces speed gains of up to a factor of five with little increase in the fog level.

(b) Hydrogen (*caution: possible explosion risk*). The plates are immersed in an atmosphere of nitrogen mixed with 2% to 10% hydrogen (known as forming gas) at room temperature for several hours. This is frequently combined with nitrogen baking for some of the commonly used astronomical emulsions, with a resulting gain in sensitivity by about a factor of twenty-five.

(c) Vacuum. Lodging the plates in a soft or hard vacuum for up to 24 hours, possibly combined with baking, can give gains of up to a factor of four.

Pre-exposure

We have seen that from two to twenty photons per grain are required to produce a developable latent image. Faint images may have too few photons to reach this point, but an additional overall uniform low intensity exposure may add sufficient photons to produce a developable image. In effect, pre-exposure brings an emulsion to the start of the toe of the characteristic curve (point *a* of figure 2.2.5). The required intensity of this fogging exposure can be found from the characteristic curve of the emulsion. An improvement in the response by a factor of up to twenty is possible, but the fog level of the emulsion is obviously increased. This is the only method of hypersensitisation (along with its equivalent; post-exposure illumination) which can be applied to colour films.

Post-exposure techniques

The treatments outlined above are all applied before exposure. A few techniques are available which may be applied after the exposure. Their use is less common, but they may be very valuable when a planned exposure is interrupted by weather or some other factor before it has been completed. A usable, if not ideal, photograph may still perhaps be obtained through their application. Most of the methods involve intensifying the latent image, and are therefore also sometimes known as latensification.

One obvious possibility is to develop the emulsion for longer than normal and/or use a more active developer. A second possibility is that the fogging exposure mentioned above may be applied after the main exposure instead of prior to it.

The main techniques in latensification, however, are chemical, and include bathing in some of the following solutions and gases: mercury vapour; gold salts (aurous thiocyanate or aurous thiosulphate); reducing agents such as hydrogen peroxide, sodium perborate, benzoyl peroxide etc; sodium bisulphate or potassium metabisulphate. Latensification must be undertaken within a few hours of the exposure and it is generally most effective for short, high intensity exposures, but some effect still occurs for the long, low intensity exposures of interest to astronomers.

Faint image contrast enhancement

Although this and the following technique are not strictly a part of hypersensitisation, they are used to obtain information from faint images and so are best discussed within the context of this section. Faint images occur in the top layer of the emulsion, while the grains which produce the fog are distributed throughout the emulsion. If a contact print is made of the negative using diffused light, then only the surface layer of the emulsion is reproduced. Thus the background fog intensity is much reduced in comparison with that of the image and the overall signal-to-noise ratio improved. The same technique can also be used to extract information from very dense negatives which have perhaps resulted from overexposure or overdevelopment, or from the intense portions of an image which has a wide range of intensities. Then the improvement relies upon the contact print's image being formed by scattered light from within the emulsion rather than the transmitted light.

Autoradiographic image retrieval

A fully developed negative with a very faint image can have some or all of its information retrieved by autoradiography. The negative is immersed in a solution containing thiourea labelled with radioactive sulphur ($^{35}_{16}$S). This combines chemically with the silver in the image. The treated negative is then placed in contact with a fresh sheet of emulsion, which by preference should be nuclear or x-ray emulsion but can be of any type. The beta decay of the radioactive sulphur then produces an image in the second emulsion whose density may be taken to almost any desired level by leaving the emulsions in contact for a sufficient length of time. Processing of the second emulsion proceeds normally after its exposure.

Film types

It is usually unhelpful to discuss specific commercial products in a book such as this, for they change rapidly. There is also the possibility of accusations of favouritism from those manufacturers who are not included to their own satisfaction! Nevertheless, for both amateur and professional (Schmidt camera) use, Kodak Technical Pan film has recently achieved great popularity. This is a very fine grain film available in 35 mm and other formats. Its sensitivity extends to 685 nm, making it suitable for solar photography at Hα. Its resolution can reach 300 lines/mm and its speed is variable from ISO 25 to 200 by varying the developing conditions. For long exposures, however, it is almost essential to hypersensitise it in forming gas or with silver nitrate (see above).

Techniques of astronomical photography

The prime difference between astronomical photography and more normal types of photography lies in the exposure length. Because of this, not only must the telescope be driven to follow the motion of the stars, but it must be continually guided in order to correct any errors in the drive. Usually guiding is by means of a second and smaller telescope attached to the main instrument. The astronomer guides the telescope by keeping cross wires in the eyepiece of the secondary telescope centred on the object, or by the use of automatic systems (section 1.1). For faint and/or diffuse objects, off-set guiding may be necessary. That is, the guide telescope is moved from its alignment with the main telescope until it is viewing a nearby bright star while the main telescope continues to point at the required object. Guiding then continues using this bright star. Sometimes the main telescope can be used as its own guide. For example in spectroscopy, the star's image may well be larger than the entrance slit to the spectrograph so that the overspill can be viewed for guiding purposes.

The determination of the required length of the exposure may be difficult; exposures may be interrupted by cloud, or the star may be a variable whose brightness is not known accurately. This problem can be overcome by using an integrating exposure meter. A small percentage of the object's light is intercepted and led to a detector. By integrating the output from this, the exposure may be continued until some specified amount of light has been accumulated as indicated by the output of the integrator, irrespective of the variations of the source. Such a system is expensive, however, and is not a viable option for smaller telescopes. On such instruments, the astronomer has to act as his own exposure meter, and use his experience to estimate the required exposure time. Fortunately the linear portion of the characteristic curve is usually sufficiently extensive to provide a considerable latitude in the range of usable exposures.

Amongst amateur astronomers, a commonly used technique is to take photographs by eyepiece projection, instead of at prime focus. A camera without its normal lens is positioned a little distance behind the eyepiece, and the focus adjusted to produce a sharp image on the film. This provides much higher plate scales so that even small objects such as the planets can be resolved with a small telescope. A guide to the extra magnification of eyepiece projection over prime focus photography, and to the change in the exposure which may be required is given by the effective focal ratio

$$\text{EFR} = f(d - e)/e \tag{2.2.10}$$

where f is the focal ratio of the objective, d is the projection distance i.e. the distance from the optical centre of the eyepiece to the film plane and e is the focal length of the eyepiece. If T is the time required for an exposure on an extended object at prime focus, then T', the exposure time for the projected image is given by

$$T' = T(\text{EFR}/f)^2. \tag{2.2.11}$$

The limit to the length of exposure times is usually imposed by the background sky brightness. This produces additional fog on the emulsion, and no improvement in the signal-to-noise ratio will occur once the total fog has reached the linear portion of the characteristic curve. Since the brightness of star images is proportional to the square of the objective's diameter, while that of an extended object, which includes the sky background, is inversely proportional to the square of the focal ratio (section 1.1), it is usually possible to obtain adequate photographs of star fields by using a telescope with a large focal ratio. But for extended objects no such improvement is possible since their brightness scales in the same manner as that of the sky background. Much of the light in the sky background comes from scattered artificial light, so that observatories are normally sited in remote spots well away from built-up areas. Even so, a fast Schmidt camera may have an exposure limit of a few tens of minutes at best. Some further improvement may sometimes then be gained by the use of filters. The scattered artificial light comes largely from sodium and mercury vapour street lighting, so that a filter which rejects the most important of the emission lines from these sources can improve the limiting magnitude by two or three stellar magnitudes at badly light-polluted sites. For the study of hot interstellar gas clouds, a narrow band filter centred on the $H\alpha$ line will almost eliminate the sky background while still allowing about half the energy from the source to pass through it.

As already mentioned, colour photography is rarely required for astronomical purposes. When, however, it is necessary, colour film is not normally the best material to use by reason of its slow speed and the differential reciprocity failure between its various constituent emulsions. Instead three separate black and white exposures through magenta, yellow and cyan filters are taken. The image is then reconstituted by colour printing each of these negatives in turn onto a single sheet of printing paper through these same filters, care of course being taken that the images are correctly aligned with each other. For non-critical work, cooling of the emulsion may give sufficiently satisfactory results as discussed earlier.

Analysis of photographic images

Photographs are used by the astronomer to provide two main types of information: the relative positions of objects, and the relative intensities of objects. The plate analysis for the first of these is straightforward, although care is needed for the highest quality results. The positions of the images are measured by means of a machine which is essentially a highly accurate travelling microscope (see sections 4.2 and 5.1 for further discussions of the techniques of position measurement on photographic plates).

To determine the relative intensities of sources from their photographs is far more of a problem. This is due to the non-linear nature of the emulsion's characteristic curve (figure 2.2.5). In order to convert back from the photographic density of the image to the intensity of the source, this curve must be known with some precision. It is obtained by means of a photometric calibration exposure, which is a photograph of a series of sources of differing and known intensities. The characteristic curve is then plotted from the measured densities of the images of these sources. To produce accurate results, great care must be exercised thus:

(a) the photometric calibration plate must be from the same batch of plates as that used for the main exposure;

(b) the treatment of the two plates, including their storage, must be identical;

(c) both plates must be processed, hypersensitised etc together in the same processing baths;

(d) the exposure lengths, intermittency effects, temperatures during exposures etc. should be as similar as possible for the two plates;

(e) if the main exposure is for a spectrum, then the calibration exposure should include a range of wavelengths as well as intensities;

(f) both plates must be measured under identical conditions.

With these precautions, a reasonably accurate characteristic curve may be plotted. It is then a simple although tedious procedure to convert the density of the main image back to intensity. The measurement of a plate is undertaken on a microdensitometer. This is a machine which shines a small spot of light through the image and measures the transmitted intensity. Such machines may readily be connected to a small computer, and the characteristic curve also fed into it, so that the conversion of the image to intensity may be undertaken automatically and in real time.

Photometry of the highest accuracy is no longer attempted from photographs for stars, but a rough guide to their brightnesses may be obtained from the diameters of their images, as well as by direct microphotometry. Scattering, halation, diffraction rings etc around star images mean that bright stars have larger images than faint stars. If a sequence of stars of known brightnesses is on the plate, then a calibration curve may be plotted, and estimates of the magnitudes of large numbers of stars then made very rapidly. Integrated magnitudes of galaxies can be obtained by deliberately moving the plate holder during the exposure. If the movement is circular and rapid enough for many cycles to occur during the exposure, then the images of both stars and galaxies will be smeared out in comparable small circular images. The magnitudes of the galaxies may then be obtained from the known magnitudes of the stars. Further discussion of photometry based on photographs will be found in sections 3.2 and 4.2.

Special applications

Special, high density emulsions are used to detect x- and gamma rays, and the electrons in electronographic cameras. These applications are reviewed in more detail in sections 1.3, 1.4 and 2.3.

A technique which has recently sprung into prominence because of its ability to show full details of images with very wide ranges of contrast is unsharp masking. A contact positive print is made of the original negative on to film, and this is then superimposed on to the negative, but with a small separation. The composite is then printed conventionally with the negative image sharply focused. The positive image is out of focus and acts as a filter to even out the density range of the negative.

There is increasing use of 'false colour' photography to highlight particular items of interest. This technique uses a colour print made from a composite of black and white negatives obtained through filters which were different from those used to produce the colour print. Frequently one of the filters used to take the original exposures will be an infrared transmission filter, so that the technique then enables temperature differences to be picked out easily.

'Pseudo' photographs are frequently produced by computers from imaging data obtained by non-photographic means. Our familiarity with photographs renders such images more acceptable than, say, intensity contour maps. Data from radar, radio telescopes, and satellite-borne instrumentation are often presented in this manner

2.3   ELECTRONIC IMAGING

Introduction

The alternatives to photography for recording images directly are almost all electronic in nature. They have two great advantages over the photographic emulsion. Firstly the image is usually produced as an electrical signal and can therefore be relayed or transmitted to a remote observer, which is vital for satellite-borne instrumentation and useful in many other circumstances, and also enables the data to be fed directly into a computer. Secondly, with many of the systems the quantum efficiency is up to a factor of a hundred or so higher than that of the photographic plate. Subsidiary advantages can include intrinsic amplification of the signal, linear response, and long wavelength sensitivity. The major disadvantages include the small number of pixels (1% or less of those available on a good photograph), low reliability, complexity of operation, high cost, short working life, and geometrical distortion of the image. Very high voltages are sometimes required and this can be a disadvantage or even a safety hazard under normal practical observing conditions. Another problem with electronic imaging is that the initial development and application is usually military in nature, so that the systems which are declassified and available for civilian use are a long way from being the most sensitive state-of-the-art devices.

Arrays

The most basic form of electronic imaging simply consists of an array of point-source detecting elements. The method is most appropriate for the intrinsically small detectors such as the solid state photovoltaic and photoconductive cells (section 1.1). Hundreds or thousands or even more of the individual elements (usually termed pixels) can then be used to give high spatial resolution. The array is simply placed at the focus of the telescope or spectroscope etc in place of the photographic plate.

The slightly more complex arrays called charge-coupled devices and charge injection devices were reviewed in detail in section 1.1. In them the individual detecting elements are cross-linked, instead of having to have separate connections to each pixel. Addressing an individual pixel is therefore less direct, but is simpler and cheaper in terms of the circuits which are required. They can also integrate the signal for long periods of time.

Television and related systems

The standard commercially available television cameras have several advantages over photographic emulsions, such as high quantum efficiency and no reciprocity failure. But their application to imaging in astronomy has been very limited. Their main use has been to provide displays for parties of visitors to the observatory! The reason for this lies in the short integration times which are available with the standard television camera systems. Even if one of these cameras is adapted to provide a slower scan, it is still of little use; at room temperature the pixel isolation rapidly breaks down, while if the tube is cooled, the response becomes very patchy because of variations in the electrical isolation of the pixels.

Low light level television systems, combined with image intensifiers and perhaps using a slower scan rate than normal are useful. They are particularly found on the guide systems of large telescopes where they enable the operator at a remote-control console to have a viewing/finding/guiding display.

For some applications the cost of developing a custom-designed and built camera may not be prohibitive. These applications are primarily for satellite instrumentation. Planetary probes make extensive use of television cameras of many varieties for direct imaging. Their scanning rate is slow compared with conventional systems, but this is more often due to the rate of transmission of data back to Earth, than to the necessity for intrinsically long integration times. Indirect imaging, such as to record spectra etc. also employs television systems, particularly for ultraviolet work. Secondary electron conduction (SEC) cameras which allow integration for many hours are especially useful in this region when allied to an ultraviolet-to-optical converter.

The SEC and the related EBS (electron bounce silicon) systems are based upon the vidicon television camera. The sensitive element in a basic vidicon

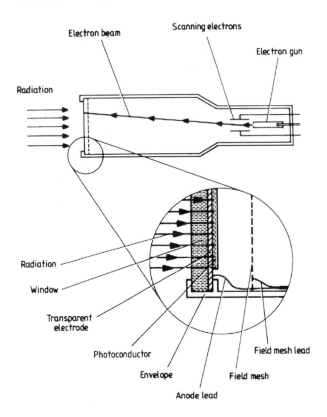

Figure 2.3.1. The vidicon television camera.

is a photoconducting layer (section 1.1). This is deposited onto a transparent conductor which is held at a slight positive potential and which faces the window of the tube. The photoconducting layer changes its conductivity in the pattern of the image which is falling upon it, becoming the most conducting where it is most intensely illuminated. Since the photoconductor is also a semiconductor, the positive charge applied by the transparent electrode will slowly leak through to its inner face. But since the rate of leakage will depend upon the conductivity, the potential on the inner surface will have a pattern similar to that of the original image. This pattern of potential is then scanned by an electron beam (figure 2.3.1) which almost instantaneously raises the inner face potential at the point of impact to that of the cathode. Capacitive coupling across the photoconducting layer to the transparent anode then produces the required video output current. The magnitude of the current is proportional to the original image intensity at the point being scanned by the electron beam. A positively charged field mesh is also necessary to protect the photoconductive layer from positive ions produced in the residual gas inside the tube. Photoconductive

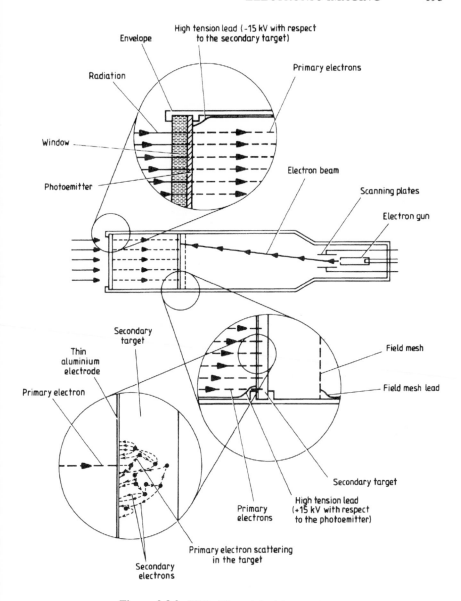

Figure 2.3.2. SEC vidicon television camera.

materials which may be used include selenium, antimony trisulphide, cadmium selenide, lead oxide and silicon. The latter two substances have low intrinsic resistivities and so must be divided into a grid of separately insulated pixels. The SEC vidicon works on a similar principle except that the image which is scanned is produced by primary electrons, and not directly by the photons. The photons impinge onto a photoemitter (section 1.1) on the window of the tube (figure 2.3.2). The ejected electrons are accelerated down the tube by a potential of 10 to 20 kV. On hitting the secondary target, the primary electrons eject many secondary electrons. These are then attracted to a positive electrode and removed, leaving a pattern of potential on the target which matches the original image. This potential image is then read off in a manner identical to that of the normal vidicon. The secondary target is formed of potassium chloride. The primary target can be any photoemitter, and can be varied to suit the wavelength region being observed (see section 1.1 for a list of suitable photoemitters). The SEC tube has two major advantages. Firstly, the primary electrons produce many times their own number of secondary electrons so that there is an intrinsic amplification of the signal by up to a factor of one hundred. Secondly, long exposures are possible since the secondary electrons are removed so that the potential pattern may continue to build up in the secondary target without being erased. Integration times of hours are easily achievable. The tube also has a disadvantage in that the image in the secondary target may not be completely wiped out by the scanning of the electron beam. Several expose and wipe sequences using uniform illuminations may be required to eliminate a previous image. The turn-round time between successive exposures may therefore be several minutes, or even several tens of minutes, with the consequent reduction in observing efficiency. The EBS tube works on the same principle as the SEC tube, but uses silicon for its secondary target. It can have intrinsic gains of up to a factor of one thousand, but requires cooling if integration times of longer than a few seconds are needed.

Image intensifiers

We have already encountered a type of image intensifier in the SEC and EBS television cameras discussed above. They have an intrinsic amplification by a factor of up to 1000 through the production of many secondary electrons by a single primary electron. The electronographic cameras discussed below also have an amplification through the use of accelerated electrons. The conventional use of the term 'image intensifier', however, is usually restricted to devices which produce amplified optical images as their output. The amplification is again achieved through the use of accelerated electrons for all the devices currently in use. There are three main designs for image intensifiers: magnetically focused, electrostatically focused, and proximity focused devices.

Magnetically focused image intensifiers

A standard photocathode (see section 1.1) for the spectral region to be observed is coated onto the inside of the window of the device (figure 2.3.3). The electrons emitted from this when it is illuminated, are then accelerated down the tube by a potential of up to 40 kV and focused onto the phosphor by the magnetic field

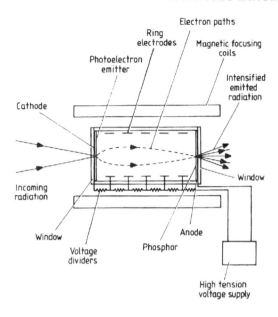

Figure 2.3.3. Cross section through a magnetically focused single-stage image intensifier.

of the surrounding solenoid and by the electric fields of the ring electrodes. The phosphor emits light through the rapid recombination of the electron–hole pairs produced in it by the primary electrons. The spectrum of the emitted light corresponds to the energy gap between the valence and the conduction bands of the phosphor. Because of the very high energy of the primary electrons, many electron–hole pairs are produced by each primary electron, and so the light emitted by the phosphor is many times brighter than that in the original image. The intensified image can be viewed directly, or recorded by almost any of the imaging systems that we have discussed. The main drawback of the device is the geometrical distortion which it can introduce into the image. Figure 2.3.4 shows a fairly typical example of this distortion for a good quality tube. Some of the other problems of the devices, such as cathode sensitivity variations, are similar to those of photomultipliers (section 1.1). The very high voltages which are used can lead to breakdown of the insulation and/or glow discharges under conditions of high humidity.

Electrostatically focused image intensifiers

These are identical to the magnetically focused systems which we have just discussed, except that the focusing of the electron paths is entirely electrostatic. A typical arrangement is shown schematically in figure 2.3.5.

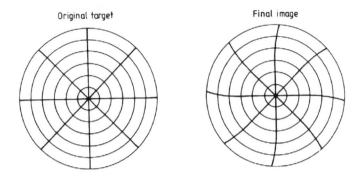

Figure 2.3.4. Rotational and 'S' distortion of the image by an image intensifier.

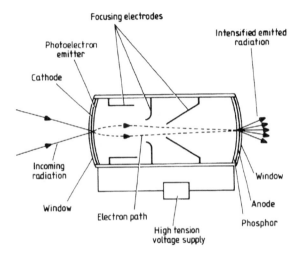

Figure 2.3.5. Cross section through an electrostatically focused single-stage image intensifier.

Proximity focused image intensifiers

In these devices the electron paths are kept very short so that they have little opportunity to become defocused, and in addition they may be mechanically restricted by channelling down small tubes. Magnetic and/or electrostatic focusing is not used so that geometrical distortion of the image is avoided; however, the spatial resolution is reduced by the spread of electron paths. The simplest system is just a vacuum tube with the photocathode close to the phosphor. More commonly used is a microchannel plate of the type discussed in section 1.3 for x-ray imaging. The only difference lies in the type of photoemitter being used; one appropriate to the observed wavelength range must be selected instead of those for x-rays. The other details of the system are covered in section 1.3 and so are not repeated here.

Cascades

A single image intensifier of any of the above types amplifies the intensity of the original image by a factor usually between fifty and a hundred. If larger amplifications are needed, then several of the basic units can be linked together. In this way amplifications by up to a factor of 10^7 can be achieved by commercially available devices. The output from one basic unit is usually connected to the photocathode of the next by a coherent fibre optic light guide. The only modification otherwise required to a basic unit is that the second and subsequent stages must have photoelectron emitters whose response is matched to the emission spectrum of the phosphor, and not to the wavelength region which is being observed.

Recording

The large amplification of the image intensifier is primarily of use in that it allows a detector with a low quantum efficiency such as the eye or a photographic emulsion to work at the quantum efficiency of the photocathode. Since this can reach 50% or more, there is a corresponding gain in the signal-to-noise ratio of the final image. Additionally with photographic plates the higher image intensity may remove or reduce the problem of reciprocity failure. Television cameras and charge-coupled devices are the other methods which are commonly used to detect the final image of an image intensifier. One subsidiary advantage of the image intensifier is that the phosphor of the final stage can be chosen so that its output spectrum matches the response of the detector, irrespective of original wavelengths being observed. The amplification of the image intensity naturally leads to shorter exposure times, but the improvement is only given by the improvement in the effective quantum efficiency of the final detector, if a given signal-to-noise ratio in the image is to be reached. The very large apparent gains of image intensifiers are therefore somewhat misleading as a guide to the savings in telescope time which their use may provide.

Photon counting imaging systems

A combination of a high-gain image intensifier and TV camera led in the 1970s to the development by Boksenberg and others of the IPCS (Image Photon Counting System). In the original IPCS, the image intensifier was placed at the telescope's (or spectroscope's etc) focal plane and produced a blip of some 10^7 photons at its output stage for each incoming photon in the original image. This output was then viewed by a relatively conventional TV camera and the video signal fed to a computer for storage. In its basic form the IPCS suffered from two major drawbacks. Firstly there was a large amount of noise from the image intensifier. Secondly, the image tube spread the cloud of electrons from a single original photon into a blob of significant physical size. The IPCS overcame these problems by using the computer to discriminate between the blips from incoming photons and those from other sources which made up the background noise. The computer also determined the centre of the output 'blob' in order to provide precise positional information. In this way a sharp, almost noise-free, picture could be built-up of the original image.

The IPCS has many advantages, apart from its high quantum efficiency. Thus, long integration times are possible since the data are stored in the computer's memory. The image can be inspected during its formation, allowing real-time judgements to be made of the required exposures etc so that telescope time is used very efficiently, and finally many of the early stages of data reduction such as correction for distortion, image enhancement, etc can be performed immediately after the exposure is completed. Despite these advantages, the IPCS is now frequently superseded by CCD detectors with their higher quantum efficiencies and lower noise levels (section 1.1), and so is becoming less widely used.

More recently a number of variants on Boksenberg's original IPCS have been developed. These include microchannel plate image intensifiers linked to CCDs, and other varieties of image intensifiers linked to CCDs or other array-type detectors such as the Reticon. The principles of the devices remain unchanged however.

Electronographic cameras

These are essentially image intensifiers in which the electrons of the final stage interact directly with the photographic emulsion. There is an overall speed gain compared with direct photography by up to a factor of ten, and a subsidiary advantage arising from the use of nuclear photographic emulsion (section 1.4) in that there is an almost linear characteristic curve. In some systems the emulsion is inside the vacuum chamber, in others there is a thin mica window through which the electrons must pass, and the emulsion is pressed against this from outside. All the devices require great care in their use. In the first type of system, only a few plates can be loaded inside the vacuum chamber at any one time, and the vacuum is destroyed when they are removed for processing. The destruction of the vacuum also damages the photocathode, and so this has to be remade every time a new batch of plates is loaded and the system pumped down again. With systems where the emulsion is outside the vacuum chamber, the mica window is only a few microns thick and so is very easily damaged or broken. Should this occur, the whole tube has to be remade. Also the mica absorbs a significant number of the electrons, so that their gain is reduced in comparison with the first type of system. These devices also suffer from most of the ills which plague image intensifiers. Despite all these problems, electronographic cameras have had quite widespread use at many observatories.

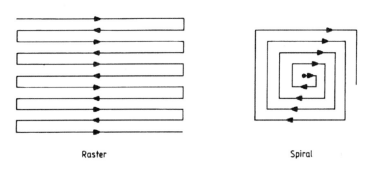

Raster   Spiral

Figure 2.4.1. Scanning patterns.

2.4 SCANNING

This is such an obvious technique as scarcely to appear to deserve a separate section, and indeed at its simplest it hardly does so. A two-dimensional image may be built up using any point source detector, if that detector is scanned over the image or vice versa. Many images at all wavelengths are obtained in this way. Sometimes the detector moves, sometimes the whole telescope or satellite etc. Occasionally it may be the movement of the secondary mirror of the telescope or of some *ad hoc* component in the apparatus which allows the image to be scanned. Scanning patterns are normally raster or spiral (figure 2.4.1). Other patterns may be encountered, however; for example, the continuously nutating roll employed by some of the early artificial satellites. The only precautions advisable are the matching of the scan rate to the response of the detector or to the integration time being used, and the matching of the separation of the scanning lines to the resolution of the system. More specialised applications of scanning occur in the spectrohelioscope (section 5.3), and the scanning spectrometer (section 4.2). Many Earth observation satellites use push-broom scanning. In this system a linear array of detectors is aligned at right angles to the spacecraft's ground track. The image is then built up as the spacecraft's motion moves the array to look at successive slices of the swathe of ground over which the satellite is passing.

A more sophisticated approach to scanning is to modulate the output of the detector by interposing a mask of some type in the light beam. Examples of this technique are discussed elsewhere and include the modulation collimator and the coded array mask used in x-ray imaging (section 1.3). A very great improvement over either these methods or the basic approach discussed above may be made by using a series of differing masks, or by scanning a single mask through the light beam so that differing portions of it are utilised. This improved method is known as Hadamard mask imaging. We may best illustrate its principles by considering one-dimensional images such as might be required for spectrophotometry. The optical arrangement is shown in figure 2.4.2. The

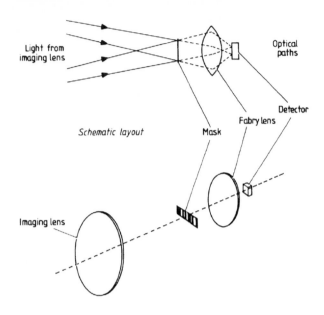

Figure 2.4.2. Schematic arrangement for a Hadamard mask imaging system.

mask is placed in the image plane of the telescope, and the Fabry lens directs all the light which passes through the mask onto the single detector. Thus the output from the system consists of a simple measurement of the intensity passed by the mask. If a different mask is now substituted for the first, then a new, and in general different intensity reading will be obtained. If the image is to be resolved into N elements, then N such different masks must be used, and N intensity readings determined. If D is the vector formed from the detector output readings, I is the vector of the intensities of the elements of the image, and $\mathbf{M}$ the $N \times N$ matrix whose columns each represent one of the masks, with the individual elements of the matrix corresponding to the transmissions of the individual segments of the mask, i.e.

$$D = [D_1 D_2 D_3 \ldots D_N] \tag{2.4.1}$$

$$I = [I_1 I_2 I_3 \ldots I_N] \tag{2.4.2}$$

$$\mathbf{M} = \begin{bmatrix} m_{11} & m_{12} & m_{13} & \cdots & m_{1N} \\ m_{21} & & & & \\ m_{31} & & & & \\ \vdots & & & & \\ m_{N1} & & & \cdots & m_{NN} \end{bmatrix}. \tag{2.4.3}$$

Then, ignoring any contribution arising from noise, we have

$$D = I\mathbf{M} \tag{2.4.4}$$

and so

$$I = D\mathbf{M}^{-1}. \tag{2.4.5}$$

Thus the original image is simply obtained by inverting the matrix representing the masks. The improvement of the method over a simple scan lies in its multiplex advantage (cf the Fourier transform spectrometer, section 4.1). The masks usually comprise segments which either transmit or obscure the radiation completely, that is

$$m_{ij} = 0 \text{ or } 1 \tag{2.4.6}$$

and on average, about half of the total image is obscured by a mask. Thus $N/2$ image segments contribute to the intensity falling onto the detector at any one time. Hence if a given signal-to-noise ratio is reached in a time T, when the detector observes a single image element, then the total time required to detect the whole image with a simple scan is

$$N \times T \tag{2.4.7}$$

and is approximately

$$\sqrt{2N}T \tag{2.4.8}$$

for the Hadamard masking system. Thus the multiplex advantage is approximately a factor of

$$\sqrt{\frac{N}{2}} \tag{2.4.9}$$

improvement in the exposure length.

In practice, the different masks are generated by moving a larger mask across the image. The matrix representing the masks must then be cyclic, or in other words, each successive column is related to the previous one by being moved down a row. The use of a single mask in this way can lead to a considerable saving in the construction costs, since $2N - 1$ segments are needed for it compared with N^2 if separate masks are used. An additional constraint on the matrix $\mathbf{M}$ arises from the presence of noise in the output. The errors introduced by this are minimised when

$$\text{Tr}\left[\mathbf{M}^{-1}\left(\mathbf{M}^{-1}\right)^{T}\right] \tag{2.4.10}$$

is minimised. There are many possible matrices which will satisfy these two constraints, but there is no completely general method for their generation. One method of finding suitable matrices and so of specifying the mask, is based upon the group of matrices known as the Hadamard matrices (hence the name for this scanning method). These are matrices whose elements are ± 1's and which have the property

$$\mathbf{H}\mathbf{H}^{T} = N\mathbf{I} \tag{2.4.11}$$

where $\mathbf{H}$ is the Hadamard matrix and $\mathbf{I}$ is the identity matrix. A typical example of the mask matrix, $\mathbf{M}$, obtained in this way might be

$$\mathbf{M} = \begin{bmatrix} 0 & 1 & 0 & 1 & 1 & 1 & 0 \\ 0 & 0 & 1 & 0 & 1 & 1 & 1 \\ 1 & 0 & 0 & 1 & 0 & 1 & 1 \\ 1 & 1 & 0 & 0 & 1 & 0 & 1 \\ 1 & 1 & 1 & 0 & 0 & 1 & 0 \\ 0 & 1 & 1 & 1 & 0 & 0 & 1 \\ 1 & 0 & 1 & 1 & 1 & 0 & 0 \end{bmatrix} \tag{2.4.12}$$

for N having a value of seven.

Variations to this scheme can include making the obscuring segments out of mirrors and using the reflected signal as well as the transmitted signal, or arranging for the opaque segments to have some standard intensity rather than zero intensity. The latter device is particularly useful for infrared work. The scheme can also easily be extended to more than one dimension, although it then rapidly becomes very complex. A two dimensional mask may be used in a straightforward extension of the one-dimensional system to provide two-dimensional imaging. A two-dimensional mask combined suitably with another one-dimensional mask can provide data on three independent variables—for example, spectrophotometry of two-dimensional images. Two two-dimensional masks could then add (for example) polarimetry to this, and so on. Except in the one-dimensional case, the use of a computer to unravel the data is obviously essential, and even in the one-dimensional case it will be helpful as soon as the number of resolution elements rises above five or six.

2.5   INTERFEROMETRY

Introduction

Interferometry is the technique of using constructive and destructive addition of radiation to determine information about the source of that radiation. Two principal types of interferometer exist: the Michelson stellar interferometer and the intensity interferometer, together with a number of subsidiary techniques such as amplitude interferometry, speckle interferometry, Fourier spectroscopy etc. The Michelson stellar interferometer is so called in order to distinguish it from the Michelson interferometer used in the 'Michelson–Morley' experiment and in Fourier spectroscopy. Since the latter type of interferometer is discussed primarily in section 4.1, we shall drop the 'stellar' qualification for this section when referring to the first of the two main types of interferometer. Both of the main types of interferometer have been used over all wavelengths from the visible to the longest radio waves, but by far the commonest combination is a Michelson interferometer used at radio frequencies, and this is the system which we discuss first.

Michelson radio interferometer

The individual elements of a radio interferometer are usually fairly conventional radio telescopes (section 1.2). Their signals may then be combined in two quite different ways to provide the interferometer output. In the simplest version, the signals are simply added together before the square law detector (figure 1.2.2), and the output will then vary with the path difference between the two signals. Such an arrangement, however, suffers from instability problems, particularly due to variations in the voltage gain.

A system that is often preferred to the simple adding interferometer is the correlation or multiplying interferometer. In this, as the name suggests, the IF signals from the receivers are multiplied together. The output of an ideal correlation inteferometer will contain only the signals from the source (which are correlated), the other components of the outputs of the telescopes will be zero as shown below.

If we take the output voltages from the two elements of the interferometer to be $(V_1 + V_1')$ and $(V_2 + V_2')$ where V_1 and V_2 are the correlated components (i.e. from the source in the sky) and V_1' and V_2' are the uncorrelated components, the product of the signals is then

$$V = \left(V_1 + V_1'\right) \times \left(V_2 + V_2'\right) \tag{2.5.1}$$
$$= V_1 V_2 + V_1' V_2 + V_1 V_2' + V_1 V_2'. \tag{2.5.2}$$

If we average the output over time, then any component of equation (2.5.2) containing an uncorrelated component will tend to zero. Thus

$$\overline{V} = \overline{V_1 V_2}. \tag{2.5.3}$$

In other words, the time-averaged output of a correlation interferometer is the product of the correlated voltages. Since most noise sources contribute primarily to the uncorrelated components, the correlation interferometer is inherently much more stable than the adding interferometer. The phase-switched interferometer (see below) is an early example of a correlation interferometer, though direct multiplying interferometers are now more common.

Radio interferometers generally use more than two antennae, and these may be in a two-dimensional array. The basic principle, however, is the same as for just two antennae, although the calculations become considerably more involved. The principle is also the same whether light or radio waves are being used. As we saw earlier, however, there are different traditions of notation etc between optical and radio astronomers which can give the impression that different phenomena are involved in the two regions even when this is not the

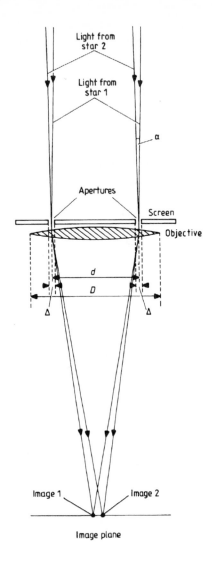

Figure 2.5.1. Optical arrangement of a Michelson interferometer.

case. Here therefore the Michelson radio interferometer is described in optical terms in order to emphasise the links between the two disciplines.

Radio aerials are generally only a small number of wavelengths in diameter (section 1.2), and the electrical signal which they output varies in phase with the received signal. In fact with most radio interferometers it is the electrical signals which are mixed to produce the interference effect rather than the radio signals themselves. The optical equivalent to a radio interferometer is therefore

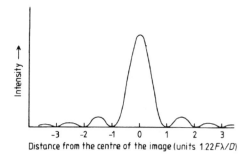

Figure 2.5.2. Image structure for one point source and the whole objective.

a pair of small apertures through which the light passes in order to be viewed. The complete optical system which we shall study is shown in figure 2.5.1. The telescope objective, which for simplicity is shown as a lens, is covered by an opaque screen with two small apertures, and is illuminated by two monochromatic point sources at infinity. The objective diameter is D, its focal length F, the separation of the apertures is d, the width of the apertures is Δ, the angular separation of the sources is α, and their emitted wavelength is λ.

Let us first consider the objective by itself, without the screen, and with just one of the sources. The image structure is then just the well known diffraction pattern of a circular lens (figure 2.5.2). With both sources viewed by the whole objective, two such patterns are superimposed. There are no interference effects between these two images, for their radiation is not mutually coherent. When the main maxima are superimposed upon the first minimum of the other pattern, we have Rayleigh's criterion for the resolution of a lens (figure 2.5.3, see also section 1.1). The Rayleigh criterion for the minimum angle between two separable

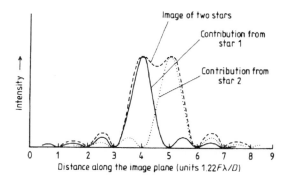

Figure 2.5.3. Image structure for two point sources and the whole objective.

sources α', as we saw in section 1.1 is

$$\alpha' = 1.22\lambda/D. \tag{2.5.4}$$

Now let us consider the situation with the screen in front of the objective, and firstly consider just one aperture looking at just one of the sources. The situation is then the same as that illustrated in figure 2.5.2, but the total intensity and the resolution are both reduced because of the smaller size of the aperture compared with the objective (figure 2.5.4). Although the image for the whole objective is also shown for comparison in figure 2.5.4, if it were truly to scale for the situation illustrated in figure 2.5.1, then it would be one seventh of the width which is shown and one thousand eight hundred times higher!

Now consider what happens when one of the sources is viewed simultaneously through both small apertures. If the apertures were infinitely small, then ignoring the inconvenient fact that no light would get through anyway, we would obtain a simple interference pattern (figure 2.5.5). The effect of the finite width of the apertures is to modulate the simple variation of figure 2.5.5 by the shape of the image for a single aperture (figure 2.5.4). Since two apertures are now contributing to the intensity and the energy 'lost' at the minima reappears at the maxima, the overall envelope of the image peaks at four times the intensity for a single aperture (figure 2.5.6). Again, for the actual situation shown in figure 2.5.1, there should be a total of 33 fringes inside the major maximum of the envelope.

Finally let us consider the case of two point sources viewed through two apertures. Each source has an image whose structure is that shown in figure 2.5.6, and these simply add together in the same manner as the case illustrated in figure 2.5.3 to form the combined image. The structure of this combined image will depend upon the separation of the two sources. When the sources are superimposed, the image structure will be identical with that shown in figure 2.5.6, except that all the intensities will have doubled. As the sources move apart, the two fringe patterns will also separate, until when the sources are

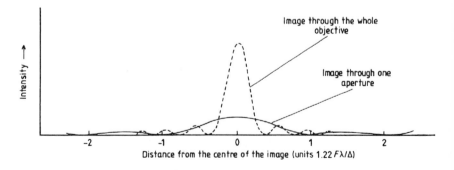

Figure 2.5.4. Image structure for one point source and one aperture.

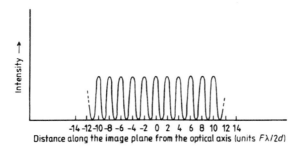

Figure 2.5.5. Image structure for a single point source viewed through two infinitely small apertures.

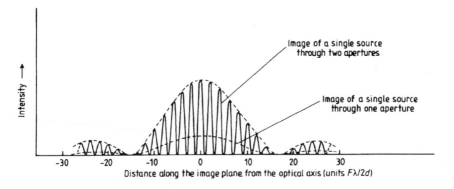

Figure 2.5.6. Image structure for one point source viewed through two apertures.

separated by an angle α'', given by

$$\alpha'' = \lambda/2d \qquad (2.5.5)$$

the maxima of one fringe pattern will be superimposed upon the minima of the other and vice versa. The fringes will then disappear and the image will be given by their envelope (figure 2.5.7). There may still be a very slight ripple on this image structure due to the incomplete filling of the minima by the maxima, but it is unlikely to be noticeable. The fringes will reappear as the sources continue to separate until the pattern is almost double that of figure 2.5.6 once again. The image patterns are then separated by a whole fringe width, and the sources by $2\alpha''$. The fringes disappear again for a source separation of $3\alpha''$, and reach yet another maximum for a source separation of $4\alpha''$, and so on. Thus the fringes are most clearly visible when the source's angular separation is given by $2n\alpha''$ (where n is an integer) and they disappear or reach a minimum in clarity for separations of $(2n + 1)\alpha''$. Applying the Rayleigh criterion for resolution, we see that the resolution of two apertures is given by the separation of the sources

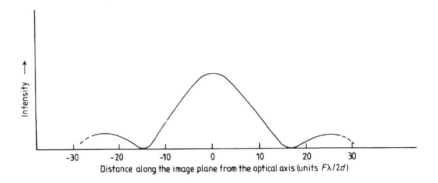

Figure 2.5.7. Image structures for two point sources separated by an angle, $\lambda/2d$, viewed through two apertures.

for which the two fringe patterns are mutually displaced by half a fringe width. This as we have seen is simply the angle α'', and so we have

$$\frac{\text{resolution through two apertures}}{\text{resolution of the objective}} = \frac{\alpha''}{\alpha'} \qquad (2.5.6)$$

$$= \frac{2.44d}{D}. \qquad (2.5.7)$$

Imagining the two apertures placed at the edge of the objective (i.e. $d = D$), we see the quite remarkable result that the resolution of an objective may be increased by almost a factor of two and a half by screening it down to two small apertures at opposite ends of one of its diameters. The improvement in the resolution is only along the relevant axis—perpendicular to this, the resolution is just that of one of the apertures. The resolution in both axes may, however, be improved by using a central occulting disc which leaves the rim of the objective clear. This also enables us to get a feeling for the physical basis for this improvement in resolution. We may regard the objective as composed of a series of concentric narrow rings, each of which has a resolution given by equation (2.5.5), with d the diameter of the ring. Since the objective's resolution is the average of all these individual resolutions, it is naturally less than their maximum. In practice some blurring of the image may be detected for separations of the sources which are smaller than the Rayleigh limit (section 1.1). This blurring is easier to detect for the fringes produced by the two apertures, than it is for the images through the whole objective. The effective improvement in the resolution may therefore be even larger than that given by equation (2.5.7).

The situation for a radio interferometer usually differs from that outlined above, in that the radiation is only rarely received perpendicularly to the line joining the antennae. Also the systematics of radio astronomy differ from those of optical work (section 1.2), so that some translation is required to relate our optical analogue to a radio interferometer. If we take the polar diagram of a single radio aerial (figure 1.2.6 for example), then this is the radio analogue of our image structure for a single source and a single aperture (figures 2.5.2 and 2.5.4). The detector in a radio telescope accepts energy from only a small fraction of this image at any given instant. Thus scanning the radio telescope across the sky corresponds to scanning this energy-accepting region through the optical image structure. The main lobe of the polar diagram is thus the equivalent of the central maximum of the optical image, and the side lobes are the equivalent of the diffraction fringes. Since an aerial is normally directed towards a source, the signal from it corresponds to a measurement of the central peak intensity of the optical image (figures 2.5.2 and 2.5.4). If two stationary aerials are now considered and their outputs combined, then when the signals arrive without any path differences, the final output from the radio system as a whole corresponds to the central peak intensity of figure 2.5.6. If a path difference does exist however, then the final output will correspond to some other point within that image. In particular when the path difference is a whole number of wavelengths, the output will correspond to the peak intensity of one of the fringes, and when it is a whole number plus half a wavelength, it will correspond to one of the minima. Now the path differences arise in two main ways: from the angle of inclination of the source to the line joining the two antennae, and from delays in the electronics and cables between the antennae and the central processing station (figure 2.5.8). The latter will normally be small and constant and may be ignored. The former will alter as the rotation of the Earth changes the angle of inclination. Thus the output of the interferometer will vary with time as the value of P changes. The output over a period of time, however, will not follow precisely the shape of figure 2.5.6 because the rate of change of path difference varies throughout the day, as we may see from the equation for P

$$P = s \cos \mu \cos(\psi - E) \qquad (2.5.8)$$

where μ is the altitude of the object and is given by

$$\mu = \sin^{-1}[\sin \delta \sin \varphi + \cos \delta \cos \varphi \cos(T - \alpha)]. \qquad (2.5.9)$$

ψ is the azimuth of the object and is given by

$$\psi = \cot^{-1}[\sin \varphi \cot(T - \alpha) - \cos \varphi \tan \delta \operatorname{cosec}(T - \alpha)]. \qquad (2.5.10)$$

In these equations E is the azimuth of the line joining the aerials, α and δ are the right ascension and declination of the object, T is the local sidereal time at the instant of observation and φ is the latitude of the interferometer. The path difference also varies because the effective separation of the aerials, d, where

$$d = s\left[\sin^2 \mu + \cos^2 \mu \sin^2(\psi - E)\right]^{1/2} \qquad (2.5.11)$$

also changes with time, thus the resolution and fringe spacing (equation (2.5.5)) are altered. Hence the output over a period of time from a radio interferometer

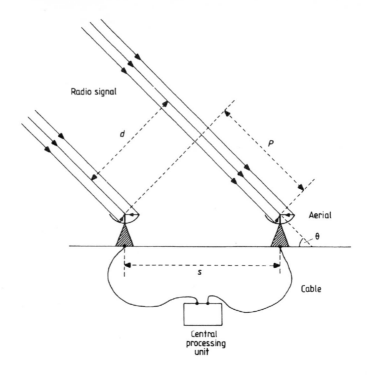

Figure 2.5.8. Schematic arrangement of a radio interferometer with fixed aerials.

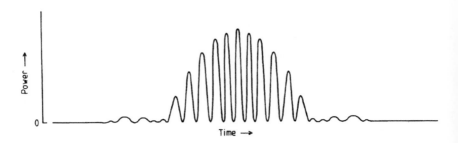

Figure 2.5.9. Output from a radio interferometer with stationary aerials, viewing a single source.

with fixed antennae is a set of fringes whose spacing varies, with maxima and minima corresponding to those of figure 2.5.6 for the instantaneous values of the path difference and effective aerial separation (figure 2.5.9).

An improved type of interferometer, which has increased sensitivity and stability is the phase switched interferometer (see earlier discussion). The phase of the signal from one aerial is periodically changed by 180° by, for example,

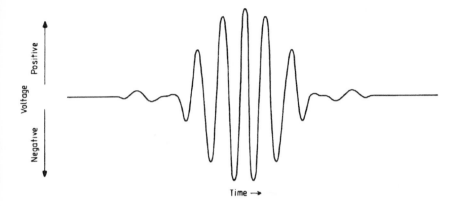

Figure 2.5.10. Output from a radio interferometer with stationary aerials and phase switching, viewing a single point source.

switching an extra piece of cable half a wavelength long into or out of the circuit. This has the effect of oscillating the beam pattern of the interferometer through half a fringe width. The difference in the signal for the two positions is then recorded. The phase switching is generally undertaken in the latter stages of the receiver (section 1.2), so that any transient effects of the switching are not amplified. The output fluctuates either side of the zero position as the object moves across the sky (figure 2.5.10).

When the aerials are driven so that they track the object across the sky (figure 2.5.11), then the output of each aerial corresponds to the central peak intensity of each image (figures 2.5.2 and 2.5.4). The path difference then causes a simple interference pattern (figure 2.5.12 and cf figure 2.5.5) whose fringe spacing alters due to the varying rate of change of path difference and aerial effective spacing as in the previous case. The maxima are now of constant amplitude since the aerial's projected effective areas are constant. In reality, many more fringes would occur than are shown in figure 2.5.12—10^4, for example, when observing at a wavelength of 0.2 m with an antenna separation of 1 km.

There is of course no advantage in observing a true point source with an interferometer, so we must now examine the effect of an extended or multiple source upon the output. In fact, only the effect of observing a double source need be considered since any extended or multiple source can be reduced to one or more pairs of point sources. Figure 2.5.13 illustrates the principle behind this. Since the resolution perpendicular to the interferometer axis is just that of a single antenna, the object is unresolved in this direction. Each resolution element along the high resolution axis is thus equivalent to a single point source whose intensity is the integrated intensity of that resolution element. The extended source thus becomes equivalent to a line of point sources and the behaviour of

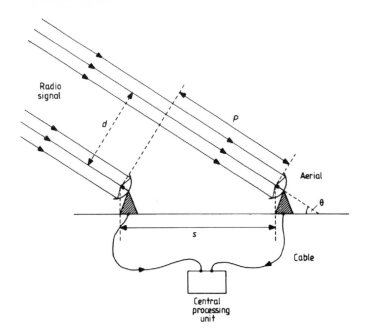

Figure 2.5.11. Schematic arrangement of a radio interferometer with tracking aerials.

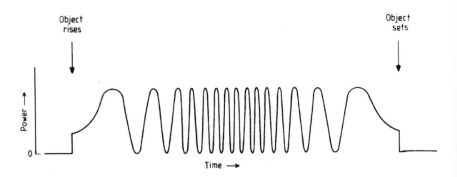

Figure 2.5.12. Output from a radio interferometer with tracking aerials viewing a single point source.

the system may simply be found by repeated application of the double-source case.

Let us consider then, two point sources of equal brightness and separated by an angle, β. Each source will produce an output from the interferometer similar to that of figure 2.5.12. Since the two sources' radiations are mutually incoherent, the two outputs will simply add together to give the total output at

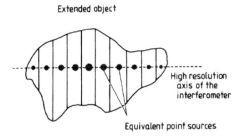

Figure 2.5.13. Extended object.

any instant. Now let us suppose that the sources are on the equator, and that the interferometer is aligned east–west. Then unless β is very large indeed, the path differences for each of the sources as they rise above the horizon will be almost identical (figure 2.5.14). Their maxima and minima will therefore be superimposed, and the total output will be twice the output for a single source. If we call the path differences between the aerials for sources 1 and 2, P_1 and P_2, then the difference between the path differences, ΔP, is simply

$$\Delta P = P_1 - P_2. \tag{2.5.12}$$

As the objects move across the sky, ΔP changes from near zero when they are on the horizon, through a maximum when they are on the meridian to zero again on the western horizon. Now if β is such that ΔP at maximum is half a wavelength (figure 2.5.15), then in this position the maximum of the output from one source will coincide with the minimum from the other and the total output will be steady. Over a 12 hour period therefore, the output will have the appearance of that shown in figure 2.5.16; the oscillations will decrease to a minimum when the source is on the meridian, and then grow again. Thus the amplitude of the fringes varies with time in contrast to the steady amplitude

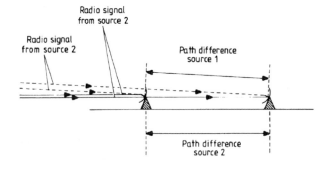

Figure 2.5.14. Interferometer viewing a close pair of sources near the horizon.

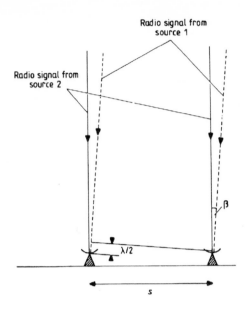

Figure 2.5.15. Interferometer viewing two sources near the meridian when the path difference is $\lambda/2$.

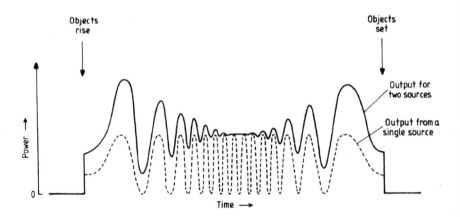

Figure 2.5.16. Output of an interferometer for two close point sources.

for a single source, and in the opposite manner to the variations of a single source viewed with fixed aerials (figure 2.5.9). Furthermore the output when the sources are transiting is equal to the maximum output for a single source so that their brightnesses may be easily found. From figure 2.5.15, we can see that the value of β for which ΔP reaches half a wavelength at maximum, β', is just

$$\beta' = \lambda/2s. \tag{2.5.13}$$

This equation, not too surprisingly, since the fringes from the two sources are mutually displaced by half a fringe width at transit, corresponds to the Rayleigh resolution of two small apertures in the optical situation (equation 2.5.5). Hence we call β' the resolution of the interferometer.

If β is larger than β', then two or more regions of steady output from the interferometer will occur, while if it is smaller, then the fringes will not die away completely. Should the sources not be of equal brightness, then some fringes will again be visible at all times. It will generally be possible to distinguish the latter situation from the case of $\beta < \beta'$, however, by the differing shapes of the envelopes of the extrema of the fringes. For widely separated and/or multiple or extended sources, the number of regions of steady output will increase until in the limit the output is constant over the whole observing period.

An interferometer works most efficiently, in the sense of returning the most information, for sources whose separation is comparable with its resolution. For objects substantially larger than the resolution, the output approaches constancy as we have just seen, and little or no useful information may be obtained. We may, however, obtain information about a larger source by using an interferometer with a *smaller* separation of its elements. The resolution is thereby degraded until it is comparable with the angular size of the source. By combining the results of two interferometers of differing separations, one might thus obtain information on both the large and small scale structure of the source. Following this idea to its logical conclusion led Ryle in the early 1960s to the invention and development of the technique of aperture synthesis.

By this technique, which also goes under the name of Earth-rotation synthesis and is closely related to synthetic aperture radar (section 2.8), observations of a stable radio source by a number of interferometers are combined to give the effect of an observation using a single very large dish. The simplest way to understand how aperture synthesis works is to take an alternative view of the operation of an interferometer. The output from a two-aperture interferometer viewing a monochromatic point source is shown in figure 2.5.5, and is a simple sine wave. This output function is just the Fourier transform (equation (2.1.3)) of the source. Such a relationship between source and interferometer output is no coincidence, but is just a special case of the general theorem:

The instantaneous output of a two-element interferometer is a measure of one component of the two-dimensional Fourier transform (equation (2.1.10)) of the objects in the field of view of one of the telescopes.

Thus, if a large number of two-element interferometers were available, so that all the components of the Fourier transform could be measured, then the inverse two-dimensional Fourier transform (equation (2.1.11)) would immediately give a map of the portion of the sky under observation.

Now the complete determination of the Fourier transform of even a small field of view would require an infinite number of interferometers. In practice, therefore, the technique of aperture synthesis is modified in several ways. The most important of these is relaxing the requirement to measure all the Fourier components *at the same instant*. However, once the measurements are spread over time, the source(s) being observed must remain unvarying over the length of time required for those measurements.

Given, then, a source which is stable at least over the measurement time, we may use one or more interferometer pairs to measure the Fourier components using different separations and angles. Of course, it is still not possible to make an infinite number of such measurements, but we may use the discrete versions of equations (2.1.10) and (2.1.11) to bring the required measurements down to a finite number (cf the one-dimensional analogues, equations (2.1.3), (2.1.4), (2.1.8) and (2.1.9)) though at the expense of losing the high-frequency components of the transform, and hence the finer details of the image (section 2.1). The problem of observing with many different base lines is eased because we are observing from the rotating Earth. Thus, if a single pair of telescopes observes a source over 24 hours, the orientation of the base line revolves through 360° (figure 2.5.17). The continuous output of such an interferometer is known as the 'visibility function'. It is a complex function whose amplitude is proportional to the amplitude of the Fourier transform and whose phase is the phase shift in the fringe pattern.

The requirement for 24 hours of observation would limit the technique to circumpolar objects. Fortunately, however, only 12 hours are actually required, the other 12 hours can then be calculated by the computer from the conjugates of the first set of observations. Hence, aperture synthesis can be applied to any object in the same hemisphere as the interferometer.

A pair of elements of an interferometer thus rotate with respect to each other as seen from the observed object. If the object is not at the north (or south) pole then the projected spacing of the interferometer will also vary and they will seem to trace out an ellipse.

Two elements arranged upon an east–west line follow a circular track perpendicular to the Earth's rotational axis. It is thus convenient to choose the plane perpendicular to the Earth's axis to work in, and this is usually termed the u–v plane. If the interferometer is not aligned east–west, then the paths of its elements will occupy a volume in the u–v–w space, and additional delays will have to be incorporated into the signals to reduce them to the u–v plane. The paths of the elements of an interferometer in the u–v plane range from circles for an object at a declination of $\pm 90°$ through increasingly narrower ellipses to a straight line for an object with a declination of $0°$ (figure 2.5.18).

A single 12 hour observation by a two-element interferometer thus samples all the components of the Fourier transform of the field of view covered by

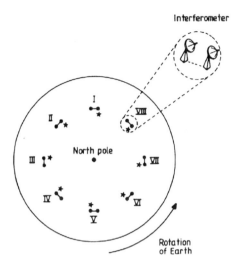

Figure 2.5.17. Changing orientation of an interferometer. The Earth is viewed from above the north pole and successive positions of the interferometer at three hour intervals are shown. Notice how the orientation of the starred aerial changes through 360° with respect to the other aerial during a day.

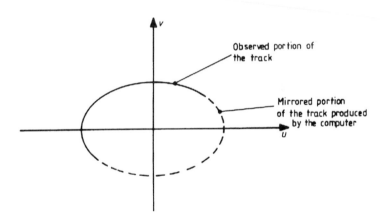

Figure 2.5.18. Track of one element of an interferometer with respect to the other in the u–v plane for an object at a declination of ±35°.

its track in the u–v plane. We require, however, the whole of the u–v plane to be sampled. Thus a series of 12 hour observations must be made, with the interferometer base line changed by the diameter of one of its elements each time (figure 2.5.19). The u–v plane is then sampled completely out to the maximum base line possible for the interferometer, and the inverse Fourier transform will

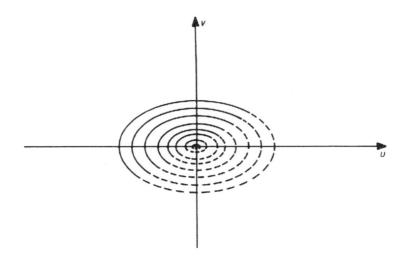

Figure 2.5.19. Successive tracks in the u–v plane of a two-element interferometer as its base line is varied.

give an image equivalent to that from a single telescope with a diameter twice that of the maximum base line.

A two-element radio interferometer with (say) 20 m diameter aerials and a maximum base line of 1 km, would need to make fifty 12 hour observations in order to synthesise a 2 km diameter telescope. By using more than 2 elements, however, the time required for the observations can be much reduced. Thus with 6 elements, there are 15 different pairs. If the spacings of these pairs of elements are all different (non-redundent spacing, something that it is not always possible to achieve) then the 50 visibility functions required by the previous example can be obtained in just four 12 hour observing sessions. If the source has a reasonably smooth spectrum, the u–v plane may be sampled even more rapidly by observing simultaneously at several nearby frequencies (multi-frequency synthesis). Since the path differences between the aerials are dependant upon the operating wavelength, this effectively multiplies the number of interferometer pairs with different separations by the number of frequencies being observed. The analysis of the results from such observations, however, will be considerably complicated by any variability of the source over the frequencies being used.

An aperture synthesis system such as we have just been considering is a filled-aperture system. That is, the observations synthesise the sensitivity and resolution of a telescope with a radius equal to the maximum available base line. Although many aperture synthesis systems are of this type, it becomes increasingly impractical and expensive to sample the whole u–v plane as the base line extends beyond 10 km or so. The larger aperture synthesis systems therefore have unfilled apertures. The Fourier transform of the sky radio emission is thus not fully sampled. Special techniques, known as hybrid mapping (see below) are then required to produce the final maps of radio sources. The MERLIN system (Multi-Element Radio Linked INterferometer) based at Jodrell Bank in the UK is typical of these larger unfilled systems. MERLIN has seven fixed aerials with base lines ranging from 6 to 230 km.

Beyond MERLIN and similar systems, we have very-long-base-line interferometry (VLBI). For VLBI, the elements of the interferometer may be separated by thousands of kilometres, over several continents. There is also the possibility of using one or more elements on board spacecraft (e.g. the proposed Quasat spacecraft) to increase the base line to more than 10 000 km and give resolutions of 10^{-5} arc seconds or better. A spacecraft with an eccentric orbit giving a high apogee and a low perigee would also provide a means of sweeping rapidly through a wide range of base lines. In VLBI, the signals from each element are separately recorded along with timing pulses from an atomic clock. The recordings are then physically brought together and processed by a computer which uses the time signals to ensure the correct registration of one radio signal with respect to another.

The extraction of the image of the sky from the visibility functions is complicated by the presence of noise and errors. Overcoming the effects of these adds additional stages to the data reduction process. In summary it becomes:

(1) Data calibration.
(2) Inverse Fourier transform.
(3) Deconvolution of instrumental effects.
(4) Self-calibration.

Data calibration is required in order to compensate for problems such as errors in the locations of the elements of the interferometer, and variations in the atmosphere. It is carried out by comparing the theoretical instrumental response function with the observed response to an isolated stable point source. These responses should be the same and if there is any difference then the data calibration process attempts to correct it by adjusting the amplitudes and phases of the Fourier components. Often ideal calibration sources cannot be found close to the observed field. Then, calibration can be attempted using special calibration signals, but these are insensitive to the atmospheric variations, and the result is much inferior to that obtained using a celestial source.

Applying the inverse Fourier transform to the visibility functions has already been discussed, and further general details are given in section 2.1.

After the inverse Fourier transformation has been completed, we are left with the 'dirty' map of the sky. This is the map contaminated by artefacts introduced by the instrumental profile of the interferometer. The primary components of the instrumental profile, apart from the central response, are the side lobes. These appear on the dirty map as a series of rings which extend widely over the sky, and which are centred on the central response of the point spread function (PSF). The instrumental profile or PSF can be calculated from interferometry theory to a high degree of precision. The deconvolution can then proceed as outlined in section 2.1.

The maximum entropy method discussed in section 2.1 has recently become widely used for determining the best source function to fit the dirty map. Another method of deconvolving the PSF, however, has long been in use, and is still used by many workers, and that is the method known as CLEAN.

The CLEAN algorithm was introduced by Hogbom in 1974. It involves the following stages:

(a) Normalise the PSF (instrumental profile or 'dirty' beam) to gI_{MAX}, where I_{MAX} is the intensity of the point of maximum intensity in the dirty map and g is the 'loop gain' and has a value between 0 and 1.
(b) Subtract the normalised PSF from the dirty map.
(c) Find the point of maximum intensity in the new map—this may or may not be the same point as before—and repeat the first two steps.
(d) Continue the process iteratively until I_{MAX} is comparable with the noise level.
(e) Produce a final clear map by returning all the components removed in the previous stages in the form of 'clean beams' with appropriate positions and amplitudes. The clean beams are typically chosen to be Gaussian with similar widths to the central response of the dirty beam.

CLEAN has proved to be a useful method, despite its lack of a substantial theoretical basis, for images made up of point sources. For extended sources MEMs are better because CLEAN may then require thousands of iterations. Though as pointed out in section 2.1, MEMs also suffer from problems, especially the variation of resolution over the image.

The final stage of self-calibration is required when the base lines become more than a few kilometres in length, since the atmospheric effects can then differ from one element to another. Under such circumstances with three or more elements we may use the closure phase which is independent of the atmospheric phase delays. The closure phase is defined as the sum of the observed phases for the three base lines made by three elements of the interferometer. It is independent of the atmospheric phase delays as we may see by defining the phases for the three base lines, *in the absence of an atmosphere* to be φ_{12}, φ_{23} and φ_{32}, and the atmospheric phase delays at each element as a_1, a_2 and a_3. The observed phases are then

$$\varphi_{12} + a_1 - a_2$$
$$\varphi_{23} + a_2 - a_3$$
$$\varphi_{31} + a_3 - a_1.$$

The closure phase is then given by the sum of the phases around the triangle of the base lines:

$$\varphi_{123} = \varphi_{12} + a_1 - a_2 + \varphi_{23} + a_2 - a_3 + \varphi_{31} + a_3 - a_1 \qquad (2.5.14)$$

$$= \varphi_{12} + \varphi_{23} + \varphi_{31}. \qquad (2.5.15)$$

From equation (2.5.15) we may see that the closure phase is independent of the atmospheric effects and is equal to the sum of the phases in the absence of an atmosphere. In a similar way an atmosphere-independent closure amplitude can be defined for four elements:

$$G_{1234} = \frac{A_{12}A_{34}}{A_{13}A_{24}} \qquad (2.5.16)$$

where A_{12} is the amplitude for the base line between elements 1 and 2 etc. Neither of these closure quantities are actually used to form the image, but they are used to reduce the number of unknowns in the procedure. At millimetre wavelengths, the principal atmospheric effects are due to water vapour. Since this absorbs the radiation, as well as leading to the phase delays, monitoring the sky brightness can provide information on the amount of water vapour along the line of sight, and so provide additional information for correcting the phase delays.

For VLBI, 'hybrid mapping' is required since there is insufficient information in the visibility functions to produce a map directly. Hybrid mapping is an iterative technique which uses a mixture of measurements and guesswork. An initial guess is made at the form of the required map. This may well actually be a lower resolution map from a smaller interferometer. From this map the visibility functions are predicted, and the true phases estimated. A new map is then generated by Fourier inversion. This map is then improved and used to provide a new start to the cycle. The iteration is continued until the hybrid map is in satisfactory agreement with the observations.

Although the majority of currently operating interferometers and aperture synthesis systems operate in the radio and microwave parts of the spectrum, there is no theoretical reason for such a restriction. Several systems to operate in the infrared and visible parts of the spectrum are therefore currently in the process of construction. These are discussed in more detail in section 1.1, and earlier in this section.

Michelson stellar interferometer

We may now turn to the consideration of the optical version of the interferometer. In fact, we have already completed the theoretical background which is required. The arrangement of the optical interferometer is essentially that of figure 2.5.1, and its image structure is given by figures 2.5.6 and 2.5.7. It merely remains to relate this earlier work to the practical observing systems.

Two basic forms of this instrument have been used. The first is identical to the arrangement in figure 2.5.1, except that the apertures are replaced by two

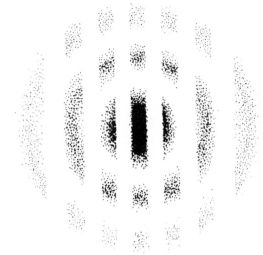

Figure 2.5.20. Appearance of the image of a source seen in a Michelson interferometer at the instant of maximum fringe clarity.

parallel slits to increase the amount of light available, and that a narrow band filter has to be included to give nearly monochromatic light; earlier the sources were assumed to be monochromatic. The resolution of the system, α'', is just that given by equation (2.5.5) for the small apertures

$$\alpha'' = \lambda/2d. \tag{2.5.17}$$

To use the instrument to measure the separation of a double star, or the width of an extended object (figure 2.5.13), the slits are aligned perpendicularly to the line joining the stars, and moved close together, so that the image fringe patterns are undisplaced and the image structure is that of figure 2.5.6. The actual appearance of the image in the eyepiece at this stage is shown in figure 2.5.20. The slits are then moved apart until the fringes disappear (figures 2.5.7 and 2.5.21). The linear separation of the slits at this instant then gives the value of d required for equation (2.5.17). Hence the separation of the source diameter may be found quite simply. Very stringent requirements on the stability and accuracy of the apparatus are required for the success of this technique. The path difference must not be greater than the coherence length, l, of the radiation which is given by

$$l = c/\Delta v = \lambda^2/\Delta\lambda \tag{2.5.18}$$

where Δv and $\Delta\lambda$ are the frequency and wavelength bandwidths of the radiation. So that for $\lambda = 500$ nm and $\Delta\lambda = 1$ nm, we have $l = 0.25$ mm. Furthermore the path difference must remain constant to considerably better than the wavelength

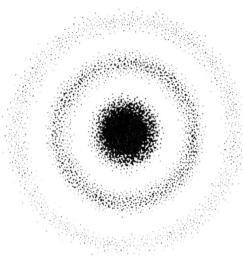

Figure 2.5.21. Appearance of the image of a double star or extended source seen in a Michelson interferometer at the time of minimum fringe clarity.

of the radiation that is being used, as the slits are separated. Vibrations and scintillation are additional limiting factors. In general, the paths will not be identical when the slits are close together and the fringes are at their most visible. But the path difference will then be some integral number of wavelengths, provided that it is less than the coherence length. Thus in practice the value which is required for d is the difference between the separations at minimum and maximum fringe visibilities. A system such as has just been described was used by Michelson to measure the diameters of the Galilean satellites. Their diameters are in the region of one second of arc, so that d has a value of a few tens of millimetres.

The improved version of the system is rather better known and is the type of interferometer which is usually intended when the Michelson stellar interferometer is mentioned. It was used by Michelson and Pease to measure the diameters of half a dozen or so of the larger stars. The slits were replaced by movable mirrors on a rigid six metre long bar which was mounted on top of the 2.5 m Hooker telescope (figure 2.5.22). By this means, d could be increased beyond the diameter of the telescope. A system of mirrors then reflected the light into the telescope. The practical difficulties with the system were very great and although larger systems have since been attempted none has been successful, and so the design is now largely of historical interest only. Recently Michelson stellar interferometers using two separate telescopes instead of two mirrors feeding a single telescope have, however, been successfully operated. Working at near infrared wavelengths, measurements have been made using base lines of ten metres and more.

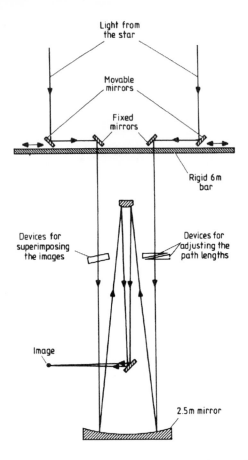

Figure 2.5.22. Schematic optical arrangement of the Michelson stellar interferometer.

A proposal for the future that would have much in common with stellar coronagraphs (section 5.3) is for a nulling interferometer based in space. This would use destructive interference to suppress the bright central object, thus allowing faint companions (e.g. planets) to be discerned. Earth-like planets around stars 10 or more parsecs away might be detected in observations of a few hours using a four-element interferometer with 50 m base lines.

A variation on the system which also has some analogy with intensity interferometry (see below) is known as amplitude interferometry. This has been used recently in a successful attempt to measure stellar diameters. At its heart is a device known as a Köster prism which splits and combines the two separate light beams (figure 2.5.23). The interfering beams are detected in a straightforward manner using photomultipliers. Their outputs may then be compared. The fringe visibility of the Michelson interferometer appears as the anticorrelated

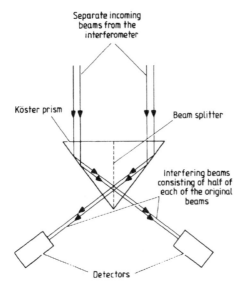

Figure 2.5.23. The Köster prism.

component of the two signals, while the atmospheric scintillation etc affects the correlated component. Thus the atmospheric effects can be screened out to a large extent, and the stability of the system is vastly improved.

Intensity interferometer

The practical difficulties of an optical Michelson interferometer have severely limited its usefulness. Most of these problems, however, may be reduced in a device which correlates intensity fluctuations. The device was originally invented by Hanbury Brown in 1949 as a radio interferometer, but it has found its main application in the optical region. The disadvantage of the system compared with the Michelson interferometer is that phase information is lost, and so the structure of a complex source cannot be reconstituted. The far greater ease of operation of the intensity interferometer, however, has led to the measurement of some hundred stellar diameters to date—a factor of fifteen or so better than the achievements of the basic Michelson interferometer.

The principle of the operation of the interferometer relies upon phase differences in the low frequency beat signals from different mutually incoherent sources at each aerial, combined with electrical filters to reject the high frequency components of the signals. The schematic arrangement of the system is shown in figure 2.5.24. We may imagine the signal from a source resolved into its Fourier components and consider the interaction of one such component from one source with another component from the other source. Let the frequencies

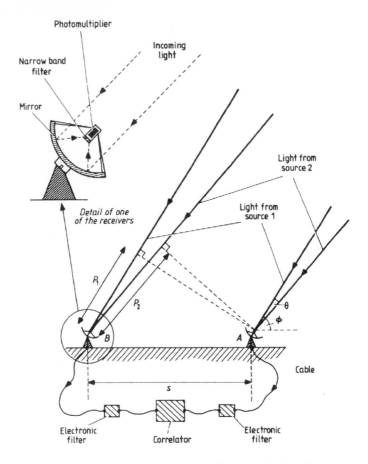

Figure 2.5.24. Schematic arrangement of an intensity interferometer.

of these two components be v_1 and v_2, then

$$\text{component from source } 1 = E_1 \sin(\pi v_1 t + \varphi_1) \qquad (2.5.19)$$

$$\text{component from source } 2 = E_2 \sin(2\pi v_2 t + \varphi_2) \qquad (2.5.20)$$

where E is the electrical amplitude, φ is the phase and t is the instant of observation. At aerial A the intensity received by the photomultiplier, I_A, will be

$$I_A = C_A \left[E_1 \sin(2\pi v_1 t + \varphi_1) + E_2 \sin(2\pi v_2 t + \varphi_2) \right]^2 \qquad (2.5.21)$$

where C_A is the constant of the system. The signals at aerial B will be delayed by times of (P_1/c) and (P_2/c), so that the intensity at aerial B will be

$$I_B = C_B \left\{ E_1 \sin\left[2\pi v_1 t + (P_1/c) + \varphi_1 \right] + E_2 \sin\left[2\pi v_2 t + (P_2/c) + \varphi_2 \right] \right\}^2. \qquad (2.5.22)$$

Expanding equation (2.5.21) we get

$$I_A = C_A\big[E_1^2 \sin^2(2\pi \nu_1 t + \varphi_1) + 2E_1 E_2 \sin(2\pi \nu_1 t + \varphi_1) \sin(2\pi \nu_2 t + \varphi_2)$$
$$+ E_2^2 \sin^2(2\pi \nu_2 t + \varphi_2)\big] \tag{2.5.23}$$

which we may rearrange into the form

$$I_A = \tfrac{1}{2}C_A\big\{(E_1^2 + E_2^2)$$
$$- \big[E_1^2 \cos(4\pi \nu_1 t + 2\varphi_1) + E_2^2 \cos(4\pi \nu_2 t + 2\varphi_2)\big]$$
$$- 2E_1 E_2 \cos\big[2\pi (\nu_1 + \nu_2)t + \varphi_1 + \varphi_2\big]$$
$$+ 2E_1 E_2 \cos\big[2\pi (\nu_1 - \nu_2)t + \varphi_1 - \varphi_2\big]\big\}. \tag{2.5.24}$$

For clarity let us put

$$E_1^2 + E_2^2 = F_a \tag{2.5.25}$$
$$E_1^2 \cos(4\pi \nu_1 t + 2\varphi_1) = F_b(\nu_1) \tag{2.5.26}$$
$$E_2^2 \cos(4\pi \nu_2 t + 2\varphi_2) = F_c(\nu_2) \tag{2.5.27}$$
$$2E_1 E_2 \cos\big[2\pi (\nu_1 + \nu_2)t + \varphi_1 + \varphi_2\big] = F_d(\nu_1 + \nu_2) \tag{2.5.28}$$
$$2E_1 E_2 \cos\big[2\pi (\nu_1 - \nu_2)t + \varphi_1 - \varphi_2\big] = F_e(\nu_1 - \nu_2). \tag{2.5.29}$$

So that equation (2.5.24) becomes

$$I_A = \tfrac{1}{2}C_A\big[F_a - F_b(\nu_1) - F_c(\nu_2) - F_d(\nu_1 + \nu_2) + F_e(\nu_1 - \nu_2)\big]. \tag{2.5.30}$$

Now we have restricted the band width of the light by means of a narrow band filter (figure 2.5.24) so that (say)

$$\nu_1 \simeq \nu_2 = \nu \tag{2.5.31}$$

and

$$0 < \nu_1 - \nu_2 \ll \nu. \tag{2.5.32}$$

We may therefore arrange for the electronic filters (figure 2.5.24) to have upper and lower frequency cut-offs, f_u and f_l, such that

$$f_l > 0 \tag{2.5.33}$$
$$f_u < \nu \tag{2.5.34}$$
$$f_l < \nu_1 - \nu_2 < f_u. \tag{2.5.35}$$

Thus of the components in equation (2.5.30), only the $F_e(\nu_1 - \nu_2)$ component will be passed by the electronic filters, F_a, $F_b(\nu_1)$, $F_c(\nu_2)$, and $F_d(\nu_1 + \nu_2)$, will all be rejected. The output, I_A', from the combination of receiver A and its filter is therefore just

$$I_A' = C_A E_1 E_2 \cos\big[2\pi (\nu_1 - \nu_2)t + \varphi_1 - \varphi_2\big]. \tag{2.5.36}$$

We may similarly obtain the output from receiver B and its filter

$$I_B' = C_B E_1 E_2 \cos\left[2\pi(\nu_1 - \nu_2)t + \varphi_1 - \varphi_2 + \frac{2\pi \nu_1 P_1}{c} - \frac{2\pi \nu_2 P_2}{c}\right]. \quad (2.5.37)$$

Comparison of equations (2.5.36) and (2.5.37) shows that both of the inputs to the correlator are functions of $(\nu_1 - \nu_2)$, i.e. the beat frequency of the original two Fourier components from each source. Thus correlation of the outputs is to be expected provided that $(P_1 - P_2)$ is fairly small. The correlation for a given separation, s, of the receivers, is measured by the product of the two inputs, denoted by $k(s)$

$$k(s) = I_A' \times I_B' \quad (2.5.38)$$

$$k(s) = C_A C_B E_1^2 E_2^2 \cos\left[2\pi(\nu_1 - \nu_2)t + \varphi_1 - \varphi_2\right]$$
$$\times \cos\left(2\pi(\nu_1 - \nu_2)t + \varphi_1 - \varphi_2 + \frac{2\pi \nu_1 P_1}{c} - \frac{2\pi \nu_2 P_2}{c}\right) \quad (2.5.39)$$

$$k(s) = \tfrac{1}{2} C_A C_B E_1^2 E_2^2 \left[\cos\left(\frac{2\pi \nu_1 P_1}{c} - \frac{2\pi \nu_2 P_2}{c}\right)\right.$$
$$\left. + \cos\left(4\pi(\nu_1 - \nu_2)t + 2\varphi_1 - 2\varphi_2 + \frac{2\pi \nu_1 P_1}{c} - \frac{2\pi \nu_2 P_2}{c}\right)\right]. \quad (2.5.40)$$

Now the time dependent term in equation (2.5.40) averages to zero and may be eliminated by integration of the correlation function over a short interval. Thus we get

$$k(s) = \tfrac{1}{2} C_A C_B E_1^2 E_2^2 \cos\left[2\pi(\nu_1 P_1 - \nu_2 P_2)/c\right]. \quad (2.5.41)$$

From figure 2.5.24 we see that

$$P_1 = s \sin \varphi \quad (2.5.42)$$

and if 0 is small

$$P_2 = s(\sin \varphi + \theta \cos \varphi). \quad (2.5.43)$$

Thus combining equations (2.5.31), (2.5.42) and (2.5.43) into equation (2.5.41), we get

$$k(s) = \tfrac{1}{2} C_A C_B E_1{}^2 E_2{}^2 \cos(2\pi \nu s \theta \cos \varphi / c) \quad (2.5.44)$$

where normally $\cos \varphi \simeq 1$. This correlation function must now be integrated over the band passes of the optical and electronic filters, and if need be over all the points comprising an extended source as well, to get the final correlation function $K(s)$ for the system. This is a non-trivial process which we will not attempt here, instead just the remarkably simple result will be quoted. The interested reader can find further details in Hanbury Brown and Twiss (1957) (*Proc. R. Soc.* A **242** 300). The result of these integrations may be expressed in terms of the visibility V of the fringes in a Michelson interferometer, where

$$V = \left(I_{\max} - I_{\min}\right) / \left(I_{\max} + I_{\min}\right) \quad (2.5.45)$$

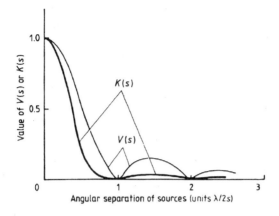

Figure 2.5.25. Comparison of the visibility function of a Michelson interferometer with the correlation function of an intensity interferometer.

where I_{max} is the intensity of a fringe maximum and I_{min} is the intensity of a fringe minimum. We then obtain

$$K(s) \propto V(s)^2 \qquad (2.5.46)$$

where $V(s)$ denotes the fringe visibility of a Michelson interferometer whose base line is s. Thus the correlation function varies with the separation of the aerials and of the sources as shown in figure 2.5.25. The resolution of the system in the sense of the angular separation required between the sources to reduce the correlation function to zero is the same as the resolution of an equivalent Michelson interferometer (equation (2.5.13)).

The intensity interferometer is normally used to measure stellar diameters. If we integrated equation (2.5.44) over a uniform circular stellar disc, then we find that $K(s)$ reaches its first zero when the angular stellar diameter, θ', is given by

$$\theta' = 1.22\lambda/s. \qquad (2.5.47)$$

So the resolution of an intensity interferometer (and also of a Michelson interferometer) for stellar discs is the same as that of a telescope (equation (2.5.1)) whose diameter is equal to the separation of the receivers.

The greater ease of construction and operation of the intensity interferometer over the Michelson interferometer arises from its dependence upon the beat frequency of two light beams of similar wavelengths, rather than upon the actual frequency of the light. A typical value of f_u (i.e. the upper limit of $\nu_1 - \nu_2$) is 100 MHz, which corresponds to a wavelength of three metres. Thus the path differences for the two receivers may vary by up to about 0.3 m during an observing sequence without ill effects. Scintillation, by the same argument, is also negligible.

Only one working intensity interferometer is in existence; built by Hanbury Brown at Narrabri in Australia. It uses two 6.5 m reflectors which are formed from several hundred smaller mirrors. There is no need for very high optical quality since the reflectors simply act as light buckets and only the brightest stars can be observed. The reflectors are mounted on trolleys on a circular track 94 m in radius. The line between the reflectors may therefore always be kept perpendicular to the line of sight to the source, and their separation can be varied from 0 to 196 m (figure 2.5.26). It operates at a wavelength of 433 nm, giving it a maximum resolution of about 0.0005 seconds of arc.

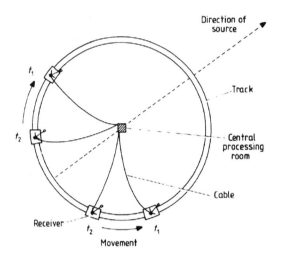

Figure 2.5.26. Schematic layout of the intensity interferometer at Narrabri, showing the positions of the receivers for observing in the same direction with different base lines. Note there are only two receivers; the diagram shows the arrangement at two separate times, superimposed.

Reflection interferometers

These are not an important class of interferometer, but they are included for the sake of completeness and for their historical interest. They are closely related to the phased array type of radio telescope discussed in section 1.2, however. The first astronomical radio interferometer measurements were made with this type of instrument and were of the Sun. Only one aerial is required, and it receives both the direct and the reflected rays (figure 2.5.27). These interfere, and the changing output as the object moves around the sky can be interpreted in terms of the structure of the object.

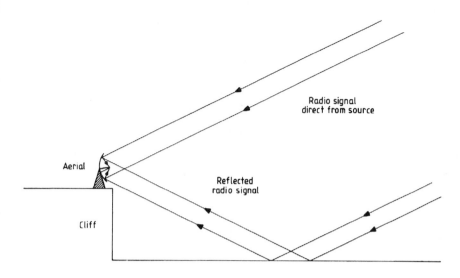

Figure 2.5.27. Schematic possible arrangement for reflection interferometry.

Exercises

2.5.1 Calculate the separation of the slits over a telescope objective in the first type of Michelson stellar interferometer, which would be required to measure the diameter of Ganymede (*a*) at opposition and (*b*) near conjunction of Jupiter. Take Ganymede's diameter to be 5000 km and the eye's sensitivity to peak at 550 nm.

2.5.2 Observations of a quasar which is thought to be at a distance of 1500 Mpc, just reveal structure to a VLBI working at 50 GHz. If its maximum base line is 9000 km, what is the linear scale of this structure?

2.5.3 Calculate the maximum distance at which the Narrabri intensity interferometer would be able to measure the diameter of a solar-type star.

2.6 SPECKLE INTERFEROMETRY

This technique is a kind of poor man's space telescope since it provides near diffraction-limited performance from Earth-based telescopes. It works by obtaining images of the object sufficiently rapidly to freeze the blurring of the image which arises from atmospheric scintillation. The total image then consists of a large number of small dots or speckles, each of which is a diffraction-limited

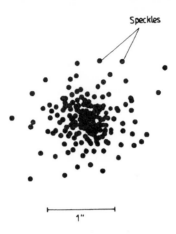

Figure 2.6.1. Schematic image of a point source obtained using a very short exposure time.

image for some objective diameter up to and including the diameter of the actual objective (figure 2.6.1). An alternative to this technique is the image sharpening telescope (section 1.1), where the speckles are recombined by adjusting the telescope optics to compensate for the atmospheric effects.

We may see how this speckled image structure arises by considering the effect of scintillation upon the incoming wavefront from the object. If we assume that above the atmosphere the wavefront is planar and coherent, then the main effect of scintillation is to introduce differential phase delays across it. The delays arise because the atmosphere is non-uniform with different cells within it having slightly different refractive indices. A typical cell size is 0.1 m, and the scintillation frequencies usually lie in the range 1 to 100 Hz. Thus an average image from a 1 m telescope will be affected by some 100 atmospheric cells at any given instant. These will be rapidly changing, and over a normal exposure time, which can range from seconds to hours, they will form an integrated image which will be large and blurred compared with the diffraction-limited image. Even under the best seeing conditions, the image is rarely less than one second of arc across. An exposure of a few milliseconds, however, is sufficiently rapid to freeze the image motion, and the observed image is then just the resultant of the contributions from the atmospheric cells across the telescope objective at that moment. Now the large number of these cells renders it highly probable that some of the phase delays will be similar to each other, and so some of the contributions to the image will be in phase with each other. These particular contributions will have been distributed over the objective when they entered the telescope in a random manner. Considering two such contributions, we have in fact a simple interferometer, and the two beams of radiation will combine in the image plane to produce results identical with those of an interferometer whose

Figure 2.6.2. Schematic image of a just-resolved double source obtained with a very short exposure time.

base line is equal to the separation of the contributions on the objective. We have already seen what this image structure might be (figure 2.5.6). If several collinear contributions are in phase, then the image structure will approach that shown in figure 4.1.9 modulated by the intensity variation due to the aperture of a single cell. The resolution of the images is then given by the maximum separation of the cells. When the in-phase cells are distributed in two dimensions over the objective, the images have resolutions in both axes given by the maximum separations along those axes at the objective. The smallest speckles in the total image therefore have the diffraction-limited resolution of the whole objective, assuming always of course that the optical quality of the telescope is sufficient to reach this limit. Similar results will be obtained for those contributions to the image which are delayed by an integral number of wavelengths with respect to each other. Intermediate phase delays will cause destructive interference to a greater or lesser extent at the point in the image plane which is symmetrical between the two contributing beams of radiation, but will again interfere constructively at other points, to produce an interference pattern which has been shifted with respect to its normal position. Thus all pairs of contributions to the final image interfere with each other to produce one or more speckles.

In its most straightforward application, speckle interferometry can produce an actual image of the object. Each speckle is a photon noise limited image of the object, so that in some cases the structure can even be seen by eye on the original photograph. A schematic instantaneous image of a double star which is just at the diffraction limit of the objective is shown in figure 2.6.2. Normally however, the details of the image are completely lost in the noise. Then, many images, perhaps several thousand, must be averaged before any improvement in the resolution is gained.

More commonly, the data are Fourier analysed, and the power spectrum obtained. This is the square of the modulus of the Fourier transform of the image

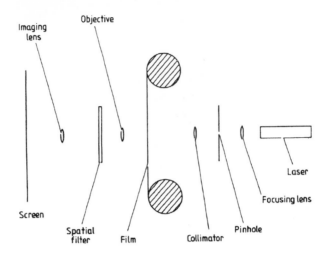

Figure 2.6.3. Arrangement for obtaining the Fourier transform of a speckle photograph by optical means.

intensity (equation (2.1.3)). The Fourier transform can be obtained directly using a large computer, or optical means can be used. In the latter case, the photograph is illuminated by collimated coherent light (figure 2.6.3). It is placed at one focus of an objective and its Fourier transform is imaged at the back focus of the objective. A spatial filter at this point can be used to remove unwanted frequencies if required, and then the Fourier transform re-imaged. However it may be obtained, the power spectrum can then be inverted to give centrisymmetric information such as diameters, limb darkening, oblateness and binarity. Non-centrisymmetric structure can only be obtained if there is a point source close enough for its image to be obtained simultaneously with that of the object and for it to be affected by the atmosphere in the same manner as the image of the object (i.e. the point source is within the same isoplanatic patch of sky as the object of interest). Then deconvolution (section 2.1) can be used to retrieve the image structure.

The practical application of the technique requires the use of large telescopes and high gain image intensifiers. Not only are very short exposures required (0.001 to 0.1 seconds), but very large plate scales (0.1 to 1 seconds of arc per millimetre) are also needed in order to separate the individual speckles. Furthermore, the wavelength range must be restricted by a narrow band filter to 20 to 30 nm. Even with such a restricted wavelength range, it may still be necessary to correct any remaining atmospheric dispersion using a low power direct vision spectroscope. Thus the limiting magnitude of the technique is currently about +18, and it has found its major applications in the study of red supergiants, the orbits of close binary stars, Seyfert galaxies, asteroids and fine details of solar structure. Speckle interferometry, however, is a recent invention and it is still developing; it is likely therefore that the limiting magnitude will be improved considerably over the next decade, particularly as photon counting techniques improve through the application of charge-coupled devices etc.

2.7 OCCULTATIONS

Background

An occultation is the more general version of an eclipse. It is the obscuration of a more distant astronomical object by the passage in front of it of a close object. Normally the term implies occultation by the Moon, or more rarely by a planet, since other types of occultation are almost unknown. It is one of the most ancient forms of astronomical observation, with records of lunar occultations stretching back some two and a half millennia. More recently interest in the events has been revived for their ability to give precise positions for objects observed at low angular resolution in, say, the x-ray region, and for their ability to give structural information about objects at better than the normal diffraction-limited resolution of a telescope. To see how the latter effect is possible, we must consider what happens during an occultation.

Firstly let us consider Fresnel diffraction at a knife edge (figure 2.7.1) for

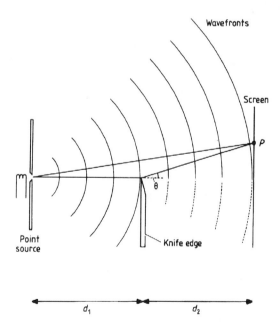

Figure 2.7.1. Fresnel diffraction at a knife edge.

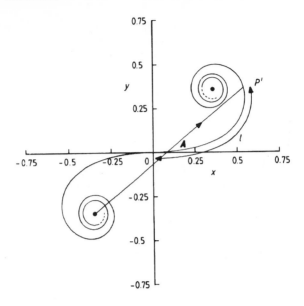

Figure 2.7.2. The Cornu spiral.

radiation from a monochromatic point source. The phase difference, δ, at a point P between the direct and the diffracted rays is then

$$\delta = 2\pi\lambda^{-1}\left\{d_1 + \left[d_2^2 + \left(d_2\tan\theta\right)^2\right]^{1/2} - \left[\left(d_1 + d_2\right)^2 + \left(d_2\tan\theta\right)^2\right]^{1/2}\right\}$$

(2.7.1)

which, since θ is very small, simplifies to

$$\delta = \frac{\pi d_1 d_2}{\lambda(d_1 + d_2)}\theta^2.$$

(2.7.2)

The intensity at a point in a diffraction pattern is obtainable from the well known Cornu spiral (figure 2.7.2), by the square of the length of the vector, $\boldsymbol{A}$. P' is the point whose distance along the curve from the origin, l, is given by

$$l = \left(\frac{2d_1 d_2}{\lambda(d_1 + d_2)}\right)^{1/2}\theta$$

(2.7.3)

and the phase difference at P' is the angle that the tangent to the curve at that point makes with the x axis, or from equation (2.7.2)

$$\delta = \tfrac{1}{2}\pi l^2.$$

(2.7.4)

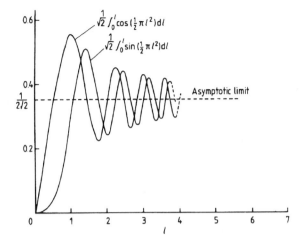

Figure 2.7.3. The Fresnel integrals.

The coordinates of P', which is the point on the Cornu spiral giving the intensity at P, x and y, are obtainable from the Fresnel integrals

$$x = \frac{1}{\sqrt{2}} \int_0^l \cos\left(\tfrac{1}{2}\pi l^2\right) dl \qquad (2.7.5)$$

$$y = \frac{1}{\sqrt{2}} \int_0^l \sin\left(\tfrac{1}{2}\pi l^2\right) dl \qquad (2.7.6)$$

whose pattern of behaviour is shown in figure 2.7.3.

If we now consider a star occulted by the Moon, then we have

$$d_1 \gg d_2 \qquad (2.7.7)$$

so that

$$l \simeq \left(2d_2/\lambda\right)\theta \qquad (2.7.8)$$

but otherwise the situation is unchanged from the one we have just discussed. The edge of the Moon of course is not a sharp knife-edge, but since even a sharp knife-edge is many wavelengths thick, the two situations are not in practice any different. The shadow of the Moon cast by the star onto the Earth therefore has a standard set of diffraction fringes around its edge. The intensities of the fringes are obtainable from the Cornu spiral and equations (2.7.5) and (2.7.6), and are shown in figure 2.7.4. The first minimum occurs for

$$l = 1.22 \qquad (2.7.9)$$

so that for the mean Earth–Moon distance

$$d_2 = 3.84 \times 10^8 \ (\mathrm{m}) \qquad (2.7.10)$$

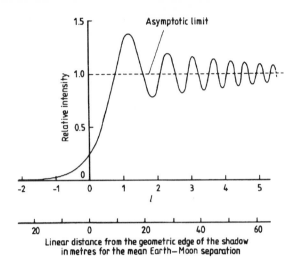

Figure 2.7.4. Fringes at the edge of the lunar shadow.

we obtain

$$\theta = 0.0064'' \tag{2.7.11}$$

at a wavelength of 500 nm. The fringes have a linear width of about 12 m therefore, when the shadow is projected onto a part of the Earth's surface which is perpendicular to the line of sight. The mean rate of angular motion of the Moon, $\dot{\theta}$, is

$$\dot{\theta} = 0.55 \quad \left('' \mathrm{s}^{-1}\right) \tag{2.7.12}$$

so that at a given spot on the Earth, the fringes will be observed as intensity variations of the star as it is occulted, with a basic frequency of up to 85 Hz. Usually the basic frequency is lower than this since the Earth's rotation can partially off-set the Moon's motion, and because the section of the lunar limb which occults the star will generally be inclined to the direction of the motion.

If the star is not a point source, then the fringe pattern shown in figure 2.7.4 becomes modified. We may see how this happens by imagining two point sources separated by an angle of about 0.0037'' in a direction perpendicular to the limb of the Moon. At a wavelength of 500 nm the first maximum of one star is then superimposed upon the first minimum of the other star, and the amplitude of the resultant fringes is much reduced compared with the single point source case. The separation of the sources parallel to the lunar limb is, within reason, unimportant in terms of its effect on the fringes. Thus an extended source can be divided into strips parallel to the lunar limb (figure 2.7.5), and each strip then behaves during the occultation as though it were a centred point source of the relevant intensity. The fringe patterns from all these point sources are then superimposed in the final tracing of the intensity variations. The precise nature

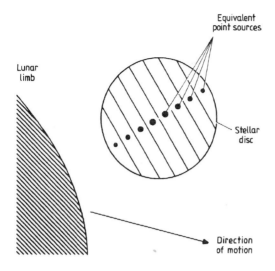

Figure 2.7.5. Schematic decomposition of a non-point source into equivalent point sources for occultation purposes.

of the alteration from the point source case will depend upon the size of the star and upon its surface intensity variations through such factors as limb darkening and gravity darkening etc. as discussed later in this section. A rough guide, however, is that changes from the point source pattern are just detectable for stellar diameters of 0.002″, while the fringes disappear completely for stellar diameters of about 0.2″. In the latter situation, the diameter may be recoverable from the total length of time which is required for the star to fade from sight. Double stars may also be distinguished from point sources when their separation exceeds about 0.002″.

Techniques

In complete contrast to almost all other areas of astronomy, the detection of occultations in the optical region is best undertaken using comparatively small telescopes. This arises from the linear size of the fringes which we have seen is about 12 m. Telescopes larger than about 1 m will therefore simultaneously sample widely differing parts of the fringe so that the detected luminosity variations become smeared. Since the signal-to-noise ratio decreases with decreasing size of telescope, the optimum size for observing occultations usually lies between 0.5 and 2 m.

Most of the photometers described in section 3.2 can be used to detect the star. However, we have seen that the observed intensity variations can have basic frequencies of 85 Hz with harmonics of several hundred hertz. The photometer and its associated electronics must therefore be capable of responding sufficiently rapidly to pick up these frequencies. Thus long integration times are not possible. Also chopping and phase-sensitive detection are ruled out unless they are operating at 10 kHz or more. The photometers used to detect occultations thus require response times of one to a few milliseconds.

Fringes are also blurred by the waveband over which the photometer is operating. We have seen from equation (2.7.8) that the monochromatic fringe pattern is wavelength dependent. A bichromatic source with wavelengths differing by a factor of 2.37 would have the first maximum for one wavelength superimposed upon the first minimum for the other, so that the fringes would almost disappear. Smaller bandwidths will still degrade the visibility of the fringes although to a lesser extent. Using the standard UBV filters (section 3.1) only five or six fringes will be detectable even in the absence of all other noise sources. A bandwidth of about 20 nm must be used if the fringes found using a medium sized telescope are not to deteriorate through this effect.

The main difference between a photometer used for occultation observations and one used for other purposes lies in the recording technique. In normal photometry the data rate is usually suffficiently slow that a chart recorder can be used, or the readings can even be read off a meter and noted down by hand. An occultation by contrast only lasts for a second or two at most, and upwards of a thousand data points must be recorded in that time. Although ultraviolet chart recorders can operate at chart speeds of metres per second and so could record this quantity of data, electronic methods are generally preferred for the ease with which the data may then be fed into a computer. Since standard commercial tape recorders have responses ranging to over 10 kHz, they can be used to record the occultation if allied to a voltage-to-frequency converter. Alternatively, that data can be fed straight into a computer memory via an analogue-to-digital converter. In the latter case a completely automatic system can be arranged by using a recycling store. Some few kilobits of information are stored and the latest data overwrite the oldest. A simple detection system to stop the recording a second or two after the occultation will then leave the desired data in the computer's memory ready for processing.

CCDs can also be used to determine the diffraction pattern of an occulted star. The diffraction pattern moves over the CCD at a calculable velocity, and the charges in the pixels are moved through the device at the same rate (cf image tracking for liquid mirrors, section 1.1). By this process of time delay and integration (TDI), high signal-to-noise ratios can be reached because each portion of the diffraction pattern is observed for a longer time than when a single element detector is used.

Since the observations are obviously always carried out within a few minutes of arc of the brightly illuminated portion of the Moon, the scattered background light is a serious problem. It can be minimised by using clean, dust free optics, precisely made light baffles, and a very small entrance aperture for the photometer, but it can never be completely eliminated. The total intensity of the background can easily be many times that of the star even when all these precautions have been taken. Thus electronic methods must be used to compensate for its effects. The simplest system is to use a differential amplifier which subtracts a preset voltage from the photometer's signal before amplification. It must be set manually a few seconds before the occultation since the background is continually changing. A more sophisticated approach uses a feedback to the differential amplifier which continually adjusts the preset voltage so that a zero output results. The time constant of this adjustment, however, is arranged to be long, so that while the slowly changing background is counteracted, the rapid changes during the occultation are unaffected. The latter system is particularly important for observing disappearances, since it does not require last minute movements of the telescope onto and off the star.

An occultation can be observed at the leading edge of the Moon (a disappearance event) or at the trailing edge (when it is a reappearance) providing that the limb near the star is not illuminated. Disappearances are by far the easiest events to observe since the star is visible and can be accurately tracked right up to the instant of the occultation. Reappearances can be observed only if a very precise off-set is available for the photometer or telescope, or if the telescope can be set onto the star before its disappearance and then kept tracking precisely on its position for tens of minutes until it reappears. The same information content is available, however, from either event. The Moon often has quite a large motion in declination, so that it is occasionally possible for both a disappearance and the following reappearance to occur at a dark limb. This is a particularly important situation since the limb angles will be different between the two events, and a complete two-dimensional map of the object can be found.

Scintillation is another major problem in observing an occultation. The frequencies involved in scintillation are similar to those of the occultation so that they cannot be filtered out. The noise level is therefore largely due to the effects of scintillation, and it is rare for more than four fringes to be detectable in practice. Little can be done to reduce the problems caused by scintillation since the usual precautions of observing only near the zenith and on the steadiest nights cannot be followed; occultations have to be observed when and where they occur.

Occultations can be used to provide information other than whether the star is a double or not, or to determine its diameter. Precise timing of the event can give the lunar position to $0.05''$ or better, and can be used to calibrate ephemeris time. Until recently when navigation satellites became available, lunar occultations were used by surveyors to determine the positions on the Earth of isolated remote sites, such as oceanic islands. Occultations of stars by planets are quite rare but are of considerable interest. Their main use is as a probe for the upper atmosphere of the planet, since it may be observed through the upper layers for some time before it is completely obscured. There has also been the recent serendipitous discovery of the rings of Uranus through their occultation of a star.

In spectral regions other than the visible, lunar occultations can find application in the determination of precise positions of objects. This is now primarily of use in the infrared and x- and gamma-ray regions where the error boxes of sources may be quite large. In the past it has also been valuable for radio observations, but has now largely been superseded by interferometer systems for the determination of precise positions of radio sources. The events can also be used to provide structural information at any wavelength by the method just discussed for optical work, provided that the detector has a fast enough response time. Extensive extended objects may still be below the resolution of the detector at some wavelengths, and their structure can then be retrieved if two occultations or more of the object can be observed. It is just the slow diminution of the flux as the object is obscured that is observed. There are no diffraction effects to worry about, and the structure is obtained by relatively straightforward computer modelling of the observations.

Analysis

The analysis of the observations to determine the diameters and/or duplicity of the sources is simple to describe, but very time consuming to carry out. It involves the comparison of the observed curve with synthesised curves for various models. The synthetic curves are obtained in the manner indicated in figure 2.7.5, taking account also of the band width and the size of the telescope. This is quite a costly exercise in terms of computer time even for a single model, and a grid of models must usually be simulated in order to provide a close fit to the observed curve. One additional factor to be taken into account is the inclination of the lunar limb to the direction of motion of the Moon. If the Moon were a smooth sphere this would be a simple calculation, but the smooth limb is distorted by lunar surface features. Sometimes the actual inclination at the point of contact may be determinable from artificial satellite photographs of the Moon. On other occasions, the lunar slope must appear as an additional unknown to be determined from the analysis. Very rarely, a steep slope may be encountered and the star may reappear briefly before its final extinction. More often the surface may be too rough on a scale of tens of metres to provide a good 'straight-edge'. Both these cases can usually be recognised from the data, and have to be discarded, at least for the purposes of determination of diameters or duplicity.

The analysis for positions of objects simply comprises determining the precise position of the edge of the Moon at the instant of occultation. One observation limits the position of the source to a semicircle corresponding to the leading (or trailing) edge of the Moon. Two observations from separate sites, or from the same site at different lunations fix the object's position to the one or two intersections of two such semicircles. This will usually be sufficient to allow an unambiguous optical identification for the object, but if not, further observations can continue to be added to reduce the ambiguity in its position and to increase the accuracy, for as long as the occultations continue. Generally a source which is occulted one month as observed from the Earth or from a close Earth-orbiting satellite will continue to be occulted for several succeeding lunations. Eventually, however, the rotation of the lunar orbit in the Saros cycle of $18\frac{2}{3}$ years will move the Moon away from the object's position, and no further occultations will occur for several years.

2.8   RADAR

Introduction

Radar astronomy and radio astronomy are very closely linked because the same equipment is often used for both purposes. Radar astronomy is less familiar to the astrophysicist however, for only one star, the Sun, has ever been studied with it, and this is likely to remain the case until other stars are visited by spacecraft, which will be a while yet. Other aspects of astronomy benefit from the use of radar to a much greater extent, so the technique is included, despite being almost outside the limits of this book, for completeness, and because its results may find applications in some areas of astrophysics.

Radar, as used in astronomy, can be fairly conventional equipment for use on board spacecraft, or for studying meteor trails, or it can be of highly specialised design and construction for use in detecting the Moon, planets, Sun etc from the Earth. Its results potentially contain information on three aspects of the object being observed—distance, surface fine scale structure, and relative velocity.

Theoretical principles

Consider a radar system as shown in figure 2.8.1. The transmitter has a power P and is an isotropic emitter. The radar cross section of the object is α, where α is defined as the cross sectional area of a perfectly isotropically scattering sphere which would return the same amount of energy to the receiver as the object. Then the flux at the receiver, f is given by

$$f = \frac{P\alpha}{4\pi R_1^2 4\pi R_2^2}. \qquad (2.8.1)$$

Normally we have the transmitter and receiver close together, if they are not actually the same antenna, so that

$$R_1 = R_2 = R. \qquad (2.8.2)$$

Furthermore the transmitter would not in practice be isotropic, but would have a gain, g (see section 1.2). If the receiver has an effective collecting area of A_e then the received signal, F, is given by

$$F = \frac{A_e \alpha P g}{16\pi^2 R^4}. \qquad (2.8.3)$$

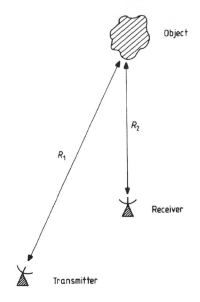

Figure 2.8.1. Schematic radar system.

For an antenna, we have the gain from equation (1.2.10)

$$g = \frac{4\pi \nu^2 A'_e}{c^2} \tag{2.8.4}$$

where A'_e is the effective area of the transmitting antenna. Much of the time in radar astronomy, the transmitting and receiving dishes are the same, so that

$$A'_e = A_e \tag{2.8.5}$$

and

$$F = \frac{P\alpha A_e^2 \nu^2}{4\pi c^2 R^4}. \tag{2.8.6}$$

This last equation is valid for objects which are not angularly resolved by the radar. For targets which are comparable with the beam width or larger, the radar cross section, α, must be replaced by an appropriate integral. Thus for a spherical target which has the radar beam directed towards its centre, we have

$$F \simeq \frac{P A_e^2 \nu^2}{4\pi c^2 R^4} \int_0^{\pi/2} 2\pi r \alpha(\varphi) \sin\varphi \left[s\left(\frac{r\sin\varphi}{R}\right) \right]^2 \mathrm{d}\varphi \tag{2.8.7}$$

where r is the radius of the target and is assumed to be small compared with the distance, R, φ is the angle at the centre of the target to a point on its surface illuminated by the radar beam, $\alpha(\varphi)$ is the radar cross section for the surface of the target when the incident and returned beams make an angle φ to the normal to the surface. The function is normalised so that the integral tends to α as the beam width increases. $s(\theta)$ is the sensitivity of the transmitter/receiver at an angle θ to its optical axis (cf equation (1.2.11)).

The amount of flux which is received is not the only criterion for the detection of objects as we saw in section 1.2 (equation (1.2.4)). The flux must also be sufficiently stronger than the noise level of the whole system if the returned pulse is to be distinguishable. Now we have seen in section 1.2 that the receiver noise may be characterised by comparing it with the noise generated in a resistor at some temperature. We may similarly characterise all the other noise sources, and obtain a temperature, T_s, for the whole system, which includes the effects of the target, and of the transmission paths as well as the transmitter and receiver. Then from equation (1.2.5), we have the noise in power terms, N

$$N = 4kT_s\Delta\nu \qquad (2.8.8)$$

where $\Delta\nu$ is the bandwidth of the receiver in frequency terms. The signal-to-noise ratio is therefore

$$\frac{F}{N} = \frac{P\alpha A_e^2\nu^2}{16\pi kc^2R^4T_s\Delta\nu} \qquad (2.8.9)$$

and must be unity or larger if a single pulse is to be detected. Since α and R are fixed by the target which is selected, the signal-to-noise ratio may be increased by increasing the power of the transmitter, the effective area of the antenna, or the frequency, or by decreasing the system temperature or the bandwidth. Only over the last of these items does the experimenter have any real control, all the others will be fixed by the choice of antenna. However, the bandwidth is related to the length of the radar pulse. Even if the transmitter emits monochromatic radiation, the pulse will have a spread of frequencies given by the Fourier transform of the pulse shape. To a first approximation

$$\Delta\nu \simeq 1/\tau \text{ (Hz)} \qquad (2.8.10)$$

where τ is the length of the transmitted pulse in seconds. Thus the signal-to-noise ratio can be increased by increasing the pulse length. Unfortunately for accurate ranging, the pulse needs to be as short as possible, and so an optimum value which minimises the conflict between these two requirements must be sought, and this will vary from radar system to radar system, and from target to target.

When ranging is not required, so that information is only being sought on the surface structure and the velocity, the pulse length may be increased very considerably. Such a system is then known as a continuous wave (CW) radar, and its useful pulse length is only limited by the stability of the transmitter frequency, and by the spread in frequencies introduced into the returned pulse through doppler shifts arising from the movement and rotation of the target. In practice CW radars work continuously even though the signal-to-noise ratio remains that for the optimum pulse. The alternative to CW radar is pulsed radar. Astronomical pulsed radar systems have a pulse length typically between 10 μs and 10 ms, with peak powers of several tens of megawatts. For both CW and pulsed systems, the receiver pass band must be matched to the returned pulse in both frequency and bandwidth. The receiver must therefore be tunable over a short range to allow for the doppler shifting of the emitted frequency.

The signal-to-noise ratio may be improved by integration. With CW radar, samples are taken at intervals given by the optimum pulse length. For pulsed radar, the pulses are simply combined. In either case the noise is reduced by a factor of $N^{1/2}$, where N is the number of samples or pulses which are combined. If the radiation in separate pulses from a pulsed radar system is coherent, then the signal-to-noise ratio improves directly with N. Coherence, however, may only be retained over the time interval given by the optimum pulse length for the system operating in the CW mode. Thus if integration is continued over several times the coherence interval, the signal-to-noise ratio for a pulsed system decreases as $(NN'^{1/2})$, where N is the number of pulses in a coherence interval, and N' the number of coherence intervals over which the integration extends.

For most astronomical targets, whether they are angularly resolved or not, there is some depth resolution with pulsed systems. The physical length of a pulse is c_τ, and this ranges from about 3 to 3000 km for most practical astronomical radar systems. Thus if the depth of the target in the sense of the distance along the line of sight between the nearest and furthest points returning detectable echoes, is larger than the physical pulse length, the echo will be spread over a greater time interval then the original pulse. This has both advantages and disadvantages. Firstly it provides information on the structure of the target with depth. If the depth information can be combined with doppler shifts arising through the target's rotation, then the whole surface may be mappable, although there may still remain a two-fold ambiguity about the equator. Secondly, and on the debit side, the signal-to-noise ratio is reduced in proportion approximately to the ratio of the lengths of the emitted and returned pulses.

Equation (2.8.9) and its variants which take account of resolved targets etc is often called the radar equation. Of great practical importance is its R^{-4} dependence. Thus if other things are equal, the power required to detect an object increases by a factor of sixteen when the distance to the object doubles. Thus Venus at greatest elongation requires thirty-seven times the power required for its detection at inferior conjunction. The high dependence upon R can be made to work in our favour, however, when using satellite-borne radar systems. The value of R can then be reduced from tens of millions of kilometres to a few hundred. The enormous gain in the sensitivity that results means that the small low power radar systems which are suitable for satellite use are sufficient to map the target in detail. The prime example of this to date is, of course, Venus. Most of the data we have on the surface features have come from the radar carried by Magellan and the various Mariner and Pioneer spacecraft.

The radar on board Magellan is of a different type from those so far

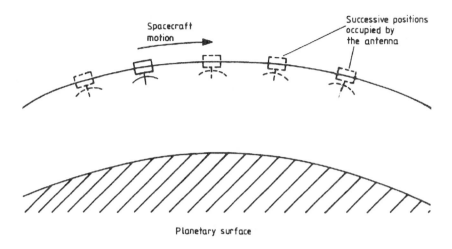

Figure 2.8.2. Synthesis of an interferometer array by a moving single antenna.

considered. It is known as Synthetic Aperture Radar (SAR) and has much in common with the technique of aperture synthesis (section 2.5). Such radars have also been used on board several remote sensing satellites, such as Seasat and ERS 1 for studying the Earth.

Synthetic aperture radar uses a single antenna which, since it is on board a spacecraft, is moving with respect to the surface of the planet below. The single antenna therefore successively occupies the positions of the multiple antennas of an array (figure 2.8.2). To simulate the observation of such an array, all that is necessary, therefore, is to record the output from the radar when it occupies the position of one of the elements of the array, and then to add the successive recordings with appropriate shifts to counteract their various time delays. In this way the radar can build up an image whose resolution, along the line of flight of the spacecraft, is many times better than that which the simple antenna would provide by itself. In practice, of course, the SAR operates continually, not just at the positions of the elements of the synthesised array.

The maximum array length that can be synthesised in this way, is limited by the period over which a single point on the planet's surface can be kept under observation. The resolution of a parabolic dish used as a radar is approximately λ/D. If the spacecraft is at a height, h, the radar 'footprint' of the planetary surface is thus about

$$L = \frac{2\lambda h}{D} \tag{2.8.11}$$

in diameter (figure 2.8.3). Since h, in general, will be small compared with the size of the planet, we may ignore the curvature of the orbit. We may then easily see (figure 2.8.4) that a given point on the surface only remains observable while the spacecraft moves a distance, L.

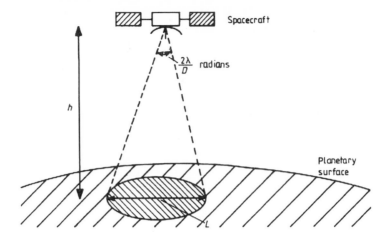

Figure 2.8.3. Radar footprint.

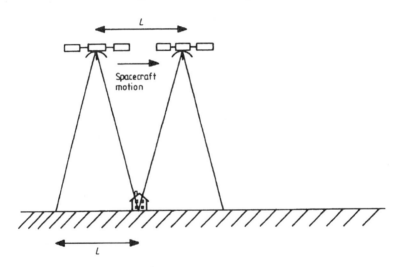

Figure 2.8.4. Maximum length of an SAR.

Now, the resolution of an interferometer (section 2.5) is given by equation (2.5.13). Thus, the angular resolution of an SAR of length L is

$$\text{Resolution} = \frac{\lambda}{2L} = \frac{D}{4h} \text{ radians.} \qquad (2.8.12)$$

The diameter of the synthesised radar footprint is then

$$\text{Footprint diameter} = 2(D/4h)h = D/2. \qquad (2.8.13)$$

Equation (2.8.13) shows the remarkable result that the linear resolution of an SAR is improved by *decreasing* the size of the radar dish! The linear resolution is just half the dish diameter. Other considerations, such as signal-to-noise ratio (equation (2.8.9)) require the dish to be as large as possible. Thus, an actual SAR system has a dish size that is a compromise, and existing systems typically have 5 to 10 m diameter dishes.

The above analysis applies to a focused SAR, that is to say, an SAR in which the changing distance of the spacecraft from the point of observation is compensated by phase shifting the echoes appropriately. In an unfocused SAR the synthesised array length is limited by the point at which the uncorrelated phase shifts due to the changing distance to the observed object reach about $\lambda/4$. The diameter of the array footprint is then about

$$\text{Unfocused SAR footprint} = \sqrt{2\lambda h}. \qquad (2.8.14)$$

Equipment

Radar systems for use in studying planets etc have much in common with radio telescopes. Almost any filled aperture radio antenna is usable as a radar antenna. Steerable paraboloids are particularly favoured because of their convenience in use. Unfilled apertures, such as collinear arrays (section 1.2) are not used since most of the transmitted power is wasted.

The receiver is also generally similar to those in use in radio astronomy, except that it must be very accurately tunable to provide compensation for doppler shifts, and the stability of its frequency must be very high. Its band pass should also be adjustable so that it may be matched accurately to the expected profile of the returned pulse in order to minimise noise.

Radar systems, however, do require three additional components which are not found in radio telescopes. Firstly, and obviously, there must be a transmitter, secondly a very high stability master oscillator and thirdly an accurate timing system. The transmitter is usually second only to the antenna as a proportion of the cost of the whole system. Different targets or purposes will generally require different powers, pulse lengths, frequencies etc from the transmitter. Thus it must be sufficiently flexible to cope with demands which might typically range from pulse emission at several megawatts for a millisecond burst to continuous wave generation at several hundred kilowatts. Separate pulses must be coherent, at least over a time interval equal to the optimum pulse length for the CW mode, if full advantage of integration to improve signal-to-noise ratio is to be gained. The master oscillator is in many ways the heart of the system. In a typical observing situation, the returned pulse must be resolved to 0.1 Hz for a frequency in the region of 1 GHz, and this must be accomplished after a delay between the emitted pulse and the echo which can range from minutes to hours. Thus the frequency must be stable to one part in 10^{12} or 10^{13}, and it is the master oscillator which provides this stability. It may be a temperature controlled crystal oscillator, or an atomic clock. In the former case, corrections for aging will be necessary. The frequency of the master oscillator is then stepped up or down as required and used to drive the exciter for the transmitter, the local oscillators for the superheterodyne receivers, the doppler compensation system, and the timing system. The final item, the timing system, is essential if the range of the target is to be found. It is also essential to know with quite a high accuracy when to expect the echo. This arises because the pulses are not broadcast at constant intervals. If this were to be the case then there would be no certain way of relating a given echo to a given emitted pulse. Thus instead, the emitted pulses are sent out at varying and coded intervals, and the same pattern searched for amongst the echoes. Since the distance between the Earth and the target is also varying, an accurate timing system is vital to allow the resulting echoes with their changing intervals to be found amongst all the noise.

Data analysis

The analysis of the data in order to obtain the distance to the target is simple in principle, being merely half the delay time multiplied by the speed of light. In practice the process is much more complicated. Corrections for the atmospheres of the Earth and the target (should it have one) are of particular importance. The radar pulse will be refracted and delayed by the Earth's atmosphere, and there will be similar effects in the target's atmosphere with in addition the possibility of reflection from an ionosphere as well as, or instead of, reflection from the surface. This applies especially to solar detection. Then reflection can often occur from layers high above the visible surface; ten thousand kilometres above the photosphere, for example, for a radar frequency of 1 GHz. Furthermore if the target is a deep one, i.e. it is resolved in depth, then the returned pulse will be spread over a greater time interval than the emitted one so that the delay length becomes ambiguous. Other effects such as refraction or delay in the interplanetary medium may also need to be taken into account.

For a deep, rotating target, the pulse is spread out in time and frequency, and this can be used to plot maps of the surface of the object (figure 2.8.5). A cross section through the returned pulse at a given instant of time will be formed from echoes from all the points over an annular region such as that shown in figure 2.8.6. For a rotating target, which for simplicity we assume has its rotational axis perpendicular to the line of sight, points such as A and A' will have the same approach velocity, and their echoes will be shifted to the same higher frequency. Similarly the echoes from points such as B and B' will be shifted to the same lower frequency. Thus the intensity of the echo at a particular instant and frequency is related to the radar properties of two points which are equidistant from the equator and are north and south of it. Hence a map of the radar reflectivity of the target's surface may be produced, though it will have a two-fold ambiguity about the equator. If the rotational axis is not perpendicular to the line of sight, then the principle is similar, but the calculations are more

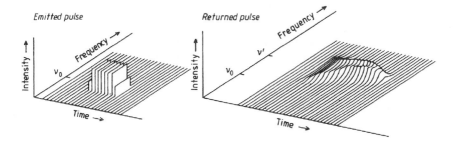

Figure 2.8.5. Schematic change in pulse shape for radar pulses reflected from a deep rotating target. The emitted pulse would normally have a Gaussian profile in frequency, but it is shown here as a square wave in order to enhance the clarity of the changes between it and the echo.

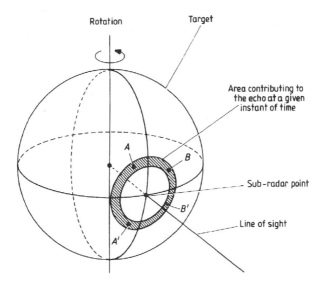

Figure 2.8.6. Region of a deep radar target contributing to the echo at a particular instant of time.

complex. However, if such a target can be observed at two different angles of inclination of its rotational axis, then the two-fold ambiguity may be removed and a genuine map of the surface produced. Again in practice the process needs many corrections, and must have the frequency and time profiles of the emitted pulse removed from the echo by deconvolution (section 2.1). The features on radar maps are often difficult to relate to more normal surface features, since they are a conflation of the effects of surface reflectivity, small scale structure, height, gradient, etc.

The relative velocity of the Earth and the target is much simpler to obtain. Provided that the transmitter is sufficiently stable in frequency (see the earlier discussion), then it may be found from the shift in the mean frequency of the echo compared with the mean frequency of the emitted pulse, by the application of the Doppler formula. Corrections for the Earth's orbital and rotational velocities are required. The former is given by equation (4.1.1), while the latter may be obtained from

$$v_c = v_o + 462\cos\delta\sin(\alpha - \text{LST})\sin(\text{LAT}) \qquad (\text{m s}^{-1}) \qquad (2.8.15)$$

where v_c is the corrected velocity (m s^{-1}), v_o is the observed velocity (m s^{-1}), α and δ are the right ascension and declination of the target, LST is the local sidereal time at the radar station when the observation is made and LAT is the latitude of the radar station.

The combination of accurate knowledge of both distance and velocity for a planet on several occasions enables its orbit to be calculated to a very high degree of precision. This then provides information on basic quantities such as the astronomical unit, and enables tests of metrical gravitational theories such as general relativity to be made, and so is of considerable interest and importance even outside the area of astronomy. Producing a precise orbit for a planet by this method requires, however, that the Earth's and the planet's orbits both be known precisely so that the various corrections etc can be applied. Thus the procedure is a 'bootstrap' one, with the initial data leading to improved orbits, and these improved orbits then allowing better interpretation of the data and so on over many iterations.

The radar reflection contains information on the surface structure of the target. Changes in the phase, polarisation and of the manner in which the radar reflectivity changes with angle, can in principle be used to determine rock types, roughness of the surface and so on. In practice, the data are rarely sufficient to allow anything more than vague possibilities to be indicated. The range of unfixed parameters describing the surface is so large, that very many different models for the surface can fit the same data. This aspect of radar astronomy has therefore been of little use to date.

Meteors

The ionised vapour along the track of a meteor reflects radar transmissions admirably. Over the last few decades, therefore, many stations have been set up to detect meteors by such means. Since they can be observed equally well during the day as at night, many previously unknown daytime meteor showers have been discovered. Either pulsed or CW radar is suitable, and a wavelength of a few metres is usually used.

The echoes from a meteor trail are modulated into a Fresnel diffraction pattern (figure 2.7.4), with the echoed power expressible in terms of the Fresnel integrals (equations (2.7.5) and (2.7.6)) as

$$F(t) = F(x^2 + y^2) \qquad (2.8.16)$$

where F is the echo strength for the fully formed trail, $F(t)$ is the echo strength at time t during the formation of the trail. The parameter l of the Fresnel integrals in this case is the distance of the meteor from the minimum range point. The modulation of the echo arises because of the phase differences between echoes returning from different parts of the trail. The distance of the meteor, and hence its height may be found from the delay time as usual. In addition to the distance however, the cross range velocity can also be obtained from the echo modulation. Several stations can combine, or a single transmitter can have several well separated receivers, in order to provide further information such as the position of the radiant, atmospheric deceleration of the meteor, etc.

Exercise

2.8.1 If a 10 MW peak power radar is used to observe Venus from the Earth at inferior conjunction, using a 300 m dish aerial, calculate the peak power required for a satellite-borne radar situated 1000 km above Venus' surface which has a 1 m dish aerial if the same signal-to-noise ratio is to be achieved. Assume that all the other parameters of the radar systems are similar, and take the distance of Venus from the Earth at inferior conjunction to be 4.14×10^7 km.

If the linear resolution of the Earth-based radar at Venus is 10^4 km, what is that of the satellite-based system?

2.9   ELECTRONIC IMAGES

Image formats

There are various ways of storing electronic images. One currently in widespread use for astronomical images is the Flexible Image Transport System, or FITS. There are various versions of FITS, but all have a header, the image in binary form, and an end section. The header must be a multiple of 2880 bytes (this number originated in the days of punched card input to computers and represents 36 cards each of 80 bytes). The header contains information on the number of bits representing the data in the image, the number of dimensions of the image (e.g. one for a spectrum, two for a monochromatic image, three for a colour image, etc), the number of elements along each dimension, the object observed, the telescope used, details of the observation and other comments. The header is then padded out to a multiple of 2880 with zeros. The image section of a FITS file is simply the image data in the form specified by the header. The end section then pads the image section with zeros until it too is an integer multiple of 2880 bytes.

Other formats may be encountered such as GIF and TIFF which are more useful for processed images than raw data.

Image compression

Most images, howsoever they may be produced and at whatever wavelength they were obtained, are nowadays stored in electronic form. This is even true of many photographic images which may be scanned using microdensitometers (section 2.2) and fed into a computer. Electronic images may be stored by simply recording the precise value of the intensity at each pixel in an appropriate two-dimensional matrix. However, this uses large amounts of memory; for example 2 megabytes for a 16-bit image from a 1000×1000 pixel CCD (and 32-bit, $10\,000 \times 10\,000$ CCDs, requiring 400 megabytes per image are being envisaged). Means whereby storage space for electronic images may be utilised most efficiently are therefore at a premium. The most widely used approach to the problem is to compress the images.

Image compression relies for its success upon some or all of the information in the image being redundant. That information may therefore be omitted or stored in shortened form without losing information from the image as a whole. Image compression which conserves all the information in the original image is known as loss-less compression. For some purposes a reduction in the amount of information in the image may be acceptable (for example, if one is only interested in certain objects or intensity ranges within the image), and this is the basis of lossy compression (see below).

There are three main approaches to loss-less image compression. The first is differential compression. This stores not the intensity values for each pixel, but the differences between the values for adjacent pixels. For an image with smooth variations, the differences will normally require fewer bits for their storage than the absolute intensity values. The second approach, known as run compression, may be used when many sequential pixels have the same intensity. This is often true for the sky background in astronomical images. Each run of constant intensity values may be stored as a single intensity combined with the number of pixels in the run, rather than storing the value for every pixel. The third approach, which is widely used on the Internet and elsewhere, where it goes by various proprietary names such as ZIP and DoubleSpace, is similar to run compression except that it stores repeated sequences whatever their nature as a single code, not just constant runs. For some images, loss-less compressions by a factor of five or ten times are possible.

Sometimes not all the information in the original images is required. Lossy compression can then be by factors of 20, 50 or more. For example, a 32-bit image to be displayed on a computer monitor which has 256 brightness levels need only be stored to 8-bit accuracy, or only the (say) 64×64 pixels covering the image of a planet, galaxy, etc need be stored from a 1000×1000 pixel image if that is all that is of interest within the image. Most workers, however, will prefer to store the original data in a loss-less form.

Compression techniques do not work effectively on noisy images since the noise is incompressible. Then a lossy compression may be of use since it will primarily be the noise that is lost. An increasingly widely used lossy compression for such images is based on the wavelet transform

$$T = \frac{1}{\sqrt{|a|}} \int f(t)\psi\left(\frac{t-b}{a}\right) dt. \qquad (2.9.1)$$

This can give high compression ratios for noisy images. The details of the technique are, however, beyond the scope of this book, and the interested reader may pursue them in sources listed in the Bibliography.

Image processing

Image processing is the means whereby images are produced in an optimum form to provide the information that is required from them. Photographic image-processing techniques are possible, such as unsharp masking (section 2.2), but most image-processing techniques are applied to electronic images.

Image processing divides into data reduction and image optimisation. Data reduction is the relatively mechanical process of correcting for known problems and faults. Many aspects of it are discussed elsewhere, such as dark frame subtraction, flat fielding, and cosmic ray spike elimination on CCD images (section 1.1), deconvolution and maximum entropy processing (section 2.1) and CLEANing aperture synthesis images (section 2.5), etc.

Image optimisation is more of an art than a science. It is the further processing of an image with the aim of displaying those parts of the image which are of interest to their best effect for the objective for which they are required. The same image may therefore be optimised in quite different ways if it is to be used for different purposes. For example, an image of Jupiter and its satellites would be processed quite differently if the surface features on Jupiter were to be studied, than if the positions of the satellites with respect to the planet were of interest. There are dozens of techniques used in this aspect of image processing, and there are few rules other than experience to suggest which will work best for a particular application. The main techniques are outlined below, but the interested reader will need to look to other sources and to obtain practical experience before being able to apply them with consistent success.

Grey scaling

This is probably the technique in most widespread use. It arises because many detectors have dynamic ranges far greater than those of the computer monitors or hard copy devices which are used to display their images. In many cases the interesting part of the image will cover only a restricted part of the dynamic range of the detector. Grey scaling then consists of stretching that section of the dynamic range over the available intensity levels of the display device.

For example, the spiral arms of a galaxy on a 16-bit CCD image (with a dynamic range from 0 to 65 535) might have CCD level values ranging from 20 100 to 20 862. On a computer monitor with 256 intensity levels, this would be grey scaled so that CCD levels 0 to 20 099 corresponded to intensity 0 (i.e. black) on the monitor, levels 20 100 to 20 103 to intensity 1, levels 20 104 to 20 106 to intensity 2, and so on up to levels 20 859 to 20 862 corresponding to intensity 254. CCD levels 20 863 to 65 535 would then all be lumped together into intensity 255 (i.e. white) on the monitor.

Other types of grey scaling may aim at producing the most visually acceptable version of an image. There are many variations of this technique whose operation may best be envisaged by its effect on a histogram of the display pixel intensities. In histogram equalisation, for example, the image pixels are mapped to the display pixels in such a way that there are equal numbers of image pixels in each of the display levels. For many astronomical images, a mapping which results in an exponentially decreasing histogram for the numbers of image pixels in each display level often gives a pleasing image since it produces a good dark background for the image. The number of other mappings available is limited only by the imagination of the operator.

Image combination

Several images of the same object may be added together to provide a resultant image with lower noise. More powerful combination techniques, such as the

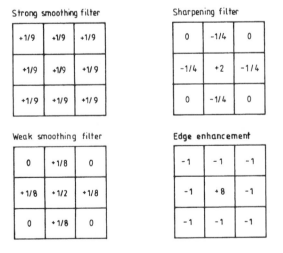

Figure 2.9.1. Examples of commonly used spatial filters.

subtraction of two images to highlight changes, or false-colour displays, are, however, probably in more widespread use.

Spatial filtering

This technique has many variations which can have quite different effects upon the image. The basic process is to combine the intensity of one pixel with those of its surrounding pixels in some manner. The filter may best be displayed by a 3×3 matrix (or 5×5, 7×7 matrices, etc) with the pixel of interest in the centre. The elements of the matrix are the weightings given to each pixel intensity when they are combined and used to replace the original value for the central pixel (figure 2.9.1). Some commonly used filters are shown in figure 2.9.1.

Chapter 3

Photometry

3.1 PHOTOMETRY

Background

Magnitudes

The system used by astronomers to measure the brightnesses of stars is a very ancient one, and for that reason is a very awkward one. It originates with the earliest of the astronomical catalogues; that due to Hipparchos in the late second century BC. The stars in this catalogue were divided into six classes of brightness, the brightest being of the first class, and the faintest of the sixth class. With the invention and development of the telescope such a system proved to be inadequate, and it had to be refined and extended. This was initially undertaken by W Herschel, but our present day system is based upon the work of Pogson in 1856. He suggested a logarithmic scale which approximately agreed with the earlier measures. As we have seen in section 1.1, the response of the eye is nearly logarithmic, so that the ancient system due to Hipparchos was logarithmic in terms of intensity. Hipparchos' class five stars were about $2\frac{1}{2}$ times brighter than his class six stars, and were about $2\frac{1}{2}$ times fainter than class four stars and so on. Pogson expressed this as a precise mathematical law

$$m_1 - m_2 = -2.5 \log(E_1/E_2) \tag{3.1.1}$$

where m_1 and m_2 are the magnitudes of stars 1 and 2 and E_1 and E_2 are the energies per unit area at the surface of the Earth for stars 1 and 2. The scale is awkward to use because of its peculiar base, and because the brighter stars have magnitudes whose absolute values are smaller than those of the fainter stars. However, it seems unlikely that astronomers will change the system, so the student must perforce become familiar with it as it stands. One aspect of the scale is immediately apparent from equation (3.1.1), and that is that the scale is a relative one—the magnitude of one star is expressed in terms of that of another, there is no direct reference to absolute energy or intensity units. The zero of

260

the scale is therefore fixed by reference to standard stars. On Hipparchos' scale, those stars which were just visible to the naked eye were of class six. Pogson's scale is thus arranged so that stars of magnitude six are just visible to a normally sensitive eye from a good observing site on a clear, moonless night. The standard stars are chosen to be non-variables, and their magnitudes are assigned so that the above criterion is fulfilled. The primary stars are known as the North Polar Sequence, and comprise stars within 2° of the pole star, so that they may be observed from almost all northern hemisphere observatories throughout the year. They are listed in Appendix II. Secondary standards may be set up from these to allow observations to be made in any part of the sky.

The faintest stars visible to the naked eye, from the definition of the scale, are of magnitude six, this is termed the limiting magnitude of the eye. For point sources the brightnesses are increased by the use of a telescope by a factor G, which is called the light grasp of the telescope (section 1.1). Since the dark-adapted human eye has a pupil diameter of about 7 mm, G is given by

$$G \simeq 2 \times 10^4 d^2 \qquad (3.1.2)$$

where d is the telescope's objective diameter in metres. Thus the limiting magnitude through a visually used telescope, m_L, is

$$m_L = 16.8 + 5 \log d. \qquad (3.1.3)$$

Photography and other detection techniques will generally improve on this by some two to four stellar magnitudes.

The magnitudes so far discussed are all apparent magnitudes and so result from a combination of the intrinsic brightness of the object and its distance. This is obviously an essential measurement for an observer, since it is the criterion determining exposure times, etc; however, it is of little intrinsic significance for the object in question. A second type of magnitude is therefore defined which is related to the actual brightness of the object. This is called the absolute magnitude and it is 'the apparent magnitude of the object if its distance from the Earth were ten parsecs'. It is usually denoted by M, while apparent magnitude uses the lower case, m. The relation between apparent and absolute magnitudes may easily be obtained from equation (3.1.1). Imagine the object moved from its real distance to ten parsecs from the Earth. Its energy per unit area at the surface of the Earth will then change by a factor $(D/10)^2$, where D is the object's actual distance in parsecs. Thus

$$M - m = -2.5 \log(D/10)^2 \qquad (3.1.4)$$
$$M = m + 5 - 5 \log D. \qquad (3.1.5)$$

The difference between apparent and absolute magnitudes is called the distance modulus and is occasionally used in place of the distance itself

$$\text{distance modulus} = m - M = 5 \log D - 5. \qquad (3.1.6)$$

Equations (3.1.5) and (3.1.6) are valid so long as the only factors affecting the apparent magnitude are distance and intrinsic brightness. However, light from the object often has to pass through interstellar gas and dust clouds, where it may be absorbed. A more complete form of equation (3.1.5) is therefore

$$M = m + 5 - 5\log D - AD \qquad (3.1.7)$$

where A is the interstellar absorption in magnitudes per parsec. A typical value for A for lines of sight within the galactic plane is 0.002 mag pc^{-1}. Thus we may determine the absolute magnitude of an object via equations (3.1.5) or (3.1.7) once its distance is known. More frequently, however, the equations are used in the reverse sense in order to determine the distance. Often the absolute magnitude may be estimated by some independent method, and then

$$D = 10^{[(m-M+5)/5]} \text{ (pc)}. \qquad (3.1.8)$$

Such methods for determining distance are known as standard candle methods, since the object is in effect acting as a standard of some known luminosity. The best known examples are the classical Cepheids with their period–luminosity relationship.

$$M = -1.9 - 2.8\log P \qquad (3.1.9)$$

where P is the period of the variable in days. Many other types of stars such as dwarf Cepheids, RR Lyrae stars, W Virginis stars and β Cepheids are also suitable. Also the brightest novae or supernovae, or the brightest globular clusters around a galaxy, or even the brightest galaxy in a cluster of galaxies can provide rough standard candles. Yet another method is due to Wilson and Bappu who found a relationship between the width of the emission core of the ionised calcium line at 393.3 nm (the Ca II K line) in late-type stars, and their absolute magnitudes. The luminosity is then proportional to the sixth power of the line width.

Both absolute and apparent magnitudes are normally measured over some well defined spectral region. While the above discussion is quite general, the equations only have a validity within a given spectral region. Since the spectra of two objects may be quite different, their relationship at one wavelength may be very different from that at another. The next section discusses the definitions of these spectral regions and their interrelationships.

Filter systems

A very wide range of filter systems and filter–detector combinations has been devised. They may be grouped into wide, intermediate and narrow band systems according to the widths of their transmission curves. Wide band filters typically have bandwidths of around 100 nm, intermediate band filters range from 10 to 50 nm, while narrow band filters range from 0.05 to 10 nm. The division is convenient for the purposes of discussion here, but is not of any real physical significance.

The earliest 'filter' system was given by the response of the human eye (figure 1.1.3) and peaks around 510 nm with a bandwidth of 200 nm or so. Since all eyes have roughly comparable responses, the importance of the spectral region within which a magnitude was measured was not realised until the application of the photographic plate to astronomical detection. Then it was discovered that the magnitudes of stars which had been determined from such plates often differed significantly from those found visually. The discrepancy arose from the differing sensitivities of the eye and the photographic emulsion. Early emulsions and today's unsensitised emulsions (section 2.2) have a response which peaks near 450 nm, little contribution is made to the image by radiation of a wavelength longer than about 500 nm. The magnitudes determined with the two methods became known as visual and photographic magnitudes respectively, and were denoted by m_v and m_p or M_v and M_p. This system is still used, for it allows the brightnesses of many stars to be found rapidly from plates taken by Schmidt cameras. The visual magnitude, however, is now found from the images on an orthochromatic plate which mimics the eye's sensitivity closely when it is combined with a yellow filter, and is then usually known as the photovisual magnitude.

With the development of photoelectric methods of detection of light, and their application to astronomy by Stebbins and others earlier this century, other spectral regions became available, and more precise definitions of them were required. There are now a very great number of different photometric systems. While it is probably not quite true that there is one system for every observer, there is certainly at least one separate system for every slightly differing purpose to which photometric measurements may be put. Many of these are developed for highly specialised purposes, and so are unlikely to be encountered outside their particular area of application. Other systems have a wide application and so are worth knowing in detail.

The most widespread of these more general systems is the *UBV* system which was defined by Johnson and Morgan in 1953. This is a wide band system with the *B* and *V* regions corresponding approximately to the photographic and visual responses, and with the *U* region in the violet and ultraviolet. The precise definition (in Johnson H L 1963 *Basic Astronomical Data* ed W A Hiltner, University of Chicago Press) requires a reflecting telescope with aluminised mirrors and uses an RCA 1P21 photomultiplier. The filters are:

Corning 3384	for the *V* region
Corning 5030 plus Schott GG 13	for the *B* region
Corning 9863	for the *U* region.

The response curves for the filter–photomultiplier combination are shown in figure 3.1.1. The absorption in the Earth's atmosphere normally cuts off the short

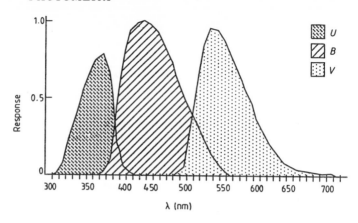

Figure 3.1.1. *UBV* response curves, excluding the effects of atmospheric absorption.

wavelength edge of the *U* response. At sea level roughly half of the incident energy at a wavelength of 360 nm is absorbed, even for small zenith angles. More importantly, the absorption can vary with changes in the atmosphere. Ideal *U* response curves can therefore only be approached at very high altitude, high quality observing sites. The effect of the atmosphere on the *U* response curve is shown in figure 3.1.2. The *B* and *V* response curves are comparatively unaffected by differential absorption across their wavebands; there is only a total reduction of the flux through these filters (see section 3.2). The primary standard stars for the *UBV* system are listed in Appendix III. The scales are arranged so that the magnitudes through all three filters are equal to each other for A0 V stars.

There are several extensions to the *UBV* system for use at longer wavelengths. The most widespread has another eight regions:

Filter name	R	I	J	K	L	M	N	Q
Central wavelength (nm)	700	900	1250	2200	3400	4900	10200	20000
Bandwidth (nm)	220	240	380	480	700	300	5000	5000

The longer wavelength regions in this system are defined by atmospheric transmission and so are very variable.

Amongst the intermediate pass band systems, the most widespread is the *uvby* system. This was proposed in the late 1960s and is now in fairly common use. Its transmission curves are shown in figure 3.1.3 and its standard stars in Appendix IV. Narrow band work concentrates on isolating spectral lines. *Hα* and *Hβ* are common choices, and their variations can be determined by measurements through a pair of filters which are centred on the line but have different bandwidths. No single system is in general use, so that a more detailed discussion is not very profitable. Other spectral features can be studied with

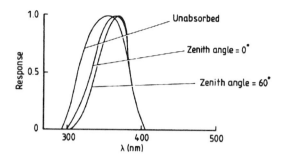

Figure 3.1.2. Effect of atmospheric absorption at sea level upon the *U* response curve (normalised).

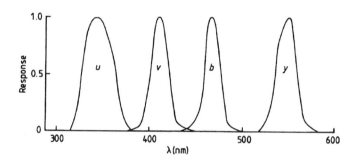

Figure 3.1.3. Normalised transmission curves for the *uvby* system, not including the effects of atmospheric absorption.

narrow band systems where one of the filters isolates the spectral feature, and the other is centred on a nearby section of the continuum. The reason for the invention of these and other photometric systems lies in the information that may be obtained from the comparison of the brightness of the object at different wavelengths. Many of the processes contributing to the final spectrum of an object as it is received on Earth preferentially affect one or more spectral regions. Thus estimates of their importance may be obtained simply and rapidly by measurements through a few filters. Some of these processes are discussed in more detail in the next section and in section 3.2. The most usual features studied in this way are the Balmer discontinuity and other ionisation edges, interstellar absorption, and strong emission or absorption lines or bands.

There is one additional photometric 'system' which has not yet been mentioned. This remaining system is the bolometric system, or rather bolometric magnitude, since it has only one pass band. The bolometric magnitude is based upon the total energy emitted by the object at all wavelengths. Since it is not possible to observe this in practice, its value is determined by modelling calculations based upon the object's intensity through one or more of the filters of a photometric system. Although x-ray, ultraviolet, infrared and radio data are now available to assist these calculations many uncertainties remain, and the bolometric magnitude is still rather imprecise especially for high temperature stars. The calculations are expressed as the difference between the bolometric magnitude and an observed magnitude. Any filter of any photometric system could be chosen as the observational basis, but in practice the *V* filter of the standard *UBV* system is always used. The difference is then known as the bolometric correction, BC,

$$BC = m_{bol} - V \qquad (3.1.10)$$

$$= M_{bol} - M_v \qquad (3.1.11)$$

and its scale is chosen so that it is zero for main sequence stars with a temperature of about 6500 K, i.e. about spectral class F5 V. The luminosity of a star is directly related to its absolute bolometric magnitude

$$L_* = 3 \times 10^{28} \times 10^{-0.4 M_{bol}} \ (W). \qquad (3.1.12)$$

Similarly, the flux of the stellar radiation just above the Earth's atmosphere, f_*, is related to the apparent bolometric magnitude

$$f_* = 2.5 \times 10^{-8} \times 10^{-0.4 m_{bol}} \ \left(W \ m^{-2} \right) \qquad (3.1.13)$$

The bolometric corrections are plotted in figure 3.1.4.

Measurements in one photometric system can sometimes be converted to another system. This must be based upon extensive observational calibrations or upon detailed calculations using specimen spectral distributions. In either case, the procedure is very much second best to obtaining the data directly in the required photometric system, and requires great care in its application. Its commonest occurrence is for the conversion of data obtained with a slightly non-standard *UBV* system to the standard system.

Stellar parameters

The purpose of making measurements of stars in the various photometric systems is to determine some aspect of the star's spectral behaviour by a simpler and more rapid method than that of actually obtaining the spectrum. The simplest approach to highlight the desired information is to calculate one or more colour indices. The colour index is just the difference between the star's magnitude through two different filters. Probably the commonest pair of filters so used are the *B* and *V* filters of the standard *UBV* system. The colour index, *C*, is then just

$$C = B - V \qquad (3.1.14)$$

where *B* and *V* are the magnitudes through the *B* and *V* filters respectively. However, it should be noted that a fairly common alternative usage is for *C*

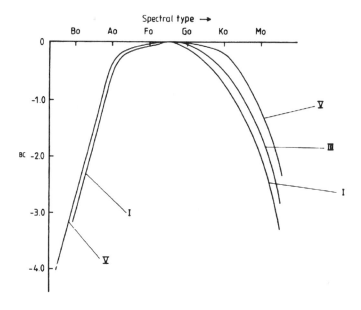

Figure 3.1.4. Bolometric corrections for main sequence stars (type V), giants (type III) and supergiants (type I).

to denote the so called international colour index, which is based upon the photographic and photovisual magnitudes. The interrelationship is

$$m_p - m_{pv} = C = B - V - 0.11. \tag{3.1.15}$$

The $B - V$ colour index is closely related to the spectral type (figure 3.1.5) with an almost linear relationship for main sequence stars. This arises from the dependence of both spectral type and colour index upon temperature. For most stars, the B and V regions are located on the long wavelength side of the maximum spectral intensity. In this part of the spectrum the intensity then varies approximately in a black body fashion for many stars. If we assume that the B and V filters effectively transmit at 440 and 550 nm respectively, then using the Planck equation

$$\mathcal{F}_\lambda = \frac{2\pi h c^2}{\lambda^5 [\exp(hc/\lambda kT) - 1]} \tag{3.1.16}$$

we obtain

$$B - V = -2.5 \log \left(\frac{(5.5 \times 10^{-7})^5 \left[\exp\left(\dfrac{6.62 \times 10^{-34} \times 3 \times 10^8}{5.5 \times 10^{-7} \times 1.38 \times 10^{-23} \times T} \right) - 1 \right]}{(4.4 \times 10^{-7})^5 \left[\exp\left(\dfrac{6.62 \times 10^{-34} \times 3 \times 10^8}{4.4 \times 10^{-7} \times 1.38 \times 10^{-23} \times T} \right) - 1 \right]} \right) \tag{3.1.17}$$

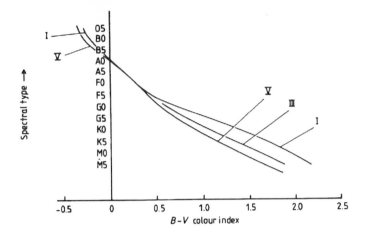

Figure 3.1.5. Relationship between spectral type and $B - V$ colour index.

which simplifies to

$$B - V = -2.5 \log\left(3.05 \frac{[\exp(2.617 \times 10^4 / T) - 1]}{[(\exp(3.27 \times 10^4 / T) - 1]}\right) \qquad (3.1.18)$$

which for $T < 10\,000$ K is approximately

$$B - V \simeq -2.5 \log\left(3.05 \frac{\exp(2.617 \times 10^4 / T)}{\exp(3.27 \times 10^4 / T)}\right) \qquad (3.1.19)$$

$$= -1.21 + (7090/T). \qquad (3.1.20)$$

Now the magnitude scale is an arbitrary one as we have seen, and it is defined in terms of standard stars, with the relationship between the B and the V magnitude scale such that $B - V$ is zero for stars of spectral type A0 (figure 3.1.5). Such stars have a surface temperature of about 10 000 K, and so a correction term of +0.5 must be added to equation (3.1.20) to bring it into line with the observed relationship. Thus we get

$$B - V = -0.71 + 7090T^{-1} \qquad (3.1.21)$$

$$T = 7090/[(B - V) + 0.71] \text{ (K)}. \qquad (3.1.22)$$

Equations (3.1.21) and (3.1.22) are still only poor approximations because the filters are very broad, and so the monochromatic approximation used in obtaining the equations is a very crude one. Furthermore, the effective wavelengths (i.e. the 'average' wavelength of the filter, taking account of the energy distribution within the spectrum) of the filters change with the stellar temperature (figure 3.1.6). However, an equation of a similar nature may be

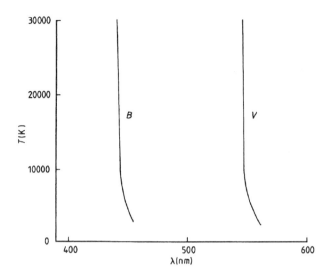

Figure 3.1.6. Change in the effective wavelengths of the standard B and V filters with changing black body temperature.

fitted empirically to the observed variation (figure 3.1.7), with great success for the temperature range 4000 to 10 000 K, and this is

$$B - V = -0.865 + 8540/T \tag{3.1.23}$$

$$T = 8540/[(B - V) + 0.865] \text{ (K)}. \tag{3.1.24}$$

At higher temperatures, the relationship breaks down, as would be expected from the approximations made to obtain equation (3.1.19). The complete observed relationship is shown in figure 3.1.8. Similar relationships may be found for other photometric systems which have filters near the wavelengths of the standard B and V filters. For example, the relationship between spectral type and the $b - y$ colour index of the $uvby$ system is shown in figure 3.1.9.

For filters which differ from the B and V filters, the relationship of colour index with spectral type or temperature may be much more complex. In many cases the colour index is a measure of some other feature of the spectrum, and the temperature relation is of little use or interest. The $U - B$ index for the standard system is an example of such a case, since the U and B responses bracket the Balmer discontinuity (figure 3.1.10). The extent of the Balmer discontinuity is measured by a parameter, D, defined by

$$D = \log\left(I_{364+}/I_{364-}\right) \tag{3.1.25}$$

where I_{364+} is the spectral intensity at wavelengths just longer than the Balmer discontinuity (which is at or near 364 nm) and I_{364-} is the spectral intensity

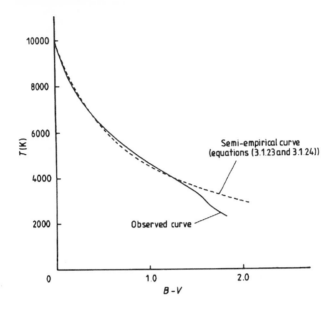

Figure 3.1.7. Observed and semi-empirical $B - V/T$ relationships over the lower part of the main sequence.

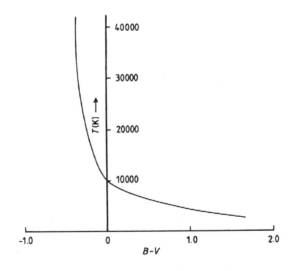

Figure 3.1.8. Observed $B - V$ versus T relationship for the whole of the main sequence.

at wavelengths just shorter than the Balmer discontinuity, and the variation of D with the $U - B$ colour index is shown in figure 3.1.11. The relationship is complicated by the overlap of the B filter with the Balmer discontinuity and

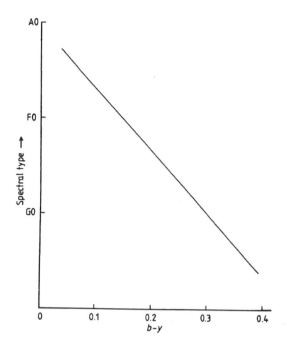

Figure 3.1.9. Relationship between spectral type and $b - y$ colour index.

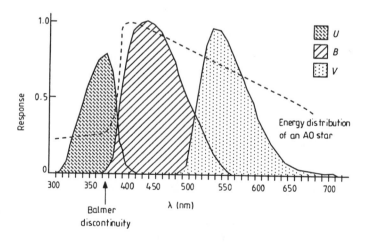

Figure 3.1.10. Position of the standard *UBV* filters with respect to the Balmer discontinuity.

the variation of the effective position of that discontinuity with spectral and luminosity class. The discontinuity reaches a maximum at about A0 for main

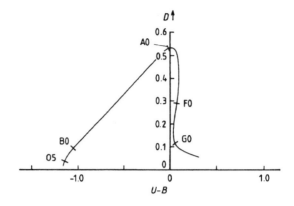

Figure 3.1.11. Variation of D with $U - B$ for main sequence stars.

sequence stars and at about F0 for supergiants. It almost vanishes for very hot stars, and for stars cooler than about G0, corresponding to the cases of too high and too low temperatures for the existence of significant populations in the hydrogen $n = 2$ level. The colour index is also affected by the changing absorption coefficient of the negative hydrogen ion (figure 3.1.12), and by line blanketing in the later spectral types. A similar relationship may be obtained for any pair of filters which bracket the Balmer discontinuity, and another commonly used index is the c_1 index of the $uvby$ system

$$c_1 = u + b - 2v. \tag{3.1.26}$$

The filters are narrower and do not overlap the discontinuity, leading to a simpler relationship (figure 3.1.13), but the effects of line blanketing must still be taken into account.

Thus the $B - V$ colour index is primarily a measure of stellar temperature, while the $U - B$ index is a more complex function of both luminosity and temperature. A plot of one against the other provides a useful classification system analogous to the Hertzsprung–Russell diagram. It is commonly called the colour–colour diagram and is shown in figure 3.1.14. The deviations of the curves from that of a black body arise from the effects which we have just mentioned: Balmer discontinuity, line blanketing, negative hydrogen ion absorption coefficient variation, and also because the radiation originates over a region of the stellar atmosphere rather than in a single layer characterised by a single temperature. The latter effect is due to limb darkening, and because the thermalisation length (the distance between successive absorptions and re-emissions) normally corresponds to a scattering optical depth many times unity. The final spectral distribution therefore contains contributions from regions ranging from perhaps many thousands of optical depths below the visible surface to well above the photosphere and so cannot be assigned a single temperature.

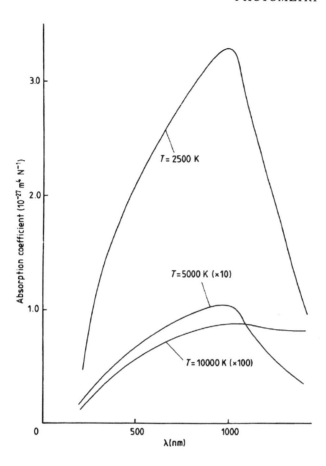

Figure 3.1.12. Change in the absorption coefficient of the negative hydrogen ion with temperature.

Figure 3.1.14 is based upon measurements of nearby stars. More distant stars are affected by interstellar absorption, and since this is strongly inversely dependent upon wavelength (figure 3.1.15), the $U - B$ and the $B - V$ colour indices are altered. The star's spectrum is progressively weakened at shorter wavelengths and so the process is often called interstellar reddening. The degree to which the spectrum is reddened is measured by the colour excesses

$$E_{U-B} = (U - B) - (U - B)_0 \qquad (3.1.27)$$

$$E_{B-V} = (B - V) - (B - V)_0 \qquad (3.1.28)$$

where the subscript 0 denotes the unreddened quantities. These are called the intrinsic colour indices and may be obtained from the spectral type and figures 3.1.5 and 3.1.16. The interstellar absorption (figure 3.1.15) varies over most of

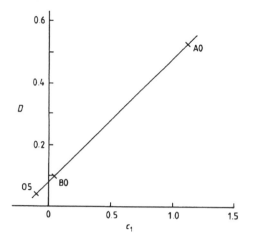

Figure 3.1.13. Variation of D with the c_1 index of the $uvby$ system for main sequence stars.

the optical spectrum in a manner which may be described by the semiempirical relationship

$$A_\lambda = 6.5 \times 10^{-10}\lambda^{-1} - 2.0 \times 10^{-4} \; \left(\text{mag pc}^{-1}\right). \tag{3.1.29}$$

Hence, approximating U, B and V filters by monochromatic responses at their central wavelengths, we have, for a distance of D parsecs,

$$\frac{E_{U-B}}{E_{B-V}} = \frac{(U - U_0) - (B - B_0)}{(B - B_0) - (V - V_0)} \tag{3.1.30}$$

$$= \big[\big[\{[(6.5 \times 10^{-10})/(3.65 \times 10^{-7})] - 2.0 \times 10^{-4}\}D$$
$$- \{[(6.5 \times 10^{-10})/(4.4 \times 10^{-7})] - 2.0 \times 10^{-4}\}D\big]$$
$$\times \big[\big[\{[(6.5 \times 10^{-10})/(4.4 \times 10^{-7})] - 2.0 \times 10^{-4}\}D$$
$$- \{[(6.5 \times 10^{-10})/(5.5 \times 10^{-7})] - 2.0 \times 10^{-4}\}D\big]^{-1} \tag{3.1.31}$$

$$= \left(\frac{1}{365} - \frac{1}{440}\right)\left(\frac{1}{440} - \frac{1}{550}\right)^{-1} \tag{3.1.32}$$

$$= 1.027. \tag{3.1.33}$$

Thus the ratio of the colour excesses is independent of the reddening. This is an important result and the ratio is called the reddening ratio. Its actual value is somewhat different from that given in equation (3.1.33), because of our monochromatic approximation, and furthermore it is slightly temperature dependent and not quite independent of the reddening. Its best empirical values are

$$\frac{E_{U-B}}{E_{B-V}} = (0.70 \pm 0.10) + (0.045 \pm 0.015)E_{B-V} \quad \text{at } 30\,000 \text{ K} \tag{3.1.34}$$

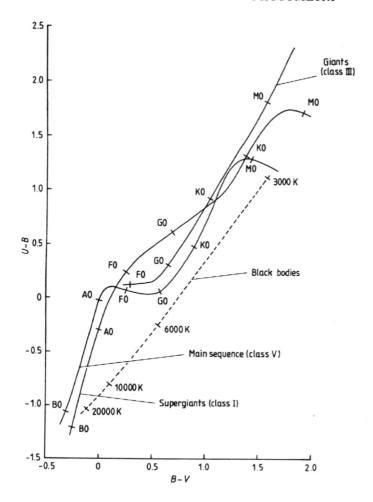

Figure 3.1.14. $(U - B)$ versus $(B - V)$ colour–colour diagram.

$$\frac{E_{U-B}}{E_{B-V}} = (0.72 \pm 0.06) + (0.05 \pm 0.01)E_{B-V} \qquad \text{at } 10\,000 \text{ K} \qquad (3.1.35)$$

$$\frac{E_{U-B}}{E_{B-V}} = (0.82 \pm 0.12) + (0.065 \pm 0.015)E_{B-V} \quad \text{at } 5\,000 \text{ K.} \qquad (3.1.36)$$

The dependence upon temperature and reddening is weak, so that for many purposes we may use a weighted average reddening ratio

$$\left(\overline{\frac{E_{U-B}}{E_{B-V}}}\right) = 0.72 \pm 0.03. \qquad (3.1.37)$$

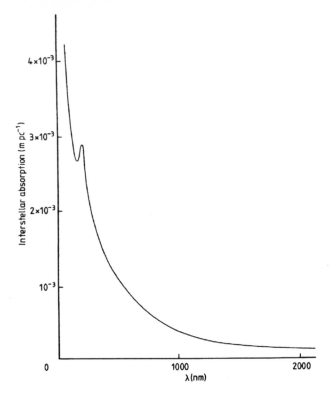

Figure 3.1.15. Average interstellar absorption.

The colour factor, Q, defined by

$$Q = (U - B) - \left(\frac{\overline{E_{U-B}}}{E_{B-V}}\right)(B - V) \qquad (3.1.38)$$

is then independent of reddening, as we may see from its expansion

$$Q = (U - B)_0 + E_{U-B} - \left(\frac{\overline{E_{U-B}}}{E_{B-V}}\right)\left[(B - V)_0 + E_{B-V}\right] \qquad (3.1.39)$$

$$= (U - B)_0 - \left(\frac{\overline{E_{U-B}}}{E_{B-V}}\right)(B - V)_0 \qquad (3.1.40)$$

$$= (U - B)_0 - 0.72(B - V)_0 \qquad (3.1.41)$$

and provides a precise measure of the spectral class of the B type stars (figure 3.1.17). Since the intrinsic $B - V$ colour index is closely related to the spectral type (figure 3.1.5), we therefore also have the empirical relation for the early spectral type

$$(B - V)_0 = 0.332Q \qquad (3.1.42)$$

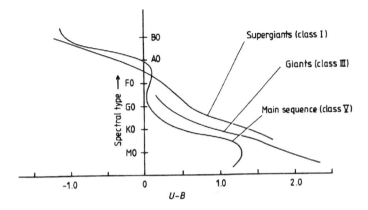

Figure 3.1.16. Relationship between spectral type and $U - B$ colour index.

shown in figure 3.1.18. Hence we have

$$E_{B-V} = (B - V) - 0.332Q = 1.4E_{U-B}. \qquad (3.1.43)$$

Thus simple UBV photometry for hot stars results in determinations of temperature, Balmer discontinuity, spectral type and reddening. The latter may also be translated into distance when the interstellar absorption in the star's direction is known. Thus we have potentially a very high return of information

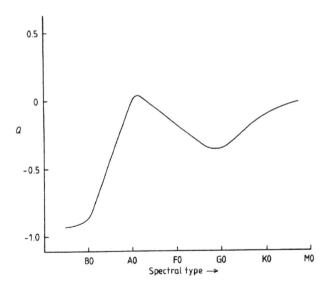

Figure 3.1.17. Variation of colour factor with spectral type.

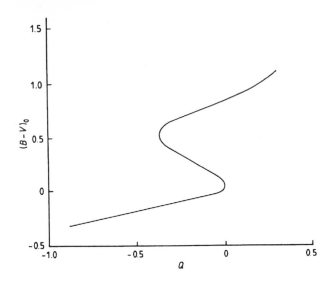

Figure 3.1.18. Relationship between colour factor and the $B - V$ intrinsic colour index.

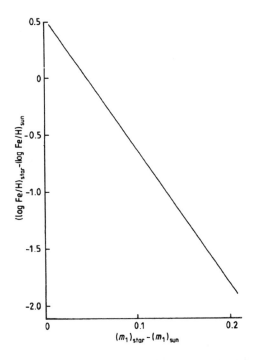

Figure 3.1.19. Relationship between metallicity index and iron abundance for solar type stars.

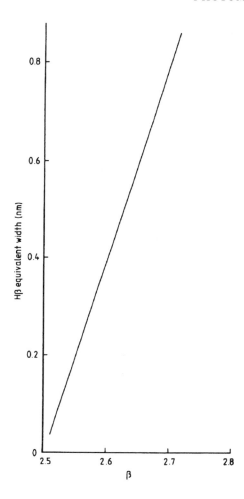

Figure 3.1.20. Relationship between β and the equivalent width of Hβ.

for a small amount of observational effort, and the reader may see why the relatively crude method of *UBV* photometry still remains popular.

Similar or analogous relationships may be set up for other filter systems. Only one further example is briefly discussed here. The reader is referred to the bibliography for further information on these and other systems. The *uvby* system plus filters centred on *Hβ* (often called the *uvbyβ* system) has several indices. We have already seen that $b - y$ is a good temperature indicator (figure 3.1.9), and that c_1 is a measure of the Balmer discontinuity for hot stars (figure 3.1.13). A third index is labelled m_1 and is given by

$$m_1 = v + y - 2b. \tag{3.1.44}$$

It is sometimes called the metallicity index, for it provides a measure of the number of absorption lines in the spectrum and hence, for a given temperature the abundance of the metals (figure 3.1.19). The fourth commonly used index of the system uses the wide and narrow $H\beta$ filters, and is given by

$$\beta = m_n - m_w \qquad (3.1.45)$$

where m_n and m_w are the magnitudes through the narrow and wide $H\beta$ filters respectively. β is directly proportional to the equivalent width of $H\beta$ (figure 3.1.20), and therefore acts as a guide to luminosity and temperature among the hotter stars (figure 3.1.21). Colour excesses may be defined as before

$$E_{b-y} = (b - y) - (b - y)_0 \qquad (3.1.46)$$

and so on. Correction for interstellar absorption may also be accomplished in a similar manner to the *UBV* case and results in

$$(c_1)_0 = c_1 - 0.20(b - y) \qquad (3.1.47)$$
$$(m_1)_0 = m_1 + 0.18(b - y) \qquad (3.1.48)$$
$$(u - b)_0 = (c_1)_0 + 2(m_1)_0. \qquad (3.1.49)$$

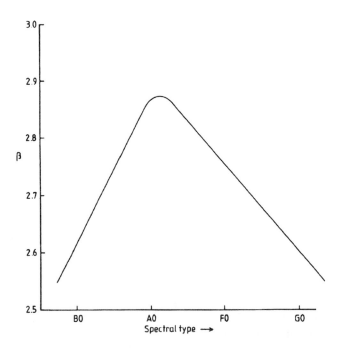

Figure 3.1.21. Relationship of β with spectral type.

By combining the photometry with model atmosphere work, much additional information such as effective temperatures and surface gravities may be deduced.

Exercises

3.1.1 Calculate the absolute magnitude of Jupiter from its apparent magnitude at opposition, -2.6, and the absolute magnitude of M 31, from its apparent magnitude, $+3.5$ (assuming it to be a point source). Their distances are respectively 4.2 AU and 670 kpc.

3.1.2 Calculate the distance of a Cepheid whose apparent magnitude is $+13$ on average, and whose absolute magnitude is -4 on average, assuming that it is affected by interstellar absorption at a rate of 1.5×10^{-3} mag pc^{-1}.

3.1.3 Standard *UBV* measures of a main sequence star give the following results: $U = 3.19$; $B = 4.03$; $V = 4.19$. Hence calculate or find: $(U - B)$, $(B - V)$; Q; $(B - V)_0$, E_{B-V}; E_{U-B}, $(U - B)_0$; spectral type; temperature; distance (assuming that the interstellar absorption shown in figure 3.1.15 may be applied); U_0, B_0, V_0 and M_V.

3.2   PHOTOMETERS

Instruments

We may distinguish three basic methods of measuring stellar brightness, based upon the eye, the photographic plate, or upon a variety of photoelectric devices. All three are still in use today. Visual photometry is the almost exclusive prerogative of the amateur astronomer. Although visual photometry suffers from large errors and lacks consistency, it is one of the few areas of astronomy wherein the amateur can still make a genuine contribution. Monitoring programmes and long term programmes, which could not be justified on a large telescope because of the pressure of other higher priority programmes, can easily be undertaken visually using small telescopes. The light curves of many stars are known only through this long term work, and one need only remember that the majority of novae, etc are discovered by amateurs to realise the value of the monitoring that they undertake. Photographic techniques have been superseded for precise work by the photoelectric methods, but they are still valuable for survey work— a Schmidt camera combined with an automatic measuring machine provides a system which can determine magnitudes at a rate of many thousands of stars per hour. Precision photometry, however, is now exclusively undertaken by means of photoelectric devices, with the photomultiplier predominating in the optical region, and solid state devices in the infrared. We now look at some of the principles and designs for instruments for each of these three techniques in more detail.

Visual photometers

The eye is comparatively sensitive to small differences in intensity between two sources, but performs poorly when judging the absolute intensity of a single source. Visual photometers are therefore comparison devices. The simplest, the extinction photometer, uses the threshold detection limit for the eye itself as the comparison. The image of the source is progressively reduced in brightness by some method such as a variable diaphragm over the telescope's objective, two polarising filters with a variable mutual angle, a variable density filter, etc. The setting of the device which is reducing the star's intensity at the instant when the star disappears from view may then be calibrated in terms of the original magnitude of the star. There are many drawbacks to this type of photometer, amongst which one might mention the variation of the eye's threshold with the physiological condition of the observer, and also with observing conditions and the imprecise nature of the extinction point itself.

Many designs of visual photometer have a comparison source, which may be a second star or an artificial source, which is visible in the field of view simultaneously with the star or source of interest. Either one or the other of the two sources is then adjusted in brightness until both are judged to be equally bright. The setting of the adjusting device can be calibrated in terms of the difference in magnitude between the two sources. Numerous designs exist for photometers of this type, but they may generally be considered to be variants of one of the basic designs shown in figure 3.2.1. Use of an artificial source is complicated by the difficulty of getting its appearance and colour to match that of the star. While the use of a comparison star improves matters in these respects, a suitable star may be lacking, and in any case the adjustments required to bring it into the field of view can be very time consuming. Furthermore, the sky background will be different for the comparison star because of the use of the filter, and this will introduce additional errors. Probably the optimum visual photometer is based upon the design for use with an artificial star, but uses a chopper in place of the beam splitter (figure 3.2.2). The images of the star and the source are then superimposed and viewed alternately at a frequency of a few hertz. Differences in the two brightnesses will show up as a flickering of the image. A very precise adjustment of the two images to equality may then be made by reducing the flickering to zero. The effects of differences in appearance, colour and background are minimised by the rapid interchange of images, which tends to result in a synthesised image whose properties are an amalgam of the properties of each of the separate images. Accuracies of a few hundredths of a magnitude can be achieved with experience with such a system.

Photographic photometry

The use of photography for stellar photometry might at first sight appear straightforward. Inspection of the characteristic curves (figure 2.2.13) of typical

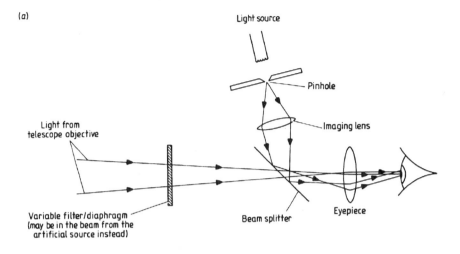

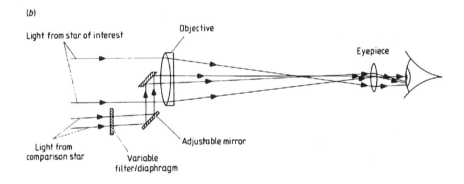

Figure 3.2.1. Visual photometers. (*a*) Schematic plan of a visual comparison photometer using an artificial source. (*b*) Schematic plan of a visual comparison photometer using a second star for comparison.

emulsions used in astronomy shows that they have a range of response by a factor of between 100 and 1000 in intensity, corresponding to a useful magnitude range of 5^m to 7.5^m. In practice the situation is very much more complicated, since the images of stars of different magnitudes will be of different sizes. This arises through the image of the star not being an ideal point. It is spread out by diffraction, scintillation, scattering within the emulsion, and by reflection off the emulsion's support (halation). The actual image structure will therefore have an intensity distribution which probably approaches a Gaussian pattern. However, the recorded image will have a size governed by the point at which the illumination increases the photographic density above the gross fog level. Images

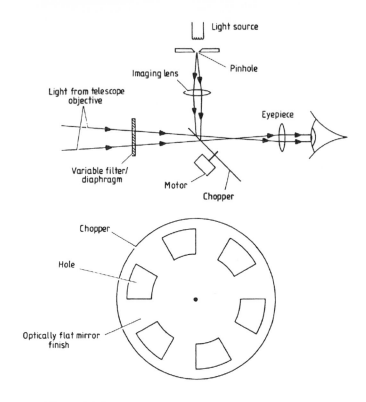

Figure 3.2.2. Schematic plan of a visual flicker photometer.

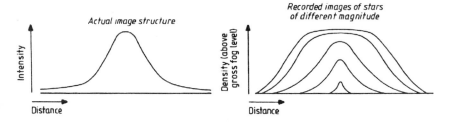

Figure 3.2.3. Schematic variation in image size and density for stars of differing magnitudes recorded on photographic emulsion.

of stars of differing magnitudes will then be recorded as shown schematically in figure 3.2.3.

One way of reducing the problem which is particularly useful for extended objects such as galaxies is to broaden the image further artificially. This may be done by moving the plate holder in some sort of raster pattern during the exposure, so that all the objects—stars and galaxies—produce square images. Such a system is sometimes called a jiggle photometer. If the resulting enlarged image is considerably larger than the normal image, then the edge effects will be negligible, and the photographic density may be converted directly back into intensity via the characteristic curve. While this method has the merit of relative simplicity at the measurement and reduction stages, the exposure required to reach any given magnitude is increased by at least a factor of ten. Normally therefore a simple photograph is used and a calibration curve produced which takes account of both the size and density variations. Ideally standard stars of known magnitude are measured in order to produce this calibration curve. If there are insufficient or none of these on the plate, then a second plate may be taken of a different area of the sky which does contain enough known stars. For this to give useful results similar precautions must be taken to those used in obtaining a calibration curve for spectroscopy (section 2.2), i.e. the two plates must come from the same batch; their treatment must have been identical; they must be processed together; the exposures must be identical and the sky conditions should be as similar as possible. Thus the plates should be obtained in quick succession, and the altitudes of the two regions of sky should be comparable. Other possibilities for determining the calibration curve include multiple exposure, with the plate moved slightly between exposures, and the exposure lengths differing by a factor 2.512 each time. Thus a series of images is obtained, with the intensities differing by one magnitude. This method suffers from differential reciprocity failure between exposures of different lengths and from the hypersensitisation effect of the pre and/or post exposure to the sky background (section 2.2).

An alternative method is to use a coarse objective grating so that each stellar image has two or more fainter satellite images with a calculable relative intensity with respect to the main image (sections 2.5, 4.1 and 5.1). Yet another possibility is to place a small low dispersion prism just before or after the objective. This is known as a Pickering–Racine prism, and it diverts a small fraction of the light into a fainter secondary image. Problems occur, however, with this method because the images have different sized Airy discs. The measurement which is made of the stellar images may simply be that of its diameter, when a semiempirical relationship of the form

$$D = A + B \log I \qquad (3.2.1)$$

(where D is the image diameter, I the intensity of the star, and A and B are constants to be determined from the data) can often be fitted to the calibration curve. A more precise measurement than the actual diameter would be the distance between points in the image which are some specified density greater than that of the gross fog, or between the half density points. Alternatively a combination of diameter and density may be measured. Some automatic measuring machines scan the image (section 5.1), so that its total density may be found by integration. This is particularly useful when searching for variable

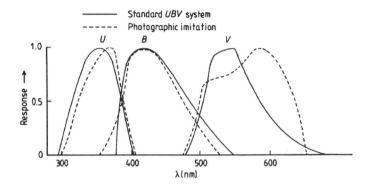

Figure 3.2.4. Comparison of the standard UBV system and its photographic simulation.

stars, since only a consistent relationship from one plate to another need be established; there is no need to find the actual magnitudes.

Another fairly common system is based upon a variable diaphragm. This is set over the image and adjusted until it just contains that image. The transmitted intensity may then be measured by some variety of microdensitometer (section 2.2), and a calibration curve plotted from known standard stars as before. A variant of this last method adjusts the diaphragm until a preset amount of light is passed through, and uses the diaphragm setting as the measure of the star's brightness. With care and luck, measurements to an accuracy of 0.01 magnitudes can be obtained from photographs, normally, however, the errors are likely to be a few times larger than this.

The magnitudes obtained from a photograph are of course not the same as those of the standard photometric system (section 3.1). Usually the magnitudes are photographic, or photovisual. However, with careful selection of the emulsion and the use of filters, a response which is close to that of some filter systems may be obtained (figure 3.2.4).

Photoelectric photometers

The most precise photometry relies upon photoelectric devices of various types. Although generally slower than the other methods, photoelectric photometry is also usually preferred even when the highest precision is not required. Designs for the instruments vary enormously, and some are discussed in more detail later in this section. The basis of the photometer, however, is similar in almost all cases (figure 3.2.5). An entrance aperture is required which is a pinhole about one and a half to two times the diameter of the stellar image. This eliminates much of the background light, while being large enough to accommodate scintillation movements of the star. Care is needed, however, not to make the pinhole too small, or light in the diffraction rings may be excluded to a significant degree. One or two finding and guiding eyepieces

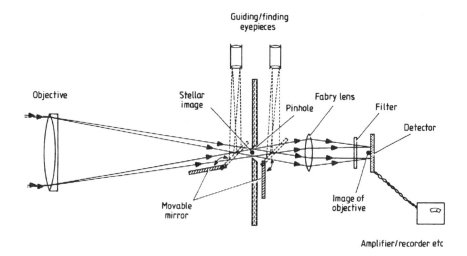

Figure 3.2.5. Schematic layout of a photoelectric photometer.

which view the star via mirrors, and which can be moved out of the light path when measurements are to be taken, are needed. The Fabry lens images the objective onto the detector. This has the advantage of producing a uniformly illuminated image which is stationary even when the stellar image moves due to scintillation or poor tracking etc. Detector non-uniformity is therefore no longer of quite such importance. Finally a filter selects the required waveband for the photometric system which is being used.

The commonest detector in use is a photomultiplier (section 1.1) because of its association with the standard *UBV* system, and because of its very large intrinsic amplification (though CCDs are also becoming popular, as discussed below). Compared with the visual and photographic systems, it has the enormous advantage of an almost linear response with intensity. It also has a low noise level especially if it is cooled (but see the discussion in section 1.1). The problems and disadvantages of photomultipliers were discussed in section 1.1 in some detail, here we just note the main ones of the danger, under observing conditions, of the high voltage supply, and the variation in sensitivity over the detector's area. The first of these can be mitigated by the use of a supply with a low current limit, while the Fabry lens is incorporated into the design (figure 3.2.5) specifically to reduce the effect of the second. For most of the photometric systems a photomultiplier with a caesium antimonide photoelectron emitter is used, for example the standard *UBV* system is based upon the 1P21 photomultiplier which has this type of photocathode. For Earth-based observations, the ultraviolet response of a normal photomultiplier is adequate, but for satellite-borne photometers special windows of lithium fluoride or sapphire can extend the response down to 150 nm. For even shorter wavelengths, down to about 20 nm, a windowless photomultiplier with a metal or metal oxide photocathode can be used. At long wavelengths the caesium antimonide response falls away, and is almost zero by 650 nm. The sensitivity can be extended to 1100 nm using tri-alkali and silver–oxygen–caesium cathodes, but for longer wavelengths some of the other detector types (section 1.1) such as bolometers must be used. The dark current of photomultipliers is much reduced by cooling as already mentioned in this section and in section 1.1. Most of the photoemissive substances reach a lower limit to the reduction in the dark current at about 220 K, so that dry ice at 195 K is an adequate coolant. However, with silver–oxygen–caesium, the dark current continues to decrease, and liquid nitrogen at 77 K may profitably be tried for cooling. In either case, the photomultiplier must be mounted inside an enclosure in an insulated box or a Dewar. These are available commercially, or may be constructed in a well-equipped workshop. The major problem is to avoid condensation on the photomultiplier. The enclosure must therefore be filled with a completely dry gas or be evacuated. This then necessitates the use of a window for the enclosure, and this must be heated to avoid condensation upon it, in turn. When dry ice is used, it is generally crushed and mixed with acetone in order to provide a good thermal contact. Thus with both dry ice and liquid nitrogen, the container must be designed so that liquid cannot spill out as the telescope moves, but which still allows the evaporating gases to escape. It must also be possible to refill the container during use, since it is unlikely to last throughout an entire observing session. The pre-amplifier may often with advantage be included inside the cooled enclosure, since its noise contribution will thereby be reduced, and for infrared work the filters and pinhole should be cooled as well. There are many detailed designs for such cooled chambers and figure 3.2.6 shows their general principles schematically.

The linear response and sensitivity of CCDs (section 1.1) has led to their increasing use for photometry. However, care must be taken if accurate results are to be obtained. The dark field subtraction and flat fielding must be applied carefully and 'bad' pixels avoided. Usually the image of the star will occupy more than one pixel. A suitable 'aperture' must therefore be used by the software processing the data to ensure that all the light from the star is utilised. In crowded fields, star images may overlap. It may then be necessary to fit a suitable intensity profile to each of the stars in order to extract accurate magnitudes for them.

Other types of detectors may be used to extend photometry to other wavelengths, or for reasons of convenience be used within the band obtainable with photomultipliers. We have already seen the types of detectors used at x- and gamma-ray wavelengths (section 1.3), and their normal mode of usage is photometric, so that the earlier discussion need not be repeated here. Similarly photometry at radio wavelengths is discussed in section 1.2.

In the visual region there are many solid state detectors available (section 1.1). Very convenient small photometers can be made using them, and some are produced commercially for the amateur market. The detector area is generally

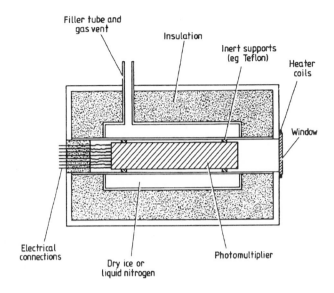

Figure 3.2.6. Schematic arrangement of a cooling system for a photomultiplier.

quite small, and sensitivity variations across it are usually less than one per cent or so. Thus with these photometers no Fabry lens is needed, and the stellar image focuses directly onto the detector. The response curves of these devices are mostly quite different from those of photomultipliers, usually having peak sensitivities in the red or near infrared. Hence it is difficult to fit their measurements into any of the photometric systems which were listed earlier. With careful selection of filters, however, it is usually possible to mimic fairly closely any given region of any of the systems, but the relative response from one region to another within the system is likely to be quite different than when using a photomultiplier, and must therefore be corrected before comparisons with other measurements can be made.

A variety of detectors are available for use at longer infrared wavelengths (section 1.1), but their incorporation into a photometer requires some modification of the basic optical system. The detectors are again fairly small in size, and so a Fabry lens can be dispensed with. The desired signal is usually much weaker than the background noise arising from radiation from the telescope structure, the atmosphere, etc. Thus sophisticated techniques must be used to extract the signal from the noise. A commonly used method employs a phase-sensitive detector, and switches between the source and either the nearby sky background or some standard source within the apparatus. Some telescopes are designed with a secondary mirror which can be oscillated to achieve this switching. Alternatively a chopper must be incorporated into the photometer (cf figure 3.2.2). Two of the many possible ways of doing this are shown in

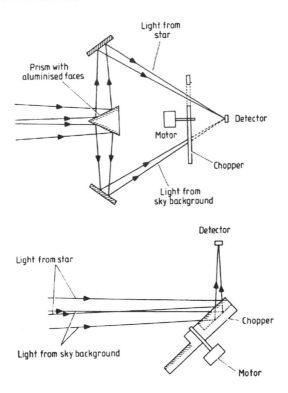

Figure 3.2.7. Two possible methods of sky background/source chopping.

figure 3.2.7. The phase-sensitive detector filters out any constant component in the resulting signal. It then integrates the remainder of the signal, but reverses its sign at the same frequency as, and in phase with, the chopper. The noise, being random, tends to cancel out in this process, while since both phases of the stellar signal now have the same sign, it accumulates. Results can be obtained in this way even when the signal is a million times fainter than the noise. The switching of the phase-sensitive detector is usually also controlled by the chopper so that its motor speed variations do not matter, and the phases of the signal and the switching cannot get out of step. Similar or improved results can be obtained with a variety of modifications to the basic phase-sensitive detector, or the signal can be fed into a dedicated microprocessor with a suitable program to perform the same task. Additional stages of amplification are also usually necessary both before and after the phase-sensitive detector, and the final output may need further integration, especially for faint sources and/or precision work. The other components of the basic photometer (figure 3.2.5) are quite straightforward. The filter is normally mounted onto a slide or a wheel so that all the filters of the photometric system may be quickly interchanged (sometimes under computer control). The pinhole may also be mounted similarly since it will need to be altered as the observing conditions change. Some sort of finding/guiding system is needed, but this is only shown very schematically in figure 3.2.5. The mirror in front of the pinhole acts as a wide field finder, while the second mirror allows the star to be viewed through the pinhole for precision alignment. When the star must be observed for a lengthy period, such as during integration of the signal, provision must also be made for guiding. This is mostly commonly accomplished by an off-set system; an additional movable eyepiece can inspect the field of view around the star of interest, and a convenient nearby object is found and centred in this eyepiece while the main object is aligned on the pinhole. Subsequent guiding can continue using this second object. A mirror with a hole in it to allow the main light beam through while reflecting the peripheral light is a simple system to enable such guiding to be undertaken, and an alternative is a dichroic mirror which reflects the wavelengths not required for the photometry, and transmits the remainder (or vice versa). The latter system is particularly suitable for infrared photometry—a mirror thinly plated with gold is transparent to visual radiation, but reflects the infrared. If neither of these guiding systems is available, then a normal, separate guiding telescope may be used, but this is very much a second choice, since flexure etc is likely to make guiding to the precision required to keep the star aligned with the pinhole almost impossible.

The basic photometer as we have just described it is used extensively. Its mode of use requires a separate observation for each filter, and for the star and its comparison standard stars. A single set of observations of a star is therefore quite a lengthy procedure (see the discussion later in this section). In order to try and speed up the observation of a star, several more specialised designs of photometer have been developed. These fall primarily into two categories— multiband and multistar photometers. In the former, measurements of a single star are made simultaneously through two or more filters of a photometric system. This requires a separate basic photometer for each filter and the light is split and fed to each of these separately. Beam splitters or dichroic filters can be used for this, or alternatively the filters may be dispensed with and a low dispersion spectrograph used to split the light into the required spectral regions. The use of a spectrograph is to be preferred over the use of dichroic filters since the latter will tend to distort the spectral regions and make comparison with 'normal' measurements made in the photometric system difficult. Using a low dispersion spectrograph is also more flexible in that the number of possible channels is limited only by finance and by the necessity of attaching the final instrument to a telescope, so that it must be fairly light and compact. A schematic arrangement for such a multichannel photometer is shown in figure 3.2.8, but other spectrograph designs may be used (section 4.2) and some simplification is likely because even the narrow band photometric systems have very low spectral resolutions by spectroscopic standards. Solid state detectors may be used with advantage in multichannel photometers. One of the arrays (sections 1.1 and 2.3)

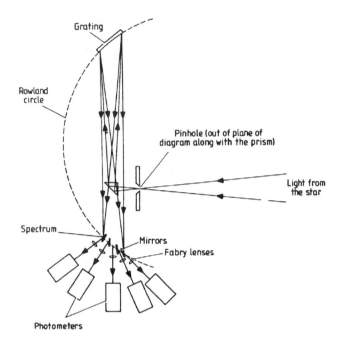

Figure 3.2.8.   Schematic plan of a multichannel photometer based on an Eagle spectroscope.

of many closely spaced small detectors is simply placed at the focal plane of the spectrograph, eliminating the need for separate photometers for each spectral region.

Multistar photometers enable observations of the star of interest to be made simultaneously with those for the standard star or the sky background. The latter is simpler to achieve, and one possible design is shown in figure 3.2.9. Simultaneous observation of two stars requires a method of guiding the light of each star to its individual photometer. One possible method is similar to that shown in the second half of figure 3.2.1 but without the filter or diaphragm, and with the photometers taking the place of the eye. Both photometers are fixed, and the telescope aligned so that the image of one star falls onto one of them, the adjustable mirror is then moved to bring the second image onto the other photometer. This is not a particularly good approach since the light from each star follows a different light path, and so the images are not strictly comparable. A better method employs one fixed and one movable photometer. The star is directed onto the fixed photometer, and then the other photometer moved around the field of view, until it receives the light from the second star (figure 3.2.10).

A problem with the multistar/multichannel photometers that we have just discussed, and with other similar designs, is that the individual photometers are

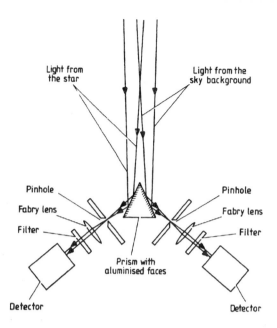

Figure 3.2.9. Schematic arrangement for a simultaneous star/sky background photometer.

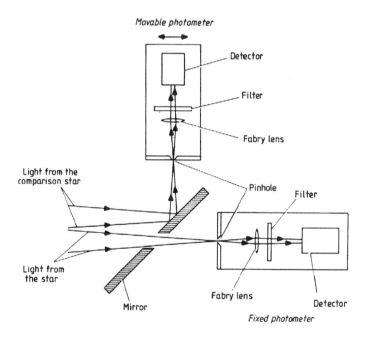

Figure 3.2.10. Schematic arrangement for a two-star photometer.

likely to vary slightly in their behaviour. The detectors in particular may have differing responses (section 1.1). Careful and frequently repeated calibrations are therefore necessary (see the next section). For multistar photometers, an additional safeguard is possible. Once the first set of observations has been completed, it is repeated with the object of interest set onto the second photometer, and the comparison source on the first photometer. This method of observing reduces the time saving aspect of the multistar photometer, but gives very precise results because both the star and its comparison are similarly affected by the observing conditions and the differences between the photometers cancel out. Accuracies of 0.001^m or even better may be achieved with this type of system.

Finally of course the multistar and the multichannel photometer designs can be combined to produce a photometer which simultaneously measures the two stars, or the star and background, in several different photometric regions. Computer control of such systems with real-time data reduction is now becoming quite common, and the telescope time can then be used most efficiently by only observing long enough to reach the required signal-to-noise ratio in the noisiest of the photometric bands being used.

Another variety of photometer which is generally of the two-star design is particularly adapted for making rapid measurements. Known as a high speed photometer, it obtains readings at a rate of between a few per second to one per few seconds. Its main problem arises through the small number of photons which are received in such a short interval from the fainter stars, and the consequent high noise level. It must therefore be as efficient as possible, and often may be used without filters in order to maximise the information obtained.

All the photometers so far discussed are spot photometers; that is, they measure the brightness of a point source. However, many objects are extended, and then there are two different possible measures of their brightness which are of interest. The first is the intensity per unit area and its variation over the object, while the second is the total integrated brightness of the object. The former measurement may simply be obtained using a spot photometer and by scanning over the object. Generally the data are then fed into a computer and a picture is built up which may be presented in many forms: contours, side-view contours, simulated photographs, etc. The integrated magnitude may then be obtained from this information by integrating it. Alternatively a specialised photometer such as the jiggle photographic photometer (see p. 285) may be used to obtain the integrated magnitude directly.

Observing techniques

Probably the single most important 'technique' for successful photometry lies in the selection of the observing site. Only the clearest and most consistent of skies are suitable for precision photometry. Haze, dust, clouds, excessive scintillation, light pollution, etc and the variations in these, all render a site unsuitable for photometry. Furthermore, for infrared work, the amount of water vapour above the site should be as low as possible. For these reasons good photometric observing sites are rather rare. Generally they are at high altitudes and are located where the weather is particularly stable. Oceanic islands, and mountain ranges with a prevailing wind from an ocean, with the site above the inversion layer, are fairly typical of the best choices. Sites which are less than ideal are still used for photometry, but their number of good nights will be reduced. The use of multichannel and/or multistar photometers which enable the data to be obtained rapidly, before the observing conditions change significantly, will allow such non-optimum sizes to be utilised most effectively. Restricting the observations to near the zenith, and certainly always above an altitude of 45°, is also likely to improve the results obtained at a mediocre observing site.

The second most vital part of photometry lies in the selection of the comparison star(s). This must be non-variable, close to the star of interest, of similar apparent magnitude and spectral class, and have its own magnitude known reliably on the photometric system which is in use. Amongst the brighter stars there is usually some difficulty in finding a suitable standard which is close enough. With fainter stars the likelihood that a suitable comparison star exists is higher, but the chances of its details being known are much lower. Thus ideal comparison stars are found only rarely. The best known variables already have lists of good comparison stars, which may generally be found from the literature. But for less well studied stars, and those being investigated for the first time, there is no such useful information to hand. It may even be necessary to undertake extensive prior investigations, such as checking back through photographic records to find non-variable stars in the region, and then measuring these to obtain their magnitudes before studying the star itself.

The observing procedure varies with the observer and with the instrument. A typical approach for a small basic photometer is outlined below. In more complex instruments, such as those discussed earlier, the procedure may well be shortened very considerably since this is the aim of introducing the increased complexity in the hardware. Essentially, however, the method is similar for all instruments even though short-cuts may be available sometimes. The comparison star is first found and centred in the photometer, and the first filter of the photometric system is selected. The telescope is then moved slightly to one side of the star, and a measurement of the sky background intensity obtained. The star is then returned to the photometer and its intensity measured. Finally a second measurement of the sky background is made. This sequence may then be repeated several times. The next filter is then selected, and the whole set of measurements repeated through this filter, and then so on through the rest of the filters. Similar measures are then obtained for the star of interest, and then the original measures repeated on the comparison star. The repetition of measurements allows averages to be taken and gives an indication of the amplitude of any fluctuations arising from changing observing conditions. A typical output for such an observing sequence which has been recorded on a

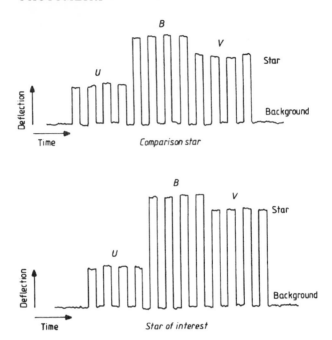

Figure 3.2.11. Typical schematic outputs of a photometric observing run.

chart recorder is shown in figure 3.2.11. If there is more than one comparison star, then each must be measured both before and after the observations of the star of interest are made.

The final technique to be discussed in this section is that of photon counting. In the normal mode of use a photomultiplier-based photometer, the final output signal is fed to a chart recorder or computer, and it is the magnitude of that signal which is the measure of the star's brightness. However, the signal is then the aggregate sum of many individual pulses each of which originates in a single interaction of a photon with the photocathode. Now not all the pulses are identical since some of the electrons can be lost during the cascade down to the anode, or there may be sensitivity variations over the dynodes etc (section 1.1). Thus the final pulses at the anode may vary by as much as a factor of ten in their strengths even though they originated in similar photons, and these variations form an additional contribution to the noise of the signal. With bright stars the number of photons is sufficiently large for these statistical fluctuations to cancel out, but for faint stars, the noise level is significantly worsened by them. However, for the faint stars the pulses are also likely to be infrequent enough for them to be distinguished as individual events. Too many photons cause the pulses to overlap, leading to errors similar to those caused by the dead time of a Geiger counter (section 1.3). Thus a much better method of determining the brightness of a faint star than that of measuring the total signal strength, is to count the number of pulses contributing to that strength. Pulses of different amplitudes are then equally weighted. Further reductions in noise can be obtained by the use of a discriminator to disregard those pulses which are too strong or too weak to be due to photons. Such pulses can arise from cosmic radiation, or from spontaneous emission of electrons from the dynode surfaces, or through the operation of other nearby equipment (section 1.1). Much fainter stars can be detected using pulse counting techniques rather than by looking at the signal strength directly, but the major advantages of the method lie elsewhere. Firstly, integration of the signal is very simple, the pulses simply being counted by the computer for as long as required, furthermore a continuous display of the state of the integration can easily be arranged so that the observations need only be continued until some desired signal-to-noise ratio has been reached. Secondly, in the multichannel and/or multistar photometers the variations between the sensitivities and amplifications of the detectors are no longer of importance. These variations result in changes in the pulse strength, but not in the number of pulses. Thus photon counting can be used directly with such photometers for observing faint stars, and also used to calibrate the responses of the photomultipliers, so that the observations of brighter stars may be made from the total signal strength.

Data reduction and analysis

The reduction of the data is performed in three stages—correction for the effects of the Earth's atmosphere, correction to a standard photometric system, and correction to heliocentric time.

The atmosphere absorbs and reddens the star's light, and its effects are expressed by Bouguer's law

$$m_{\lambda,0} = m_{\lambda,z} - a_\lambda \sec z \qquad (3.2.2)$$

where $m_{\lambda,z}$ is the magnitude at wavelength λ and at zenith distance z. a_λ is a constant which depends on λ. The law is accurate for zenith distances up to about 60°, which is usually sufficient since photometry is rarely carried out on stars whose zenith distances are greater than 45°. Correction for atmospheric extinction may be simply carried out once the value of the extinction coefficient a_λ is known. Unfortunately a_λ varies from one observing site to another, with the time of year, from day to day, and even throughout the night. Thus its value must be found on every observing occasion. This is done by observing a standard star at several different zenith distances, and by plotting its observed brightness against $\sec z$. Now for zenith angles less than 60° or 70°, $\sec z$ is just a measure of the air mass along the line of sight, to a good approximation, so that we may reduce the observations to above the atmosphere by extrapolating them back to an air mass of zero, ignoring the question of the meaning of a value of $\sec z$ of zero! (see figure 3.2.12). For the same standard star, this

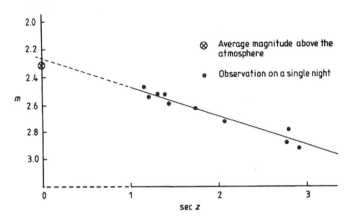

Figure 3.2.12. Schematic variation in magnitude of a standard star with zenith distance.

brightness should always be the same on all nights, so that an additional point is available to add to the observations on any given night, which is the average of all the previous determinations of the above-atmosphere magnitude of the star. The extinction coefficient is simply obtained from the slope of the line in figure 3.2.12. The coefficient is strongly wavelength dependent (figure 3.2.13), and so it must be determined separately for every filter which is being used. Once the extinction coefficient has been found, the above-atmosphere magnitude of the star, m_λ, is given by

$$m_\lambda = m_{\lambda.z} - a_\lambda(1 + \sec z). \qquad (3.2.3)$$

Thus the observations of the star of interest and its standards must all be multiplied by a factor k_λ,

$$k_\lambda = 10^{0.4a_\lambda(1+\sec z)} \qquad (3.2.4)$$

to correct them to their unabsorbed values. When the unknown star and its comparisons are very close together in the sky, the differential extinction will be negligible and this correction need not be applied. But the separation must be very small; for example at a zenith distance of $45°$, the star and its comparisons must be within ten minutes of arc of each other if the differential extinction is to be less than a thousandth of a magnitude. If E_λ and E'_λ are the original average signals for the star and its comparison, respectively, through the filter centred on λ then the corrected magnitude is given by

$$m_\lambda = m'_\lambda - 2.5 \log\left(\frac{E_\lambda 10^{0.4a_\lambda(1+\sec z)}}{E'_\lambda 10^{0.4a_\lambda(1+\sec z')}}\right) \qquad (3.2.5)$$

$$= m'_\lambda + a_\lambda(\sec z' - \sec z) - 2.5 \log\left(E_\lambda/E'_\lambda\right) \qquad (3.2.6)$$

where m_λ and m'_λ are the magnitudes of the unknown star and its comparison respectively, and z and z' are similarly their zenith distances. The zenith angle

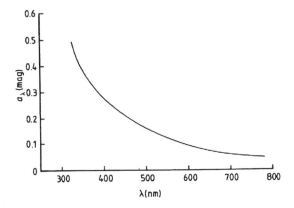

Figure 3.2.13. A typical dependence of the extinction coefficient with wavelength for a good observing site.

is given by

$$\cos z = \sin \varphi \sin \delta + \cos \varphi \cos \delta \cos(\text{LST} - \alpha) \qquad (3.2.7)$$

where φ is the latitude of the observatory, α and δ are the right ascension and declination of the star and LST is the local sidereal time of the observation.

If the photometer is working in one of the standard photometric systems, then the magnitudes obtained through equation (3.2.6) may be used directly, and colour indices, colour excesses, etc may be obtained as discussed in section 3.1. Often, however, the filters or the detector may not be of the standard type. Then the magnitudes must be corrected to a standard system. This can only be attempted if the difference is small, since absorption lines, ionisation edges, etc will make any large correction exceedingly complex. The required correction is best determined empirically by observing a wide range of the standard stars of the photometric system. Suitable correction curves may then be plotted from the known magnitudes of these stars. However, if the filters are standard, but the detector has a non-standard response, for example if a photomultiplier with a tri-alkali photocathode should be used for the standard *UBV* system in place of the usual caesium antimonide one, then the correction may be obtainable from the published data on the detector. In the example just cited, the improved sensitivity of the tri-alkali photocathode may easily be compensated within the *U* and *B* regions. However, the long wavelength cut-off of the *V* region is not due to the filter, but to the decreasing sensitivity of the caesium antimonide photocathode. Since the tri-alkali photocathode responds out to a wavelength of 850 nm, it will be almost impossible to correct a reading through a standard *V* filter to give a normal *V* magnitude. If a tri-alkali-based photomultiplier has to be used for standard *UBV* photometry, then an additional filter must be added to eliminate this long wave sensitivity.

Finally, and especially for variable stars, the time of the observation should be expressed in heliocentric Julian days. The geocentric Julian date is tabulated in Appendix VII, and if the time of the observation is expressed on this scale, then the heliocentric time is obtained by correcting for the travel time of the light to the Sun,

$$T_\odot = T + 5.787 \times 10^{-3} \left[\sin \delta_* \sin \delta_\odot - \cos \delta_* \cos \delta_\odot \cos(\alpha_\odot - \alpha_*) \right] \quad (3.2.8)$$

where T is the actual time of observation in geocentric Julian days, $T_\odot$ is the time referred to the Sun, $\alpha_\odot$ and $\delta_\odot$ are the right ascension and declination of the Sun at the time of the observation, and α_* and δ_* are the right ascension and declination of the star. For very precise work the varying distance of the Earth from the Sun may need to be taken into account as well.

Further items in the reduction and analysis of the data such as the corrections for interstellar reddening, calculation of colour index, temperature, etc were covered in section 3.1.

Exercise

3.2.1 Show that if the differential extinction correction is to be less than Δm magnitudes, then the zenith distances of the star and its comparison must differ by less than Δz, where

$$\Delta z = \frac{\Delta m}{a_\lambda} \cot z \operatorname{cosec} z \quad \text{(radians)}.$$

Chapter 4

Spectroscopy

4.1 SPECTROSCOPY

Optics

Practical spectroscopes are usually based upon one or other of two quite separate optical principles—interference or differential refraction. The former produces instruments based upon diffraction gratings or interferometers, while the latter results in prism-based spectroscopes. There are also a few hybrid designs, but these are quite uncommon. The details of the spectroscopes themselves are considered in the next section, here we discuss the basic optical principles which underlie their designs.

Prisms

When monochromatic light passes through an interface between two transparent isotropic media at a fixed temperature, then we can apply the well known Snell's law relating the angle of incidence, i, to the angle of refraction, r, at that interface

$$\mu_1 \sin i = \mu_2 \sin r \tag{4.1.1}$$

where μ_1 and μ_2 are constants which are characteristic of the two media. When $\mu_1 = 1$, which strictly only occurs for a vacuum, but which holds to a good approximation for most gases, including air, we have

$$\sin i \,/\, \sin r = \mu_2 \tag{4.1.2}$$

and μ_2 is known as the refractive index of the second medium. Now we have already seen that the refractive index varies with wavelength for many media (section 1.1, equation (1.1.32) etc). The manner of this variation may, over a restricted wavelength interval, be approximated by the Hartmann dispersion formula, in which A, B and C are known as the Hartmann constants:

$$\mu_\lambda = A + B/(\lambda - C). \tag{4.1.3}$$

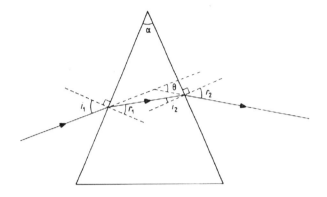

Figure 4.1.1. Optical path in a prism.

If the refractive index is known at three different wavelengths, then we can obtain three simultaneous equations for the constants, from equation (4.1.3), giving

$$C = \frac{\{[(\mu_1 - \mu_2)/(\mu_2 - \mu_3)]\lambda_1(\lambda_2 - \lambda_3)\} - \lambda_3(\lambda_1 - \lambda_2)}{\{[(\mu_1 - \mu_2)/(\mu_2 - \mu_3)](\lambda_2 - \lambda_3)\} - (\lambda_2 - \lambda_2)} \quad (4.1.4)$$

$$B = (\mu_1 - \mu_2)\left(\frac{1}{\lambda_1 - C} - \frac{1}{\lambda_2 - C}\right)^{-1} \quad (4.1.5)$$

$$A = \mu_1 - \frac{B}{\lambda_1 - C}. \quad (4.1.6)$$

The values for the constants for the optical region for typical optical glasses are:

	A	B	C
Crown glass	1.477	3.2×10^{-8}	-2.1×10^{-7}
Dense flint glass	1.603	2.08×10^{-8}	1.43×10^{-7}

Thus white light refracted at an interface is spread out into a spectrum with the longer wavelengths refracted less than the shorter ones. This phenomenon was encountered in section 1.1 as chromatic aberration, and there we were concerned with eliminating or minimising its effects; for spectroscopy by contrast we are interested in maximising the dispersion.

Consider, then, a prism with light incident upon it as shown in figure 4.1.1. For light of wavelength λ, the deviation, θ, is given by

$$\theta = i_1 + r_2 - \alpha. \quad (4.1.7)$$

So that using equation (4.1.3) and using the relations

$$\mu_\lambda = \frac{\sin i_1}{\sin r_1} = \frac{\sin r_2}{\sin i_2} \quad (4.1.8)$$

and

$$\alpha = r_1 + i_2 \tag{4.1.9}$$

we get

$$\theta = i_1 - \alpha + \sin^{-1}\left\{ \left(A + \frac{B}{\lambda - C} \right) \sin\left[\alpha - \sin^{-1}\left(\frac{\sin i_1}{A + [B/(\lambda - C)]} \right) \right] \right\}. \tag{4.1.10}$$

Now we wish to maximise $\partial\theta/\partial\lambda$, which we could study by differentiating equation (4.1.10), but which is far easier to obtain from

$$\frac{\Delta\theta}{\Delta\lambda} = \frac{\theta_{\lambda_1} - \theta_{\lambda_2}}{\lambda_1 - \lambda_2} \tag{4.1.11}$$

so that

$$\frac{\Delta\theta}{\Delta\lambda} = (\lambda_1 - \lambda_2)^{-1}\left[\!\!\left[\sin^{-1}\left\{ \mu_{\lambda_1} \sin\left[\alpha - \sin^{-1}\left(\frac{\sin i_1}{\mu_{\lambda_1}} \right) \right] \right\}\right.\right.$$
$$\left.\left. - \sin^{-1}\left\{ \mu_{\lambda_2} \sin\left[\alpha - \sin^{-1}\left(\frac{\sin i_1}{\mu_{\lambda_2}} \right) \right] \right\} \right]\!\!\right]. \tag{4.1.12}$$

The effect of altering the angle of incidence or the prism angle is now most simply followed by an example. Consider a dense flint prism for which

$$\lambda_1 = 4.86 \times 10^{-7}\ \text{m} \qquad \mu_{\lambda_1} = 1.664 \tag{4.1.13}$$
$$\lambda_2 = 5.89 \times 10^{-7}\ \text{m} \qquad \mu_{\lambda_2} = 1.650 \tag{4.1.14}$$

and

$$\frac{\Delta\theta}{\Delta\lambda} = 9.71 \times 10^6 \left[\!\!\left[\sin^{-1}\left\{ 1.664 \sin\left[\alpha - \sin^{-1}\left(\sin i_1/1.664 \right) \right] \right\}\right.\right.$$
$$\left.\left. - \sin^{-1}\left\{ 1.650 \sin\left[\alpha - \sin^{-1}\left(\sin i_1/1.650 \right) \right] \right\} \right]\!\!\right] \tag{4.1.15}$$

(in $^\circ$ m^{-1}). Figure 4.1.2 shows the variation of $\Delta\theta/\Delta\lambda$ with angle of incidence for a variety of apex angles as given by equation (4.1.15). From this figure it is reasonably convincing to see that the maximum dispersion of 1.02×10^8 $^\circ$ m^{-1} occurs for an angle of incidence of 90° and an apex angle of 73.8776°. This represents the condition of glancing incidence and exit from the prism (figure 4.1.3), and the ray passes symmetrically through the prism.

The symmetrical passage of the ray through the prism is of importance apart from being one of the requirements for maximum $\Delta\theta/\Delta\lambda$. It is only when this condition applies that the astigmatism introduced by the prism is minimised. The condition of symmetrical ray passage for any prism is more normally called the position of minimum deviation. Again an example quickly illustrates why this is so. For a dense flint prism with an apex angle of 30°, and at a wavelength of 500

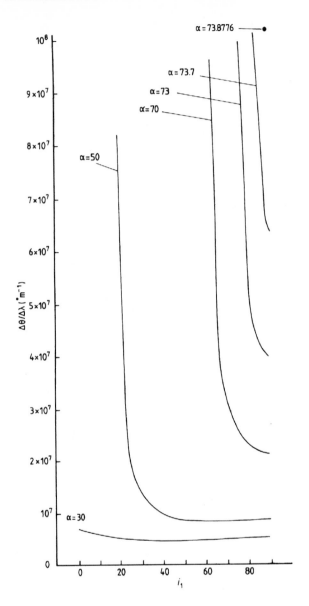

Figure 4.1.2. Variation of $\Delta\theta/\Delta\lambda$ with angle of incidence for a dense flint prism, for various apex angles, α.

nm, figure 4.1.4 shows the variation of the deviation with angle of incidence. The minimum value of θ occurs for $i_1 = 25.46°$, from which we rapidly find that $r_1 = 15°$, $i_2 = 15°$ and $r_2 = 25.46°$, and so the ray is passing through the

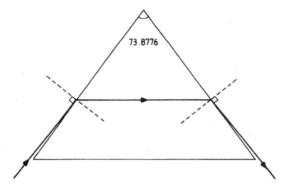

Figure 4.1.3. Optical path for maximum dispersion in a dense flint prism.

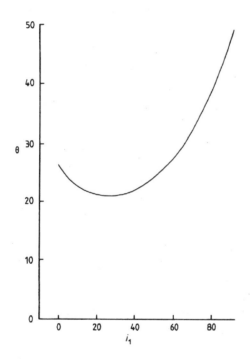

Figure 4.1.4. Deviation of dense flint prism with an apex angle of 30° at a wavelength of 500 nm.

prism symmetrically when its deviation is a minimum. More generally we may write θ in terms of r_1,

$$\theta = \sin^{-1}\left(\mu_\lambda \sin r_1\right) + \sin^{-1}\left[\mu_\lambda \sin\left(\alpha - r_1\right)\right] - \alpha \tag{4.1.16}$$

so that

$$\frac{\partial\theta}{\partial r_1} = \frac{\mu_\lambda \cos r_1}{\sqrt{1 - \mu_\lambda^2 \sin^2 r_1}} - \frac{\mu_\lambda \cos(\alpha - r_1)}{\sqrt{1 - \mu_\lambda \sin^2(\alpha - r_1)}}. \tag{4.1.17}$$

Since $\partial\theta/\partial r_1 = 0$ for θ to be an extremum, we obtain at this point after some manipulation

$$\left[\cos(2r_1) - \cos(2\alpha - 2r_1)\right](1 - 2\mu_\lambda^2) = 0 \tag{4.1.18}$$

giving

$$r_1 = \tfrac{1}{2}\alpha \tag{4.1.19}$$

and so also

$$i_2 = \tfrac{1}{2}\alpha \tag{4.1.20}$$

and the minimum deviation always occurs for a symmetrical passage of the ray through the prism.

In practice the maximum dispersion conditions of glancing incidence and exit are unusable because most of the light will be reflected at the two interactions and not refracted through the prism. Anti-reflection coatings (section 1.1) cannot be employed because these can only be optimised for a single wavelength and so the spectral distribution would be disturbed. Thus the apex angle must be less than the optimum, but the minimum deviation condition must be retained in order to minimise astigmatism. A fairly common compromise, then, is to use a prism with an apex angle of 60°, which has advantages from the manufacturer's point of view in that it reduces the amount of waste material if the initial blank is cut from a billet of glass. For an apex angle of 60°, the dense flint prism considered earlier has a dispersion of 1.39×10^7 ° m^{-1} for the angle of incidence of 55.9° which is required to give minimum deviation ray passage. This is almost a factor of ten lower than the maximum possible value.

With white light, it is obviously impossible to obtain the minimum deviation condition for all the wavelengths, and the prism is usually adjusted so that this condition is preserved for the central wavelength of the region of interest. To see how the deviation varies with wavelength, we consider the case of a prism with a normal apex angle of 60° . At minimum deviation we then have

$$r_1 = 30°. \tag{4.1.21}$$

Putting these values into equation (4.1.16) and using equation (4.1.3), we get

$$\theta = 2\sin^{-1}\left\{\tfrac{1}{2}[A + B/(\lambda - C)]\right\} - 60 \quad \text{(degrees)} \tag{4.1.22}$$

so that

$$\frac{d\theta}{d\lambda} = \frac{-180B}{\pi(\lambda - C)^2\{1 - \tfrac{1}{4}[A + B/(\lambda - C)]^2\}^{1/2}} \quad (°m^{-1}). \tag{4.1.23}$$

Now

$$A + B/(\lambda - C) = \mu_\lambda \simeq 1.5 \tag{4.1.24}$$

and so

$$\left[1 - \frac{1}{4}\left(A + \frac{B}{\lambda - C}\right)^2\right]^{-1/2} \simeq A + \frac{B}{\lambda - C}. \tag{4.1.25}$$

Thus

$$\frac{\mathrm{d}\theta}{\mathrm{d}\lambda} \simeq \frac{-180AB}{\pi(\lambda - C)^2} - \frac{180B^2}{(\lambda - C)^3}. \tag{4.1.26}$$

Now the first term on the right-hand side of equation (4.1.26) has a magnitude about thirty times larger than that of the second term for typical values of λ, A, B and C. Hence

$$\frac{\mathrm{d}\theta}{\mathrm{d}\lambda} \simeq \frac{-180AB}{\pi(\lambda - C)^2} \quad (^\circ\mathrm{m}^{-1}). \tag{4.1.27}$$

and hence

$$\frac{\mathrm{d}\theta}{\mathrm{d}\lambda} \propto (\lambda - C)^{-2} \tag{4.1.28}$$

and so the dispersion of a prism increases rapidly towards shorter wavelengths. For the example we have been considering involving a dense flint prism, the dispersion is nearly five times larger at 400 nm than at 700 nm.

To form a part of a spectrograph, a prism must be combined with other elements. The basic layout is shown in figure 4.1.5, practical designs are discussed in section 4.2. The prism is illuminated by parallel light which is usually obtained by placing a slit at the focus of a collimating lens, but sometimes may be obtained simply by allowing the light from a very distant object to fall directly onto the prism. After passage through the prism the light is focused by the imaging lens to form the required spectrum, and this may then be photographed, observed through an eyepiece, projected onto a screen, etc as desired. The collimator and imaging lenses may be simple lenses as shown, in which case the spectrum will be tilted with respect to the optical axis because of chromatic aberration, or they may be achromats or mirrors (see the next section).

The angular dispersion of a prism is not normally used as a parameter of a spectroscopic system. Instead it is combined with the focal length of the imaging element to give either the linear dispersion or the reciprocal linear dispersion. If x is the linear distance along the spectrum from some reference point, then we have for an achromatic imaging element of focal length f_2,

$$\frac{\mathrm{d}x}{\mathrm{d}\lambda} = f_2 \frac{\mathrm{d}\theta}{\mathrm{d}\lambda} \tag{4.1.29}$$

where θ is small and is measured in radians. Thus from equation (4.1.27)

$$\frac{\mathrm{d}x}{\mathrm{d}\lambda} = \frac{-ABf_2}{\pi(\lambda - C)^2}. \tag{4.1.30}$$

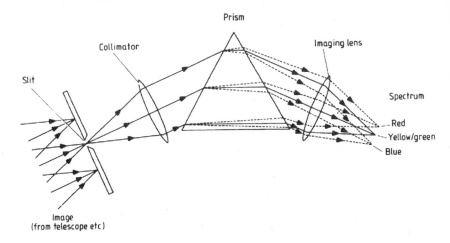

Figure 4.1.5. Basic optical arrangement of a prism spectroscope.

More commonly, the reciprocal linear dispersion, $d\lambda/dx$, is quoted and used. For practical astronomical spectrometers, this usually has values in the range

$$10^{-7} < d\lambda/dx < 5 \times 10^{-5}. \tag{4.1.31}$$

The commonly used units are nanometres change of wavelength per millimetre along the spectrum so that the above range is from 0.1 to 50 nm mm^{-1}. The use of Å mm^{-1} is still fairly common practice amongst astronomers, the magnitude of the dispersion is then a factor of ten larger than the conventional measure.

   The resolving power of a spectrometer is limited by the resolving power of its optics (section 1.1), and by the projected slit width. The spectrum is formed from an infinite number of monochromatic images of the entrance slit. It is easy to see that the width of one of these images, W, is given by

$$W = s(f_2/f_1) \tag{4.1.32}$$

where s is the slit width, f_1 is the collimator's focal length and f_2 is the imaging element's focal length. In wavelength terms, the slit width, $W(d\lambda/dx)$, is sometimes called the spectral purity of the spectroscope. If the optics of the spectroscope are well corrected then we may ignore their aberrations and consider only the diffraction limit of the system. When the prism is fully illuminated, the imaging element will intercept a rectangular beam of light. The height of the beam is just the height of the prism, and has no effect upon the spectral resolution. The width of the beam, D, is given by

$$D = L\left[1 - \mu_\lambda^2 \sin^2(\alpha/2)\right]^{1/2} \tag{4.1.33}$$

where L is the length of a prism face, and the diffraction limit is just that of a rectangular slit of width D. So that from figure 1.1.26, the linear Rayleigh limit of resolution, W', is given by

$$W' = f_2\lambda/D \tag{4.1.34}$$

$$= \frac{f_2\lambda}{L[1 - \mu^2 \sin^2(\alpha/2)]^{1/2}}. \tag{4.1.35}$$

If the beam is limited by some other element of the optical system, and/or is of circular cross section, then D must be evaluated as may be appropriate, or the Rayleigh criterion for the resolution through a circular aperture (equation 1.1.28) used in place of that for a rectangular aperture. Optimum resolution occurs when

$$W = W' \tag{4.1.36}$$

i.e.

$$s = \frac{f_1\lambda}{D} \tag{4.1.37}$$

$$= \frac{f_1\lambda}{L[1 - \mu_\lambda^2 \sin^2(\alpha/2)]^{1/2}} \tag{4.1.38}$$

The ability of a spectroscope to separate two wavelengths is called the spectral resolution and is denoted by W_λ, and it may now be found from equations (4.1.29) and (4.1.34)

$$W_\lambda = W' \frac{d\lambda}{dx} \tag{4.1.39}$$

$$= \frac{\lambda}{D} \frac{d\lambda}{d\theta} \tag{4.1.40}$$

$$\simeq \frac{\lambda(\lambda - C)^2}{ABL\{1 - [A + B/(\lambda - C)]^2 \sin^2(\alpha/2)\}^{1/2}}. \tag{4.1.41}$$

More commonly the resolution is expressed as

$$R = \lambda/W_\lambda \tag{4.1.42}$$

$$\simeq \frac{ABL\{1 - [A + B/(\lambda - C)]^2 \sin^2(\alpha/2)\}^{1/2}}{(\lambda - C)^2} \tag{4.1.43}$$

instead of the actual spectral resolution. For a dense flint prism with an apex angle of 60° and a side length of 0.1 m, we then obtain in the visible

$$R \simeq 1.5 \times 10^4 \tag{4.1.44}$$

and this is a fairly typical value for the resolution of a prism-based spectroscope. We may now see another reason why the maximum dispersion (figure 4.1.2) is not used in practice. Working back through the equations, we find that the term $L\{1 - [A + B/(\lambda - C)]^2 \sin^2(\alpha/2)\}^{1/2}$ which is involved in the numerator of

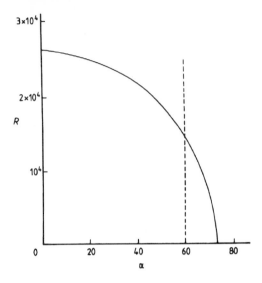

Figure 4.1.6. Resolution of a dense flint prism with a side length of 0.1 m at a wavelength of 500 nm, with changing apex angle α.

the right-hand side of equation (4.1.43) is just the width of the emergent beam from the prism. Now for maximum dispersion, the beam emerges at 90° to the normal to the last surface of the prism. Thus, however large the prism may be, the emergent beam width is zero, and thus R is zero as well. The full variation of R with α is shown in figure 4.1.6. A 60° apex angle still preserves 60% of the maximum resolution and so is a reasonable compromise in terms of resolution as well as dispersion. The truly optimum apex angle for a given type of material will generally be close to but not exactly 60°. Its calculation will involve a complex trade-off between resolution, dispersion, and throughput of light, assessed in terms of the final amount of information available in the spectrum, and is not usually attempted unless absolutely necessary.

The resolution varies slightly across the width of the spectrum, unless cylindrical lenses or mirrors are used for the collimator, and these have severe disadvantages of their own. The variation arises because the light rays from the ends of the slit impinge on the first surface of the prism at an angle to the optical axis of the collimator (figure 4.1.7). The peripheral rays therefore encounter a prism whose effective apex angle is larger than that for the paraxial rays. The deviation is also increased for such rays, and so the ends of the spectrum lines are curved towards shorter wavelengths. Fortunately astronomical spectra are mostly so narrow that both these effects can be neglected.

A quantity known variously as throughput, etendu, or light gathering power, is useful as a measure of the efficiency of the optical system. It is the amount of energy passed by the system when its entrance aperture is illuminated by unit

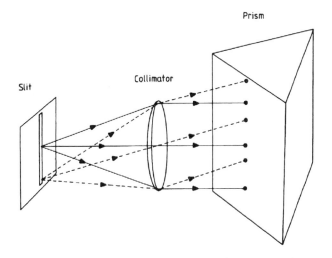

Figure 4.1.7. Light paths of rays from the centre and edge of the slit of a spectroscope.

intensity per unit area per unit solid angle, and it is denoted by u

$$u = \tau A \Omega \tag{4.1.45}$$

where Ω is the solid angle accepted by the instrument, A is the area of its aperture and τ is the fractional transmission of its optics, incorporating losses due to scattering, absorption, imperfect reflection, etc. For a spectroscope, Ω is the solid angle subtended by the entrance slit at the collimator or, for slitless spectroscopes, the solid angle accepted by the telescope–spectroscope combination. A is the area of the collimator or the effective area of the dispersing element, whichever is the smaller, τ will depend critically upon the design of the system, but as a reasonably general rule it may be taken to be the product of the transmissions for all the surfaces. These will usually be in the region of 0.8 to 0.9 for each surface, so that τ for the design illustrated in figure 4.1.5 will have a value of about 0.4. The product of resolution and throughput, P, is a useful figure for comparing the performances of different spectroscope systems

$$P = Ru. \tag{4.1.46}$$

Normally it will be found that, other things being equal, P will be largest for Fabry–Perot spectroscopes (see p. 320), of intermediate values for grating-based spectroscopes (see below), and lowest for prism-based spectroscopes.

The material used to form prisms depends upon the spectral region which is to be studied. In the visual region, the normal types of optical glass may be used, but these mostly start to absorb in the near ultraviolet. Fused silica and crystalline quartz can be formed into prisms to extend the limit down to

200 nm. Crystalline quartz, however, is optically active (section 5.2), and therefore must be used in the form of a Cornu prism. This has the optical axis parallel to the base of the prism so that the ordinary and extraordinary rays coincide, and is made in two halves cemented together. The first half is formed from a right-handed crystal, and the other half from a left-handed crystal, the deviations of left- and right-hand circularly polarised beams are then similar. If required, calcium fluoride or lithium fluoride can extend the limit down to 140 nm or so, but astronomical spectroscopes working at such short wavelengths are normally based upon gratings (see below). In the infrared, quartz can again be used for wavelengths out to about 3.5 μm. Rock salt can be used at even longer wavelengths, but it is extremely hygroscopic which makes it difficult to use. More commonly Fourier spectroscopy (see p. 327) is applied when high resolution spectroscopy is required in the far infrared.

Diffraction gratings

We have already seen in section 2.5 the structure of the image of a single source viewed through two apertures (figure 2.5.6). The angular distance of the first fringe from the central maximum is λ/d, where d is the separation of the apertures. Thus the position of the first fringe, and of course the positions of all the other fringes, is a function of wavelength. If such a pair of apertures were illuminated with white light, all the fringes apart from the central maximum would thus be short spectra with the longer wavelengths furthest from the central maximum. In an image such as that of figure 2.5.6, the spectra would be of little use since the fringes are so broad that they would overlap each other long before a useful dispersion could be obtained. However, if we add a third aperture in line with the first two, and separated from the nearer of the original apertures by a distance, d, again, then we find that the fringes remain stationary, but become

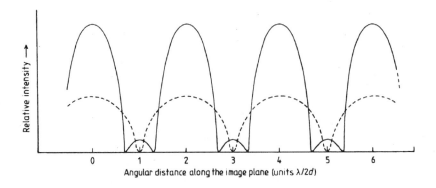

Figure 4.1.8. Small portion of the image structure for a single point source viewed through two apertures (broken curve) and three apertures (full curve).

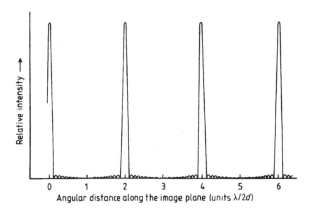

Figure 4.1.9. Small portion of the image structure for a single point source viewed through twenty apertures.

narrower and more intense. Weak secondary maxima also appear between the main fringes (figure 4.1.8). The peak intensities are of course modulated by the pattern from a single slit when looked at on a larger scale, in the manner of figure 2.5.6. If further apertures are added in line with the first three and with the same separations, then the principal fringes continue to narrow and intensify, and further weak maxima appear between them (figure 4.1.9). The intensity of the pattern at some angle, θ, to the optical axis is given by (cf equation (1.1.6))

$$I(\theta) = I(0) \times \frac{\sin^2\left(\dfrac{\pi D \sin\theta}{\lambda}\right)}{\left(\dfrac{\pi D \sin\theta}{\lambda}\right)^2} \times \frac{\sin^2\left(\dfrac{N\pi d \sin\theta}{\lambda}\right)}{\sin^2\left(\dfrac{\pi d \sin\theta}{\lambda}\right)} \qquad (4.1.47)$$

where D is the width of the aperture and N the number of apertures. The term

$$\sin^2\left(\frac{\pi D \sin\theta}{\lambda}\right)\bigg/\left(\frac{\pi D \sin\theta}{\lambda}\right)^2 \qquad (4.1.48)$$

represents the modulation of the image by the intensity structure for a single aperture, while the term

$$\sin^2\left(\frac{N\pi d \sin\theta}{\lambda}\right)\bigg/\sin^2\left(\frac{\pi d \sin\theta}{\lambda}\right) \qquad (4.1.49)$$

represents the result of the interference between N apertures. We may write

$$\Delta = (\pi D \sin\theta)/\lambda \qquad (4.1.50)$$

and

$$\delta = (\pi d \sin\theta)\lambda \qquad (4.1.51)$$

and equation (4.1.47) then becomes

$$I(\theta) = I(0)\frac{\sin^2 \Delta}{\Delta^2} \frac{\sin^2(N\delta)}{\sin^2 \delta}. \tag{4.1.52}$$

Now consider the interference component as δ tends to $m\pi$, where m is an integer. Putting

$$P = \delta - m\pi \tag{4.1.53}$$

we have

$$\lim_{\delta \to m\pi}\left(\frac{\sin(N\delta)}{\sin \delta}\right) = \lim_{p \to 0}\left(\frac{\sin[N(P + m\pi)]}{\sin(P + m\pi)}\right) \tag{4.1.54}$$

$$= \lim_{P \to 0}\left(\frac{\sin(NP)\cos(Nm\pi) + \cos(NP)\sin(Nm\pi)}{\sin P \cos(m\pi) + \cos P \sin(m\pi)}\right) \tag{4.1.55}$$

$$= \lim_{P \to 0}\left(\pm\frac{\sin(NP)}{\sin P}\right) \tag{4.1.56}$$

$$= \pm N \lim_{P \to 0}\left(\frac{\sin(NP)}{NP}\frac{P}{\sin P}\right) \tag{4.1.57}$$

$$= \pm N. \tag{4.1.58}$$

Hence integer multiples of π give the values of δ for which we have a principal fringe maximum. The angular positions of the principal maxima are given by

$$\theta = \sin^{-1}\left(\frac{m\lambda}{d}\right) \tag{4.1.59}$$

and m is usually called the order of the fringe. The zero intensities in the fringe pattern will be given by

$$N\delta = m'\pi \tag{4.1.60}$$

where m' is an integer, but excluding the cases where $m' = mN$, which are the principal fringe maxima. Their positions are given by

$$\theta = \sin^{-1}\left(\frac{m'\lambda}{Nd}\right). \tag{4.1.61}$$

The angular width of a principal maximum, W, between the first zeros on either side of it is thus given by

$$W = 2\lambda/(Nd\cos\theta). \tag{4.1.62}$$

The width of a fringe is therefore inversely proportional to the number of apertures, while its peak intensity, from equations (4.1.52) and (4.1.58), is proportional to the square of the number of apertures. Thus for a bichromatic source observed through a number of apertures we obtain the type of image structure shown in figure 4.1.10. The angular separation of fringes of the same order for the two wavelengths, for small values of θ, can be seen from equation (4.1.59) to be proportional to both the wavelength and to the order of the fringe, while the fringe width is independent of the order (equation (4.1.62)). For a white light source, by a simple extension of figure 4.1.10, we may see that the image will consist of a series of spectra on either side of a white central image. The Rayleigh resolution within this image is obtained from equation (4.1.62)

$$W' = \lambda/(Nd\cos\theta) \tag{4.1.63}$$

and is independent of the fringe order, the spectral resolution however increases directly with the fringe order because of the increasing dispersion of the spectra. Thus

$$W_\lambda = W'\frac{d\lambda}{d\theta} \tag{4.1.64}$$

but from equation (4.1.59)

$$\frac{d\lambda}{d\theta} = \frac{d}{m}\cos\theta \tag{4.1.65}$$

so that

$$W_\lambda = \lambda/Nm \tag{4.1.66}$$

and the resolution of the system, R, is therefore

$$R = \frac{\lambda}{W_\lambda} = Nm. \tag{4.1.67}$$

The resolution for a series of apertures is thus just the product of the number of apertures and the order of the spectrum. It is independent of the width and spacing of the apertures.

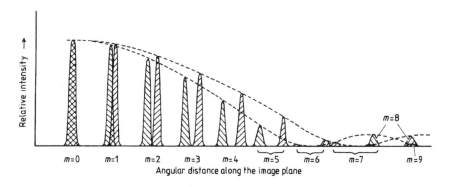

Figure 4.1.10. A portion of the image structure for a single bichromatic point source viewed through several apertures.

From figure 4.1.10 we may see that at higher orders the spectra are overlapping. This occurs at all orders when white light is used. The difference in wavelength between two superimposed wavelengths from adjacent spectral orders is called the free spectral range, Σ. From equation (4.1.59) we may see that if λ_1 and λ_2 are two such superimposed wavelengths, then

$$\sin^{-1}\left(m\lambda_1/d\right) = \sin^{-1}\left[(m+1)\lambda_2/d\right] \tag{4.1.68}$$

that is

$$\Sigma = \lambda_1 - \lambda_2 = \lambda_2/m. \tag{4.1.69}$$

For small values of m, Σ is therefore large, and the unwanted wavelengths in a practical spectroscope may be rejected by the use of filters. Some spectroscopes, such as those based on Fabry–Perot etalons (see p. 320) and echelle gratings (see p. 319), operate, however, at very high spectral orders, and both of the overlapping wavelengths may be desired. Then it is necessary to use a cross disperser so that the final spectrum consists of a two-dimensional array of short sections of the spectrum (see the discussion later in this section).

A practical device for producing spectra by diffraction uses a large number of closely spaced, parallel, narrow slits or grooves, and is called a diffraction grating. Typical gratings for astronomical use have between 100 and 1000 grooves per millimetre, and 1000 to 50 000 grooves in total. They are used at orders ranging from one up to two hundred or so. Thus the resolutions range from 10^3 to 10^5. Although the earlier discussion was based upon the use of clear apertures, each aperture can be replaced by a small plane mirror without altering the results. Thus diffraction gratings can be used either in transmission or reflection modes. Most astronomical spectroscopes are in fact based upon reflection gratings. Often the grating is inclined to the incoming beam of light, but this changes the discussion only marginally. There is a constant term, $d \sin i$, added to the path differences, where i is the angle made by the incoming beam with the normal to the grating. The whole image (figure 4.1.10) is shifted an angular distance i along the image plane. Equation (4.1.59) then becomes

$$\theta = \sin^{-1}[(m\lambda/d) - \sin i] \tag{4.1.70}$$

and in this form is often called the grating equation.

The optical layout of a grating spectroscope is similar to that of a prism spectroscope (figure 4.1.5), with the grating replacing the prism. The linear dispersion within each spectrum is thus given by

$$\frac{\mathrm{d}x}{\mathrm{d}\lambda} = \pm \frac{mf_2}{d\cos\theta}. \tag{4.1.71}$$

Now since θ varies very little over an individual spectrum, we may write

$$\frac{\mathrm{d}x}{\mathrm{d}\lambda} \simeq \text{constant.} \tag{4.1.72}$$

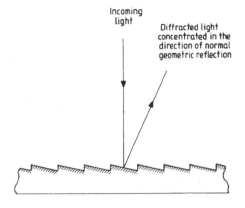

Figure 4.1.11. Enlarged section through a blazed reflection grating.

The dispersion of a grating spectroscope is thus roughly constant compared with the strong wavelength dependence of a prism spectroscope (equation (4.1.30)). The entrance slit must have a physical width of s or less, if it is not to degrade the spectral resolution, where

$$s = \lambda f_i / (N d \cos \theta) \tag{4.1.73}$$

(cf equation (4.1.38)). The other optical components of the spectroscope must also be sufficiently well corrected for their aberrations to be negligible.

The major disadvantage of a grating as a dispersing element is immediately obvious from figure 4.1.10; the light from the original source is spread over a large number of spectra. The grating's efficiency in terms of the fraction of light concentrated into the spectrum of interest is therefore very low. This disadvantage, however, may be largely overcome with reflection gratings through the use of the technique of blazing the grating. In this technique, the individual mirrors which comprise the grating are angled so that they concentrate the light into a single narrow solid angle (figure 4.1.11). For instruments based upon the use of gratings at low orders, the angle of the mirrors is arranged so that the light is concentrated into the spectrum to be used, and by this means up to 90% efficiency can be achieved. For those spectroscopes which use gratings at high orders, the grating can still be blazed, but then the light is concentrated into short segments of many different orders of spectra. By a careful choice of parameters, these short segments can be arranged so that they overlap slightly at their ends, and so coverage of a much wider spectral region may be obtained by producing a montage of the segments. Transmission gratings can also be blazed although this is less common. Each of the grooves then has the cross section of a small prism, the apex angle of which defines the blaze angle. Blazed transmission gratings for use at infrared wavelengths can be produced by etching the surface of a block of silicon in a similar manner to the way in which integrated circuits are produced.

Each inclined surface is then actually formed like a stairway—a series of small flat segments with risers between. However, as long as the height of the risers is small compared with the operating wavelength, the stairway behaves like a smooth inclined mirror.

Another problem which is intrinsically less serious, but which is harder to counteract is that of shadowing. If the incident and/or reflected light makes a large angle to the normal to the grating, then the step-like nature of the surface (figure 4.1.11) will cause a significant fraction of the light to be intercepted by the vertical portions of the grooves, and so lost to the final spectrum. There is little that can be done to eliminate this problem except either to accept the light loss, or to design the system so that large angles of incidence or reflection are not needed.

Curved reflection gratings are frequently produced. By making the curve that of an optical surface, the grating itself can be made to fulfil the function of the collimator and/or the imaging element of the spectroscope, thus reducing light losses and making for greater simplicity of design and reduced costs. The grooves should be ruled so that they appear parallel and equally spaced when viewed from infinity. The simplest optical principle employing a curved grating is that due to Rowland. The slit, grating and spectrum all lie on a single circle which is called the Rowland circle (figure 4.1.12). This has a diameter equal to the radius of curvature of the grating. The use of a curved grating at large angles to its optical axis introduces astigmatism, and spectral lines may also be curved due to the varying angles of incidence for rays from the centre and ends of the slit (cf figure 4.1.7). Careful design, however, can reduce or eliminate these defects, and there are several practical designs for spectroscopes based upon the Rowland circle (section 4.2). Aspherical curved gratings are also possible and can be used to provide very highly corrected designs with few optical components.

A grating spectrum generally suffers from unwanted additional features superimposed upon the desired spectrum. Such features are usually much fainter than the main spectrum and are called ghosts. They arise from a variety of causes. They may be due to overlapping spectra from higher or lower orders, or to the secondary maxima associated with each principal maximum (figure 4.1.9). The first of these is usually simple to eliminate by the use of filters since the overlapping ghosts are of different wavelengths from the overlapped main spectrum. The second source is usually unimportant since the secondary maxima are very faint when more than a few tens of apertures are used, though they still contribute to the wings of the instrumental profile. Of more general importance are the ghosts which arise through errors in the grating. Such errors most commonly take the form of a periodic variation in the groove spacing. A variation with a single period gives rise to Rowland ghosts, which appear as faint lines close to and on either side of strong spectrum lines. Their intensity is proportional to the square of the order of the spectrum. Thus echelle gratings (see below) must be of very high quality since they may use spectral orders in the region of several hundred. If the error is multiply-periodic, then Lyman ghosts

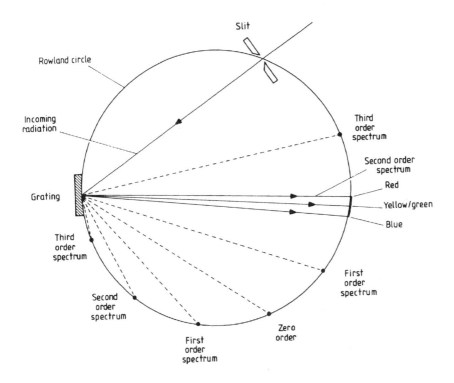

Figure 4.1.12. Schematic diagram of a spectroscope based upon a Rowland circle, using a curved grating blazed for the second order.

of strong lines may appear. These are similar to the Rowland ghosts, except that they can be formed at large distances from the line which is producing them. Some compensation for these errors can be obtained through deconvolution of the instrumental profile (section 2.1), but for critical work, the only real solution is to use a grating without periodic errors, such as a holographically produced grating (see p. 348).

Wood's anomalies may also sometimes occur. These do not arise through grating faults, but are due to light that should go into spectral orders behind the grating (were that possible) reappearing within lower order spectra. The anomalies have a sudden onset and a slower decline towards longer wavelengths and are almost 100% plane polarised. They are rarely important in efficiently blazed gratings.

By increasing the angle of a blazed grating, we obtain an echelle grating (figure 4.1.13). This is illuminated more or less normally to the groove surfaces and therefore at a large angle to the normal to the grating. It is usually a very coarse grating—ten lines per millimetre is not uncommon—so that the separation

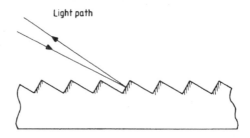

Figure 4.1.13. Enlarged view of an echelle grating.

of the apertures, d, is very large. The reciprocal linear dispersion

$$\frac{d\lambda}{dx} = \pm\frac{d\cos^3\theta}{mf_2} \tag{4.1.74}$$

is therefore also very large. Such gratings concentrate the light into many overlapping high-order spectra, and so from equation (4.1.67), the resolution is very high. A spectroscope which is based upon an echelle grating requires a second low dispersion grating or prism whose dispersion is perpendicular to that of the echelle in order to separate out each of the orders (section 4.2). The echelon is a similar device but with even coarser spacings. It was first constructed by Michelson, and consisted of a stack of flat plates, each about 10 mm in thickness, and displaced with respect to each other at their edges by about 1 mm. The edge of the stack is therefore a very coarse version of the shape of the echelle grating (figure 4.1.13). The echelon concentrates light into spectra with orders of 10^4 or higher, thus very few plates are needed in order to achieve very high degrees of resolution.

Interferometers

We only consider here in any detail the two main types of spectroscopic interferometry which are of importance in astronomy: the Fabry–Perot interferometer or etalon, and the Michelson interferometer or Fourier-transform spectrometer. Other systems exist but at present are of little importance for astronomy. We could have included the diffraction grating in this class of spectroscope, since it is also based upon interference effects, but by common consent it is deemed sufficiently different to be in a class of its own.

Fabry–Perot interferometer

Two parallel, flat, partially reflecting surfaces are illuminated at an angle θ (figure 4.1.14). The light undergoes a series of transmissions and reflections as shown, and pairs of adjoining emergent rays differ in their path lengths by ΔP,

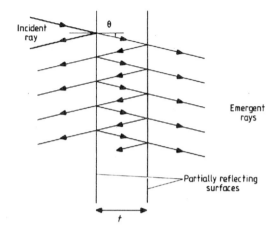

Figure 4.1.14. Optical paths in a Fabry–Perot interferometer.

where

$$\Delta P = 2t \cos \theta. \tag{4.1.75}$$

Constructive interference between the emerging rays will then occur at those wavelengths for which

$$\mu \Delta P = m\lambda \tag{4.1.76}$$

where m is an integer, i.e.

$$\lambda = (2t\mu \cos \theta)/m. \tag{4.1.77}$$

If such an interferometer is used in a spectroscope in place of the prism or grating (figure 4.1.15), then the image of a point source is still a point. However, the image is formed from only those wavelengths for which equation (4.1.77) holds. If the image is then fed into another spectroscope it will be broken up into a

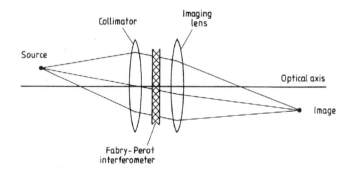

Figure 4.1.15. Optical paths in a Fabry–Perot spectroscope.

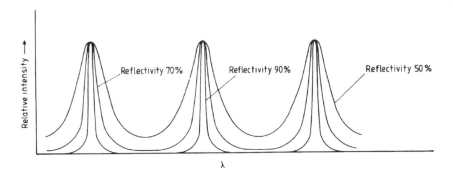

Figure 4.1.16. Intensity versus wavelength in the image of a white light point source in a Fabry–Perot spectroscope, assuming negligible absorption.

series of monochromatic images. If a slit is now used as the source, the rays from the different points along the slit will meet the etalon at differing angles, and the image will then consist of a series of superimposed short spectra. If the second spectroscope is then set so that its dispersion is perpendicular to that of the etalon (a cross-disperser), then the final image will be a rectangular array of short parallel spectra. The widths of these spectra depend upon the reflectivity of the surfaces. For a high reflectivity, we get many multiple reflections, while with a low reflectivity the intensity becomes negligible after only a few reflections. The monochromatic images of a point source are therefore not truly monochromatic but are spread over a small wavelength range in a similar, but not identical, manner to the intensity distributions for several collinear apertures (figures 4.1.8 and 4.1.9). The intensity distribution varies from that of the multiple apertures since the intensities of the emerging beams decrease as the number of reflections required for their production increases. Examples of the intensity distribution with wavelength are shown in figure 4.1.16. In practice reflectivities of about 90% are usually used.

In the absence of absorption the emergent intensity at a fringe peak is equal to the incident intensity at that wavelength. This often seems a puzzle to readers, who on inspection of figure 4.1.14, might expect there to be radiation emerging from the left as well as on the right, such that at the fringe peak the total emergent intensity would appear to be greater than the incident intensity. If we examine the situation more closely, however, we find that when at a fringe peak for the light emerging on the right, there is zero intensity in the beam emerging on the left. If the incident beam has intensity I, and amplitude a ($I = a^2$), then the amplitudes of the successive beams on the left in figure 4.1.14 are

$$-aR^{1/2}, \ aR^{1/2}T, \ aR^{3/2}T, \ aR^{5/2}T, \ aR^{7/2}T, \ \dots$$

where the first amplitude is negative because it results from an internal reflection. It has therefore an additional phase delay of 180° compared with the other reflected beams. T is the fractional intensity transmitted and R the fractional intensity reflected by the reflecting surfaces (note that $T + R = 1$ in the absence of absorption).

Summing these terms (assumed to go on to infinity) gives a zero amplitude and therefore a zero intensity for the left-hand emergent beam. Similarly the beams emerging on the right have amplitudes

$$aT, \ aTR, \ aTR^2, \ aTR^3, \ aTR^4, \ \ldots \tag{4.1.78}$$

Summing these terms to infinity gives, at a fringe peak, the amplitude on the right as a. This is the amplitude of the incident beam, and so at a fringe maximum the emergent intensity on the right equals the incident intensity (see also equation (4.1.86)).

The dispersion of an etalon may easily be found by differentiating equation (4.1.77)

$$\frac{d\lambda}{d\theta} = -\frac{2t\mu}{m}\sin\theta. \tag{4.1.79}$$

Since the material between the reflecting surfaces is usually air, and the etalon is used at small angles of inclination, we have

$$\mu \simeq 1 \tag{4.1.80}$$

$$\sin\theta \simeq \theta \tag{4.1.81}$$

and from equation (4.1.77)

$$2t/m \simeq \lambda \tag{4.1.82}$$

so that

$$\frac{d\lambda}{d\theta} \simeq \lambda\theta. \tag{4.1.83}$$

Thus the reciprocal linear dispersion for a typical system, with $\theta = 0.1°$, $f_2 = 1$ m, and used in the visible, is 0.001 nm mm^{-1}, which is a factor of 100 or so larger than that often achievable with more common dispersing elements.

The resolution of an etalon is rather more of a problem to estimate. Our usual measure—the Rayleigh criterion—is inapplicable since the minimum intensities between the fringe maxima do not reach zero, except for a reflectivity of 100%. However, if we consider the image of two equally bright point sources viewed through a telescope at its Rayleigh limit (figure 2.5.3) then the central intensity is 81% of that of either of the peak intensities. We may therefore replace the Rayleigh criterion by the more general requirement that the central intensity of the envelope of the images of two equal sources falls to 81% of the peak intensities. Consider, therefore, an etalon illuminated by a monochromatic slit source perpendicular to the optical axis (figure 4.1.15). The image will be a strip, also perpendicular to the optical axis, and the intensity will vary along the strip accordingly as the emerging rays are in or out of phase with each other.

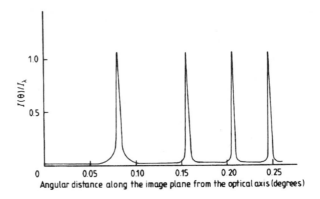

Figure 4.1.17. Image structure in a Fabry–Perot spectroscope viewing a monochromatic slit source, with $T = 0.1$, $R = 0.9$, $t = 0.1$ m, $\mu = 1$ and $\lambda = 550$ nm.

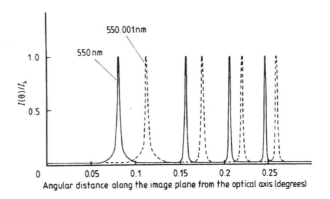

Figure 4.1.18. Image structure in a Fabry–Perot spectroscope viewing a bichromatic slit source, with $T = 0.1$, $R = 0.9$, $t = 0.1$ m, $\mu = 1$, $\lambda = 550$ nm (full curve) and $\lambda = 550.001$ nm (broken curve).

The intensity variation is given by

$$I(\theta) = T^2 I_\lambda \left[(1 - R)^2 + 4R \sin^2 \left(\frac{2\pi t \mu \cos \theta}{\lambda} \right) \right]^{-1} \qquad (4.1.84)$$

where I_λ is the incident intensity at wavelength λ. The image structure will resemble that shown in figure 4.1.17. If the source is now replaced with a bichromatic one, then the image structure will be of the type shown in figure 4.1.18. Consider just one of these fringes, its angular distance, θ_{max}, from the optical axis is, from equation (4.1.77),

$$\theta_{max} = \cos^{-1}\left(\frac{m\lambda}{2t\mu}\right) \tag{4.1.85}$$

and so the peak intensity from equation (4.1.83) is

$$I(\theta_{max}) = T^2 I_\lambda/(1-R)^2 \tag{4.1.86}$$
$$= I_\lambda \quad \text{(when there is no absorption).} \tag{4.1.87}$$

Let the angular half width of a fringe at half intensity be $\Delta\theta$, then a separation of twice the half-half width of the fringes gives a central intensity of 83% of either of the peak intensities. So that if α is the resolution by the extended Rayleigh criterion, we may write

$$\alpha \simeq 2\Delta\theta = \frac{\lambda(1-R)}{2\pi\mu t\sqrt{R}\theta_{max}\cos\theta_{max}}. \tag{4.1.88}$$

Hence from equation (4.1.82) we obtain the spectral resolution

$$W_\lambda = \alpha\frac{d\lambda}{d\theta} \tag{4.1.89}$$

$$= \frac{\lambda^2(1-R)}{2\pi\mu t\sqrt{R}\cos\theta_{max}} \tag{4.1.90}$$

and so the resolution of the system (previously given the symbol R) is

$$\frac{\lambda}{\Delta\lambda} = \frac{2\pi t\mu\sqrt{R}\cos\theta_{max}}{\lambda(1-R)} \tag{4.1.91}$$

or, since θ_{max} is small and μ is usually close to unity,

$$\frac{\lambda}{\Delta\lambda} \simeq \frac{2\pi t\sqrt{R}}{\lambda(1-R)}. \tag{4.1.92}$$

Thus for typical values of $t = 0.1$ m, $R = 0.9$ and for visible wavelengths we have

$$\frac{\lambda}{\Delta\lambda} \simeq 10^7 \tag{4.1.93}$$

which is almost two orders of magnitude higher than typical values for prisms and gratings, and is comparable with the resolution for a large echelon grating while the device is much less bulky. An alternative measure of the resolution that may be encountered is the finesse. This is the reciprocal of the half-width of a fringe measured in units of the separation of the fringes from two adjacent orders. It is given by

$$\text{Finesse} = \frac{\pi\sqrt{R}}{1-R} = \frac{\lambda}{2t} \times \text{resolution.} \tag{4.1.94}$$

For a value of R of 0.9, the finesse is therefore about 30.

The free spectral range of an etalon is small since it is operating at very high spectral orders. From equation (4.1.84) we have

$$\Sigma = \lambda_1 - \lambda_2 = \lambda_2/m \qquad (4.1.95)$$

where λ_1 and λ_2 are superimposed wavelengths from adjacent orders (cf equation (4.1.69)). Thus the device must be used with a cross disperser as already mentioned and/or the free spectral range increased. The latter may be achieved by combining two or more different etalons, then only the maxima which coincide will be transmitted through the whole system, and the intermediate maxima will be suppressed.

Practical etalons are made from two pieces of glass or quartz whose surfaces are flat to one or two per cent of their operating wavelength, they are held accurately parallel to each other by low thermal expansion spacers, with a spacing in the region of 10 to 200 mm. The inner faces are mirrors, which are usually produced by a metallic or dielectric coating. The outer faces are inclined by a very small angle to the inner faces so that the plates are the basal segments of very low angle prisms, thus any multiple reflections other than the desired ones are well displaced from the required image. The limit to the resolution of the instrument is generally imposed by departures of the two reflecting surfaces from absolute flatness. This limits the main use of the instrument to the visible and infrared regions. The absorption in metallic coatings also limits the shortwave use, so that 200 nm represents the shortest practicable wavelength even for laboratory usage. Etalons are commonly used as scanning instruments. By changing the air pressure by a few per cent, the refractive index of the material between the plates is changed and so the wavelength of a fringe at a given place within the image is altered (equation (4.1.84)). The astronomical applications of Fabry–Perot spectroscopes are comparatively few for direct observations. However, the instruments are used extensively in determining oscillator strengths and transition probabilities upon which much of the more conventional astronomical spectroscopy is based.

Another important application of etalons, and one which does have some direct applications for astronomy is in the production of narrow band filters. These are usually known as interference filters and are etalons in which the separation of the two reflecting surfaces is very small. For materials with refractive indices near 1.5, and for near normal incidence, we see from equation (4.1.77) that if t is 167 nm (say) then the maxima will occur at wavelengths of 500, 250, 167 nm and so on, accordingly as n is 1, 2, 3, etc. While from equation (4.1.88), the widths of these transmitted regions will be 8.4, 2.1, 0.9 nm, etc for 90% reflectivity of the surfaces. Thus a filter centred upon 500 nm with a bandwidth of 8.4 nm can be made by combining such an etalon with a simple dye filter to eliminate the shorter wavelength transmission regions. Other wavelengths and bandwidths can easily be chosen by changing t, μ and R. Such a filter would be constructed by evaporating a partially reflective layer onto a sheet of glass. A second layer of an appropriate dielectric material such as magnesium fluoride or cryolite is then evaporated on top of this to the desired thickness, followed by a second partially reflecting layer. A second sheet of glass is then added for protection. The reflecting layers may be silver or aluminium, or they may be formed from a double layer of two materials with very different refractive indices in order to improve the overall filter transmission. In the far infrared, pairs of inductive meshes can be used in a similar way for infrared filters.

Michelson interferometer

This Michelson interferometer should not be confused with the Michelson stellar interferometer which was discussed in section 2.5. The instrument discussed here is similar to the device used by Michelson and Morley to try and detect the Earth's motion through the aether. Its optical principles are shown in figure 4.1.19. The light from the source is split into two beams by the beam splitter, and then recombined as shown. For a particular position of the movable mirror and with a monochromatic source, there will be a path difference, ΔP, between the two beams at their focus. The intensity at the focus is then

$$I_{\Delta P} = I_{\mathrm{m}} \left[1 + \cos\left(\frac{2\pi \Delta P}{\lambda} \right) \right] \tag{4.1.96}$$

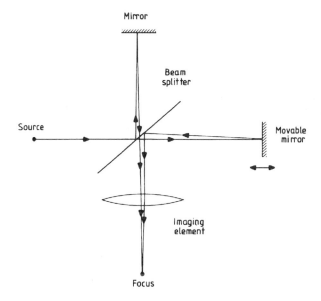

Figure 4.1.19. Optical pathways in a Michelson interferometer.

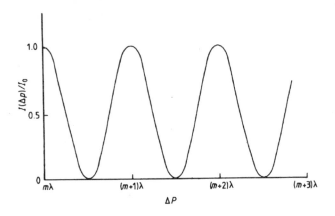

Figure 4.1.20. Variation of a fringe intensity with mirror position in a Michelson interferometer.

Figure 4.1.21. Output of a Michelson interferometer observing a bichromatic source.

where I_m is a maximum intensity. If the mirror is moved, then the path difference will change, and the final intensity will pass through a series of maxima and minima (figure 4.1.20). If the source is bichromatic, then two such variations will be superimposed with slightly differing periods and the final output will then have a beat frequency (figure 4.1.21). The difference between the two outputs (figures 4.1.20 and 4.1.21) gives the essential principle of the Michelson interferometer when it is used as a spectroscope. Neither output in any way resembles an ordinary spectrum, yet it would be simple to recognise the first as due to a monochromatic source, and the second as due to a bichromatic source. Furthermore the spacing of the fringes could be related to the original wavelength(s) through equation (4.1.95). More generally of course, sources emit a broad band of wavelengths, and the final output will vary in a complex manner. To find the spectrum of an unknown source from such an output therefore requires a rather different approach than this simple visual inspection.

Let us consider therefore a Michelson interferometer in which the path difference is ΔP, and which is observing a source whose intensity at wavelength

λ is $I(\lambda)$. The intensity in the final image due to the light of a particular wavelength, $I'_{\Delta P}(\lambda)$ is then

$$I'_{\Delta P}(\lambda) = K I(\lambda)\left[1 + \cos\left(\frac{2\pi \Delta P}{\lambda}\right)\right] \tag{4.1.97}$$

where K is a constant which takes account of the losses at the various reflections, transmissions, etc. Thus the total intensity in the image for a given path difference is just

$$I'_{\Delta P} = \int_0^\infty I'_{\Delta P}(\lambda)\mathrm{d}\lambda \tag{4.1.98}$$

$$= \int_0^\infty K I(\lambda)\mathrm{d}\lambda + \int_0^\infty K I(\lambda)\cos\left(\frac{2\pi \Delta P}{\lambda}\right)\mathrm{d}\lambda. \tag{4.1.99}$$

Now the first term on the right-hand side of equation (4.1.98) is independent of the path difference, and is simply the mean intensity of the image. We may therefore disregard it and concentrate instead on the deviations from this average level, $I(\Delta P)$. Thus

$$I(\Delta P) = K \int_0^\infty I(\lambda)\cos\left(\frac{2\pi \Delta P}{\lambda}\right)\mathrm{d}\lambda \tag{4.1.100}$$

or in frequency terms

$$I(\Delta P) = K'\int_0^\infty I(v)\cos\left(\frac{2\pi \Delta P v}{c}\right)\mathrm{d}v. \tag{4.1.101}$$

Now the Fourier transform, $F(u)$, of a function, $f(t)$, (see also section 2.1) is defined by

$$\mathcal{F}[f(t)] = F(u) = \int_{-\infty}^\infty f(t)\mathrm{e}^{-2\pi i u t}\,\mathrm{d}t \tag{4.1.102}$$

$$= \int_{-\infty}^\infty f(t)\cos(2\pi u t)\,\mathrm{d}t - i\int_{-\infty}^\infty f(t)\sin(2\pi u t)\,\mathrm{d}t. \tag{4.1.103}$$

Thus we see that the output of the Michelson interferometer is akin to the real part of the Fourier transform of the spectral intensity function of the source. Furthermore by defining

$$I(-v) = I(v) \tag{4.1.104}$$

we have

$$I(\Delta P) = \tfrac{1}{2}K'\int_{-\infty}^\infty I(v)\cos\left(\frac{2\pi \Delta P v}{c}\right)\mathrm{d}v \tag{4.1.105}$$

$$= K''\times \mathrm{Re}\left\{\int_{-\infty}^\infty I(v)\exp\left[-i\left(\frac{2\pi \Delta P}{c}\right)v\right]\mathrm{d}v\right\} \tag{4.1.106}$$

where K^{II} is an amalgam of all the constants. Now by inverting the transformation

$$\mathcal{F}^{-1}[F(u)] = f(t) = \int_{-\infty}^{\infty} F(u)e^{2\pi iut}\, du \qquad (4.1.107)$$

and taking the real part of the inversion, we may obtain the function that we require—the spectral energy distribution, or as it is more commonly known, the spectrum. Thus

$$I(v) = K^{III} \times \text{Re}\left\{\int_{-\infty}^{\infty} I\left(\frac{2\pi\, \Delta P}{c}\right) \exp\left[i\left(\frac{2\pi\, \Delta P}{c}\right)v\right] d\left(\frac{2\pi\, \Delta P}{c}\right)\right\} \qquad (4.1.108)$$

or

$$I(v) = K^{IV} \int_{-\infty}^{\infty} I\left(\frac{2\pi\, \Delta P}{c}\right) \cos\left(\frac{2\pi\, \Delta Pv}{c}\right) d(\Delta P) \qquad (4.1.109)$$

where K^{III} and K^{IV} are again amalgamated constants. Finally by defining

$$I\left(\frac{-2\pi\, \Delta P}{c}\right) = I\left(\frac{2\pi\, \Delta P}{c}\right) \qquad (4.1.110)$$

we have

$$I(v) = 2K^{IV} \int_{0}^{\infty} I\left(\frac{2\pi\, \Delta P}{c}\right) \cos\left(\frac{2\pi\, \Delta Pv}{c}\right) d(\Delta P) \qquad (4.1.111)$$

and so the spectrum is obtainable from the observed output of the interferometer as the movable mirror scans through various path differences. We may now see why a Michelson interferometer when used as a scanning spectroscope is often called a Fourier transform spectroscope.

In practice of course, it is not possible to scan over path differences from zero to infinity, and also measurements are usually made at discrete intervals rather than continuously, requiring the use of the discrete Fourier transform equations (section 2.1). These limitations are reflected in a reduction in the resolving power of the instrument. To obtain an expression for the resolving power, we may consider the Michelson interferometer as equivalent to a two-aperture interferometer (figure 4.1.8) since its image is the result of two interfering beams of light. We may therefore write equation (4.1.66) for the resolution of two wavelengths by the Rayleigh criterion as

$$W_\lambda = \lambda^2/2\Delta P. \qquad (4.1.112)$$

However, if the movable mirror in the Michelson interferometer moves a distance x, then ΔP ranges from 0 to $2x$, and we must take the the the average value of ΔP rather than the extreme value for substitution into equation (4.1.111). Thus we obtain the spectral resolution of a Michelson interferometer as

$$W_\lambda = \lambda^2/2x \qquad (4.1.113)$$

so that the system's resolution is

$$\lambda/W_\lambda = 2x/\lambda. \tag{4.1.114}$$

Since x can be as much as two metres, we obtain a resolution of up to 4×10^6 for such an instrument used in the visible region.

The sampling intervals must be sufficiently frequent to preserve the resolution, but not more frequent than this, or time and effort will be wasted. If the final spectrum extends from λ_1 to λ_2, then the number of useful intervals, n, into which it may be divided is given by

$$n = (\lambda_1 - \lambda_2)/W_\lambda \tag{4.1.115}$$

so that if λ_1 and λ_2 are not too different then

$$n \simeq \frac{8x(\lambda_1 - \lambda_2)}{(\lambda_1 + \lambda_2)^2}. \tag{4.1.116}$$

However, the inverse Fourier transform gives both $I(\nu)$ and $I(-\nu)$, so that the total number of separate intervals in the final inverse transform is $2n$. Hence we must have at least $2n$ samples in the original transformation, and therefore the spectroscope's output must be sampled $2n$ times. Thus the interval between successive positions of the movable mirrors, Δx, at which the image intensity is measured is given by

$$\Delta x = \frac{(\lambda_1 + \lambda_2)^2}{16(\lambda_1 - \lambda_2)}. \tag{4.1.117}$$

A spectrum between 500 and 550 nm therefore requires step lengths of one micron, while between 2000 and 2050 nm a spectrum would require step lengths of twenty microns. This relaxation in the physical constraints required on the accuracy of the movable mirror for longer wavelength spectra, combined with the availability of other methods of obtaining visible spectra, has led to the major applications of Fourier transform spectroscopy to date being in the infrared and far infrared.

The basic instrumental profile of the Fourier transform spectroscope is of the form $\sin(\Delta\lambda)/\Delta\lambda$ (figure 4.1.22), where $\Delta\lambda$ is the distance from the central

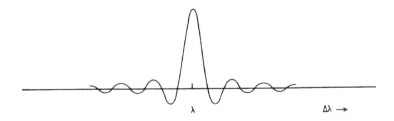

Figure 4.1.22. Basic instrumental profile of a Fourier transform spectroscope.

wavelength, λ, of a monochromatic source. This is not a particularly convenient form for the profile, and the secondary maxima may be large enough to be significant especially where the spectral energy undergoes abrupt changes, such as at ionisation edges or over molecular bands. The effect of the instrumental profile may be reduced at some cost to the theoretical resolution by a technique known as apodisation (literally 'removal of the feet', see also sections 2.1 and 2.5). The transform is weighted by some function, ω, known as the apodisation function. Many functions can be used, but the commonest is probably a triangular weighting function, i.e.

$$\omega(\Delta P) = 1 - (\Delta P/2x). \qquad (4.1.118)$$

The resolving power is halved, but the instrumental profile becomes that for a single rectangular aperture (figure 1.1.26, equation (1.1.6)), and has much reduced secondary maxima.

A major advantage of the Michelson interferometer over the etalon when the latter is used as a scanning instrument lies in its comparative rapidity of use when the instrument is detector-noise limited. Not only is the total amount of light gathered by the Michelson interferometer higher (the Jacquinot advantage), but even for equivalent image intensities the total time required to obtain a spectrum is much reduced. This arises because all wavelengths are contributing to every reading in the Michelson interferometer, whereas a reading from the etalon gives information for just a single wavelength. The gain of the Michelson interferometer is called the multiplex advantage or the Fellget advantage, and is similar to the gain of the Hadamard masking technique over simple scanning (section 2.4). If t is the integration time required to record a single spectral element, then the etalon requires a total observing time of nt. The Michelson interferometer, however, requires a time of only t/n to record each sample since it has contributions from n spectral elements, and it must obtain $2n$ samples. The total observing time for the same spectrum is therefore $2t$, and it has an advantage over the etalon of a factor of $n/2$ in observing time.

Michelson interferometers have another advantage in that they require no entrance slit in order to preserve their spectral resolution. This is of considerable significance for large telescopes where the stellar image may have a physical size of a millimetre or more due to atmospheric turbulence, while the spectroscope's slit may be only a few tenths of a millimetre wide. Thus either much of the light is wasted or complex image dissectors (section 4.2) must be used. The increased image brightness is another reason for the popularity of the Michelson interferometer for infrared work, since that is generally limited by the detector efficiency at the moment.

The inversion of the Fourier transform is carried out on large computers using the fast Fourier transform algorithm. Hence the technique has only become a feasible one in the last decade or so as such computers have developed. It is likely to find increasing applications in the future, particularly at shorter wavelengths, as its advantages become more widely known.

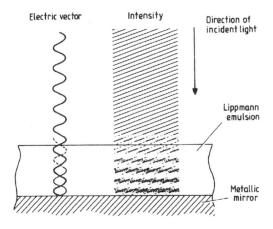

Electric vector Intensity Direction of incident light

Lippmann emulsion

Metallic mirror

Figure 4.1.23. Image structure in a thick emulsion with a reflective backing.

Holography

Another interferometric system which is worth mentioning in passing, produces a Fourier transform of the spectrum in a manner analogous to a hologram. The light interferes within a thick photographic emulsion, and a latent image is produced only where there is constructive interference. Unlike a conventional hologram, which uses two incident coherent light beams, this system is based upon the interference of one beam and its reflection. It has yet to be applied to astronomy, but potentially it could provide a superior alternative to objective prism or grating spectra.

The emulsion is thick compared to the wavelength of light, is semi-transparent, and is very fine grained (sometimes called a Lippmann emulsion). Its backing is a reflective metallic film. The incident light passes through the emulsion and is reflected back. The two beams then interfere so that the emulsion is exposed at the points of constructive interference. With monochromatic radiation there is then a layered structure in the emulsion within the image of the object (figure 4.1.23). The intensity varies sinusoidally, and so therefore will the photographic density, if the exposure is on the linear part of the characteristic curve (section 2.2). Thus the vertical structure of the image is the Fourier transform of the monochromatic source, while its structure in the plane of the emulsion is still a normal photographic image. If the source is not monochromatic, then the vertical image structure will be the appropriate Fourier transform of its actual spectrum. Thus in a single photograph, all the information available in a normal direct photograph is present, and in addition low resolution spectra which are comparable with objective prism spectra in their detail, are recorded for each image in the field of view as a Fourier transform.

The spectra can be retrieved by optically inverting the Fourier transform (cf figure 2.6.3). The developed emulsion is illuminated by collimated white light, and the reflected beam duplicates the original spectrum. This may then be rephotographed or observed directly in a visual spectroscope.

Exercises

4.1.1 Calculate the heliocentric radial velocity for a star in which the Hα line (laboratory wavelength 656.2808 nm) is observed to have a wavelength of 656.1457 nm. At the time of the observation, the solar celestial longitude was 100°, and the star's right ascension and declination were 13^h 30^m, and +45° respectively. The obliquity of the ecliptic is 23° 27'.

4.1.2 A prism spectroscope is required with a reciprocal linear dispersion of 1 nm mm^{-1} or better at a wavelength of 400 nm. The focal length of the imaging element is limited to 1 m through the need for the spectroscope to fit onto the telescope. Calculate the minimum number of 60° crown glass prisms required to achieve this, and the actually resulting reciprocal linear dispersion at 400 nm.

Note: Prisms can be arranged in a train so that the emergent beam from one forms the incident beam of the next in line. This increases the dispersion in proportion to the number of prisms, but does not affect the spectral resolution, which remains at the value appropriate for a single prism.

4.1.3 Show that the reciprocal linear dispersion of a Fabry–Perot etalon is given by $d\lambda/dx = \lambda l/2f_2^2$ where l is the length of the slit, and the slit is symmetrical about the optical axis.

4.2 SPECTROSCOPES

Basic design considerations

The specification of a spectroscope usually begins from just three parameters. One is the focal ratio of the telescope upon which the spectroscope is to operate, the second is the required spectral resolution, and the third is the required spectral range. Thus in terms of the notation used in section 4.1, we have f', W_λ and λ specified (where f' is the effective focal ratio of the telescope at the entrance aperture of the spectroscope), and we require the values of f_1, f_2, s, R, $d\lambda/d\theta$, L and D. We may immediately write down the resolution required of the dispersion element

$$R = \lambda/W_\lambda. \tag{4.2.1}$$

Now for a 60° prism from equation (4.1.43) we find that at 500 nm

$$R = 6 \times 10^4 L \qquad \text{(crown glass)} \tag{4.2.2}$$

$$R = 15 \times 10^4 L \qquad \text{(flint glass)} \tag{4.2.3}$$

where L is the length of a side of a prism, in metres. So we may write

$$R \simeq 10^5 L \qquad (4.2.4)$$

when the dispersing element is a prism. Now for a grating, the resolution depends upon the number of lines and the order of the spectrum (equation (4.1.67)). A reasonably typical astronomical grating (ignoring echelle gratings etc) will operate in its third order and have some 500 lines mm^{-1}. Thus we may write the resolution of a grating as

$$R \simeq 1.5 \times 10^6 L \qquad (4.2.5)$$

where L is the width of the ruled area of the grating, in metres. The size of the dispersing element will thus be given approximately by equations (4.2.4) and (4.2.5). It may be calculated more accurately in second and subsequent iterations through this design process, using equations (4.1.43) and (4.1.67).

The diameter of the exit beam from the dispersing element, assuming that it is fully illuminated, can then be found from

$$D = L \cos \varphi \qquad (4.2.6)$$

where φ is the angular deviation of the exit beam from the perpendicular to the exit face of the dispersing element. For a prism φ is typically 60°, while for a grating used in the third order with equal angles of incidence and reflection, it is 25°. Thus we get

$$D = 0.5L \quad \text{(prism)} \qquad (4.2.7)$$

$$D = 0.9L \quad \text{(grating).} \qquad (4.2.8)$$

The dispersion can now be determined by setting the angular resolution of the imaging element equal to the angle between two just resolved wavelengths from the dispersing element

$$\frac{\lambda}{D} = W_\lambda \frac{d\theta}{d\lambda} \qquad (4.2.9)$$

giving

$$\frac{d\theta}{d\lambda} = \frac{R}{D}. \qquad (4.2.10)$$

Since the exit beam is rectangular in cross section, we must use the resolution for a rectangular aperture of the size and shape of the beam (equation 1.1.7) for the resolution of the imaging element, and not its actual resolution, assuming that the beam is wholly intercepted by the imaging element, and that its optical quality is sufficient not to degrade the resolution below the diffraction limit.

The final parameters now follow easily. The physical separation of two just resolved wavelengths on the photographic plate (or other imaging detector, see section 2.3) must be greater than or equal to the resolution of the emulsion.

A photographic emulsion can typically resolve 50 lines mm^{-1} and CCD pixels are around 15 to 20 μm in size. So the focal length in metres of the imaging element must be at least

$$f_2 = \frac{2 \times 10^{-5}}{W_\lambda} \frac{d\lambda}{d\theta}. \tag{4.2.11}$$

The diameter of the imaging element, D_2, must be sufficient to contain the whole of the exit beam. Thus for a square cross section exit beam

$$D_2 = \sqrt{2}D. \tag{4.2.12}$$

The diameter of the collimator, D_1, must be similar to that of the imaging element in general if the dispersing element is to be fully illuminated. Thus again

$$D_1 = \sqrt{2}D. \tag{4.2.13}$$

Now in order for the collimator to be fully illuminated in its turn, its focal ratio must equal the effective focal ratio of the telescope. Hence the focal length of the collimator, f_1, is given by

$$f_1 = \sqrt{2}Df'. \tag{4.2.14}$$

Finally from equation (4.1.37) we have the slit width

$$s = \frac{f_1 \lambda}{D} \tag{4.2.15}$$

and a first approximation has been obtained to the design of the spectroscope.

These various stages in the process of specifying a spectroscope are summarised for convenience in table 4.2.1.

The low light levels involved in astronomy usually require the focal ratio of the imaging elements to be small, so that it is fast in photographic terms. Satisfying this requirement usually means a compromise in some other part of the design, so that an optimally designed system is rarely achievable in practice. The slit may also need to be wider than specified in order to use a reasonable fraction of the star's light.

The limiting magnitude of a telescope–spectroscope combination is the magnitude of the faintest star for which a useful spectrum may be obtained. This is a very imprecise quantity, for it depends upon the type of spectrum and the purpose for which it is required, as well as the properties of the instrument. For example, if strong emission lines in a spectrum are the features of interest, then fainter stars may be studied than if weak absorption lines are desired. Similarly spectra of sufficient quality to determine radial velocities may be obtained for fainter stars than if line profiles are wanted. A guide, however, to the limiting magnitude may be gained through the use of Bowen's formula

$$m = 12 + 2.5 \log\left(\frac{sD_1 T_D gqt (d\lambda/d\theta)}{f_1 f_2 \alpha H}\right) \tag{4.2.16}$$

Table 4.2.1. Stages in the specification of a spectroscope.

Stage	Equation	Notes
1	f', W_λ, λ	Initial requirements
2	$R = \lambda / W_\lambda$	
3a	$L = 10^{-5} R$ (m)	Prism—approximate value for 500 nm wavelength. Use equation (4.1.43) for more accurate determinations.
3b	$L = 6.7 \times 10^{-7} R$ (m)	Grating—used in the third order, 500 grooves mm^{-1}, wavelength 500 nm. Use equation (4.1.67) for more accurate determinations.
4a	$D = 0.5L$ (m)	Prism—reasonable approximation.
4b	$D = 0.9L$ (m)	Grating—details as for 3b, equal angles of incidence and reflection. More precise determinations must take account of the exact angles and the blaze angle.
5	$\dfrac{\mathrm{d}\theta}{\mathrm{d}\lambda} = \dfrac{R}{D}$ (rad m^{-1})	
6	$f_2 = \dfrac{2 \times 10^{-5}}{W_\lambda} \dfrac{\mathrm{d}\lambda}{\mathrm{d}\theta}$ (m)	For a typical photographic emulsion or CCD.
7	$D_2 = \sqrt{2}D$ (m)	Exit beam of square cross section.
8	$D_1 = D_2$	Only necessarily true if the incident and exit angles are equal. This is often not the case for a grating.
9	$f_1 = D_1 f'$	
10	$s = f_1 \lambda / D$	
11	Return to stage 1	Repeated iterations through the design process are usually required in order to obtain precise and optimised values for the parameters.

where m is the faintest B magnitude which will give a usable spectrum in t seconds of exposure, T_D is the telescope objective's diameter and g is the optical efficiency of the system, i.e. the ratio of the usable light at the focus of the spectroscope to that incident upon the telescope. Typically it has a value of 0.2. Note, however, that g does not include the effect of the curtailment of the image by the slit. q is the quantum efficiency of the detector. Typical values are 0.4 to 0.8 for CCDs (section 1.1) and 0.002 to 0.005 for a long exposure on photographic emulsion (section 2.2). α is the angular size of the stellar image at the telescope's focus, typically 5×10^{-6} to 2×10^{-5} radians. H is the height of the spectrum. This formula gives quite good approximations for spectroscopes in which the star's image is larger than the slit, and it is trailed along the length of the slit to broaden the spectrum. This is probably the commonest mode of use for astronomical spectroscopes. Other situations such as an untrailed image, or an image smaller than the slit, require the formula to be modified. Thus when the slit is wide enough for the whole stellar image to pass through it, the exposure varies inversely with the square of the telescope's diameter, while for extended sources, it varies inversely with the square of the telescope's focal ratio (cf equations (1.1.68) and (1.1.71) etc).

The slit is quite an important part of the spectroscope since in most astronomical spectroscopes it fulfils two functions. Firstly it acts as the entrance aperture of the spectroscope. For this purpose, its sides must be accurately parallel to each other and perpendicular to the direction of the dispersion. It is also usual for the slit width to be adjustable so that different detectors may be used and/or changing observing conditions allowed for. Although we have seen how to calculate the optimum slit width, it is usually better in practice to find the best slit width empirically so that local peculiarities in the processing or changes in the emulsion from one batch to another are taken into account. The slit width is optimised by taking a series of exposures of a sharp emission line in the comparison spectrum through slits of different widths. As the slit width decreases, the image width should also decrease at first, but should eventually become constant. The changeover point occurs when some other part of the spectroscope system starts to limit the resolution, and the slit width at changeover is the required optimum value. As well as allowing the desired light through, the slit must reject unwanted radiation. The jaws of the slit are therefore usually of a knife-edge construction with the chamfering on the inside of the slit so that light is not scattered or reflected into the spectroscope from the edges of the jaws. The secondary purpose of the slit is to assist in the guiding of the telescope on the object. The stellar image is normally larger than the slit width, and so overlaps onto the slit jaws. By polishing the front of the jaws to an optically flat mirror finish, these overlaps can be observed via an inspection microscope, and the telescope driven and guided to keep the image bisected by the slit. For extended sources, especially emission nebulae, several parallel slits can be used as the entrance aperture of the spectroscope. Then, all the information which is derivable from a single-slit spectrogram is still available, but the whole source can be covered in a fraction of the time.

For some purposes the slit may be dispensed with. Some specific designs are considered later in this section. Apart from the Fourier transform spectroscope (section 4.1), they fall into two main categories. In the first, the projected image size on the spectrum is smaller than some other constraint on the system's resolution. The slit and the collimator may be discarded and parallel light from the source allowed to impinge directly onto the dispersing element. In the second type of slitless spectroscope, the source is producing a nebular type of spectrum (i.e. a spectrum consisting almost entirely of emission lines with little or no continuum). If the slit alone is then eliminated from the telescope–spectroscope combination, the whole of the image of the source passes into the spectroscope. The spectrum then consists of a series of monochromatic images of the source in the light of each of the emission lines. Slitless spectroscopes are difficult to calibrate so that radial velocities can be found from their spectra, but they may be very much more optically efficient than a slit spectroscope. In the latter, perhaps 1 to 10% of the incident light is eventually used in the image, but some types of slitless spectroscope can use as much as 75% of the light. Furthermore some designs, such as the objective prism, can photograph as many as 10^5 stellar spectra in one exposure.

Spectroscopes, as we have seen, contain many optical elements which may be separated by large distances and arranged at large angles to each other. In order for the spectroscope to perform as expected, the relative positions of these various components must be stable to within very tight limits. The two major problems in achieving such stability arise through flexure and thermal expansion. Flexure primarily affects the smaller spectroscopes which are attached to telescopes at, for example, Newtonian or Cassegrain foci, and so move around with the telescope. Their changing attitudes as the telescope moves cause the stresses within them to alter, so that if in correct adjustment for one position, they will be out of adjustment in other positions. Early spectroscopes were simply attached at a focus with the collimator's optical axis coinciding with that of the telescope. They thus tended to be long, thin instruments which were supported only at their attachment to the telescope, and so they suffered very badly from flexure. More recently it has become fairly common practice to fold the light beam from the telescope so that it is perpendicular to the telescope's optical axis. The spectroscope is then laid out in a plane which is parallel to the back of the main mirror. Hence the spectroscope components can be rigidly mounted onto a stout metal plate which in turn is bolted flat onto the back of the telescope. Such a design can be made very rigid, and the flexure reduced to acceptable levels. In some modern spectroscopes, active supports are used to compensate for flexure along the lines of those used for telescope mirrors (section 1.1). Temperature changes affect all spectroscopes, but the relatively short light paths in the small instruments which are attached directly to telescopes mean that generally the effects are unimportant. If this is not the case, then a simple heating jacket and thermostat will usually suffice to eliminate the problem. However, this can then introduce problems of its own by causing convection and turbulence close to the light paths in the telescope. Thus ideally the thermostat should merely stabilise the temperature of the spectroscope a degree or two above the ambient temperature. On nights with rapidly changing ambient temperatures this may then require frequent resetting of the thermostat with consequent readjustment of the spectroscope settings. Of more importance in these small spectroscopes is the change in the dispersion of a prism with temperature. The prism's temperature needs to be constant to about a tenth of a degree if its spectral resolution is not to be degraded. Thus they will normally need a separate and more critical control system than that of the instrument as a whole. Gratings are far less affected in this way, especially if they are formed upon a low expansion base such as Pyrex or Cervit. The large fixed spectroscopes which operate at Coudé foci are obviously unaffected by changing flexure, and usually there is little difficulty other than that of cost in making them as rigid as desired. They experience much greater problems, however, from thermal expansion. The size of the spectrographs may be very large indeed (see exercise 4.2.1 for example) with optical path lengths measured in tens of metres. Thus the temperature control must be correspondingly strict. A major problem is that the thermal inertia of the system may be so large that it may be impossible to stabilise the spectroscope at ambient temperature before the night has ended! Thus many Coudé spectrographs are in sealed rooms with the whole room temperature-controlled to a constant temperature over the entire year. The light then has to be piped in through an optically flat window.

Any spectroscope except the Michelson interferometer can be used as a monochromator. That is, a device to observe the object in a very restricted range of wavelengths. Most scanning spectroscopes are in effect monochromators whose waveband may be varied. The most important use of the devices in astronomy, however, is in the spectrohelioscope. This builds up a picture of the Sun in the light of a single wavelength, and this is usually chosen to be coincident with a strong absorption line. Further details are given in section 5.3. A related instrument for visual use is called a prominence spectroscope. This has the spectroscope off-set from the telescope's optical axis so that the (quite wide) entrance slit covers the solar limb. A small direct-vision (see p. 342) prism then produces an image in Hα light, allowing prominences and other solar features to be discerned.

Spectroscopy is undertaken throughout the entire spectrum. In the infrared and ultraviolet regions, techniques, designs, etc are almost identical to those for visual work except that different materials may need to be used. Some indication of these has already been given in section 4.1. Generally diffraction gratings and reflection optics are preferred since there is then no worry over absorption in the optical components. The technique of Fourier transform spectroscopy, as previously mentioned, has so far had its main applications in the infrared region. At very short ultraviolet, and at x-ray wavelengths, glancing optics (section 1.3) and diffraction gratings can be used to produce spectra using the same designs as for visual spectroscopes, the appearance and layout, however, will look very different even though the optical principles are the same, because of the off-axis optical elements.

Radio spectroscopes operate quite differently and have been described in section 1.2. A radio telescope detects radiation only over a very small wavelength range. Thus all that is needed to make a radio spectrograph is to scan the detected wavelength over the required spectral range. Such scanning may be accomplished in several ways. The simplest is to alter the local oscillator of the heterodyne receiver, and the received frequency may then be scanned over almost any desired waveband. This is a fairly unstable method however, so that it is often preferable to use a number of local oscillators operating at fixed frequencies, and so obtain the spectrum by discrete rather than continuous measurements. Usually radio spectroscopes are concerned with only a single line such as the 21 cm line of hydrogen, and so the discrete approach to radio spectroscopy does not require an impossible number of oscillators and receivers. The acousto-optical radio spectrograph (section 1.2) has a much wider bandwidth and is used for spectroscopy of solar radio bursts.

Prism-based spectroscopes

The basic layout of a prism-based spectroscope is shown in figure 4.1.5. Many instruments are constructed to this design with only slight modifications, the most important of which is the use of several prisms. If several identical prisms are used with the light passing through each along minimum deviation paths, then the total dispersion is just that of one of the prisms multiplied by the number of prisms. The resolution is unchanged and remains that for a single prism. Thus such an arrangement is of use when the resolution of the system is limited by some element of the spectroscope other than the prism. A rather more compact system than that shown in figure 4.1.5 can be made by replacing the 60° prism by one with a 30° apex angle which is aluminised on one surface (figure 4.2.1). The light therefore passes twice through the prism making its effect the equivalent of a single 60° prism. However, the minimum deviation path is no longer possible, so that some astigmatism is introduced into the image, but by careful design this can be kept lower than the resolution of the system as a whole.

Another similar arrangement, which is widely used for long-focus spectroscopes in laboratory and solar work, is called the Littrow spectroscope, or autocollimating spectroscope. A single lens, or occasionally a mirror, acts as both the collimator and imaging element (figure 4.2.2) thus saving on both the cost and size of the system.

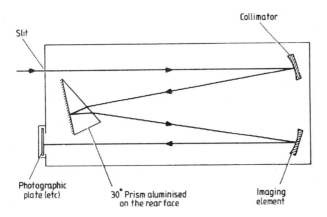

Figure 4.2.1. Compact design for the basic prism spectroscope.

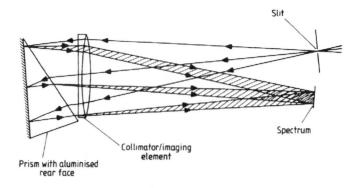

Figure 4.2.2. Light paths in a Littrow spectroscope.

The deviation of the optical axis caused by the prism can be a disadvantage for some purposes. Direct-vision spectroscopes overcome this problem and have zero deviation for some selected wavelength. There are several designs but most consist of combinations of prisms made in two different types of glass with the deviations arranged so that they cancel out while some remnant of the dispersion remains. This is the inverse of the achromatic lens discussed in section 1.1, and therefore usually uses crown and flint glasses for its prisms. The condition for zero deviation, assuming that the light passes through the prisms at minimum deviation is

$$\sin^{-1}\left(\mu_1 \sin \frac{\alpha_1}{2}\right) - \frac{\alpha_1}{2} = \sin^{-1}\left(\mu_2 \sin \frac{\alpha_2}{2}\right) - \frac{\alpha_2}{2} \qquad (4.2.17)$$

where α_1 is the apex angle of prism number 1, α_2 is the apex angle of prism number 2, μ_1 is the refractive index of prism number 1 and μ_2 is the refractive index of prism number 2. In practical designs the prisms are cemented together so that the light does not pass through all of them at minimum deviation.

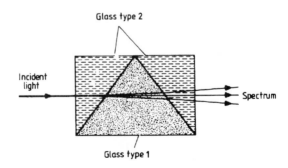

Figure 4.2.3. Treanor's direct-vision prism.

Nonetheless, equation (4.2.17) still gives the conditions for direct vision to a good degree of approximation. Applications of most direct-vision spectroscopes are non-astronomical since they are best suited to visual work. There is one ingenious astronomical application however, due to Treanor, which enables many approximate stellar radial velocities to be found rapidly. The method is based upon a direct-vision prism which is formed from one prism with an apex angle of α, and two prisms in a different glass with apex angles of $\alpha/2$. The combination forms a block of glass whose incident and exit faces are perpendicular to the light paths (figure 4.2.3), so that there is no displacement of the undeviated ray. The two glasses are chosen so that their refractive indices are identical for the desired undeviated wavelength, but their refractive index gradients against wavelength differ. One possible example which gives an undeviated ray at 530 nm combines crown glass and borosilicate glass:

λ (nm)	μ (crown glass)	μ (borosilicate glass)
400	1.530	1.533
530	1.522	1.522
600	1.519	1.517

The prism is placed without a slit in a collimated beam of light from a telescope. After passing through the prism, the spectrum is then focused to produce an image as usual. Since no slit is used all the stars in the field of view of the telescope are imaged as short spectra. If the collimated beam is slightly larger then the prism, then the light not intercepted by the prism will be focused to a stellar image, and this image will be superimposed upon the spectrum at a position corresponding to the undeviated wavelength. Thus many stellar spectra may be photographed simultaneously, each with a reference point from which to measure the wavelengths of lines in the spectrum so enabling the radial velocities to be found. The dispersion of the system is low, but the efficiency is high compared with a slit spectroscope. Higher dispersions can be achieved through the use of exotic materials such as organic liquids in place of the prism, but these introduce many of their own problems and so are rarely encountered in practice.

An alternative form of the direct-vision spectroscope exists which has applications in simple solar work. This is a combination of a prism and a transmission grating, and glories in the name 'grism'. The deviation of the prism is equal and opposite to that of one of the spectral orders of the grating, so that this order is undeviated as a whole. It can form part of a prominence spectroscope (see earlier discussion and section 5.3), but otherwise has little astronomical use since so little of the incident light is channeled into the observed spectrum.

The simplest spectroscope of all is the objective prism. This is simply a thin prism which is large enough to cover completely the telescope's objective, and it is positioned in use immediately before the objective. The starlight is

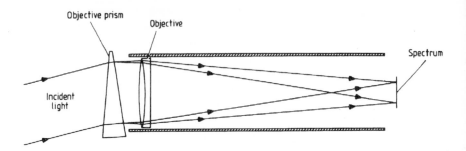

Figure 4.2.4. An objective prism spectroscope.

already parallel so that a collimator is unnecessary, while the slit is replaced by the scintillation disc of the star. The telescope acts as the imaging element (figure 4.2.4). Like Treanor's prism this has the enormous advantage that a spectrum is obtained for every star in the normal field of view. Thus if the telescope is a Schmidt camera, up to 10^5 spectra may be obtainable in a single exposure. The system has three main disadvantages. Firstly the dispersion is low, secondly the observed star field is at an angle to the telescope axis, and finally there is no reference point for wavelength measurements. Additionally, a simple prism introduces distortion into the image. The first of these problems is usually unimportant because the objective prism spectroscope is mostly used for survey work. The low dispersion is sufficient to enable interesting objects to be identified, which may then be studied in detail with other spectroscopes. The second problem can be overcome quite easily, at a cost of lowering the dispersion, by using two prisms of different glasses to form a low power direct-vision prism. The third problem is more serious, and much effort has been expended in efforts to provide reference points in the spectra. While results comparable with those from slit spectroscopes cannot be matched, some useful systems have been devised. The simplest method of all is to use the absorption lines arising from terrestrial atmospheric gases. Unfortunately, these are strongest in the red and infrared regions where the low dispersion of a prism renders the spectra of little real use. Absorption lines can be introduced artificially through the use of filters with narrow absorption bands. One substance which has had some success in this way is neodymium chloride. This has a narrow absorption feature near 427 nm; however, it is not ideal since the precise position of the feature is temperature-dependent, and it is asymmetrically shaped. Bright lines can be introduced into the spectrum through the use of a narrow-band interference filter (section 5.3). If the band pass dye filters are omitted from such a filter then it is transparent to a series of very narrow wavebands over the whole spectrum. Placing it in the light beam from an objective prism so that a small fraction of the light beam is intercepted leads to bright lines in the spectrum at these same wavebands. This system also has the disadvantage of the line positions being temperature dependent, and they also vary with the angle of inclination of the filter to the light beam. Furthermore the physical structure of the filter is much larger than its clear aperture so that a much larger fraction of the light beam is obstructed than is strictly necessary. With some photographic emulsions the long wavelength cut-off point may be sufficiently sharp and its position sufficiently consistent, that it may be used as the standard. The accuracy is fairly low with this method but it can be useful for very faint high velocity objects such as quasars and distant galaxies. A sharp cut-off point can be induced at the long or short wavelength end of the spectrum if the emulsion does not provide one, through the use of an appropriate filter. A quite different approach is to use two low angle 'grisms' oriented so that their first-order spectra are side by side and with the directions of the dispersions reversed. The separation of the same spectrum line in the two spectra can then simply be related to radial velocity provided that some stars with known radial velocities are available in the field of view to enable the photograph to be calibrated. In an improved variant of this system due to Fehrenbach, two direct-vision prisms are used which give better efficiency, and with careful design enable the field distortion caused by the use of a simple prism to be reduced. The same principle for the determination of radial velocities can still be applied even if only a single objective prism is available. Then two separate exposures are made on the same plate with the prism rotated through 180° between them in order to produce the two spectra. Yet another approach uses a direct image of the same field, and by assuming the stars to have zero radial velocities, predicts the positions of the Ca II H and K lines within the spectra on the objective prism image. The difference between the predicted and actual positions of the lines then gives the actual radial velocity. The measurement of objective prism spectra is still usually undertaken by hand, but it is now possible for some of the automatic plate measuring machines (section 5.1) to be used on the plates and, with an appropriate program for their computer, to produce large numbers of radial velocities very rapidly.

With the larger Schmidt cameras, a single objective prism becomes very large and heavy and can absorb a significant fraction of the light. A mosaic of identical co-aligned smaller prisms can then be used in place of the single larger one. The adjustment of such an array can pose problems, but with care these can be overcome, and several such systems have been used successfully.

Grating spectroscopes

With only two exceptions all the gratings used in astronomical spectroscopes are of the reflection type. This is because the light can be concentrated into the desired order by blazing quite easily, whereas for transmission gratings, blazing is much more difficult and costly. The two exceptions are the 'grism' referred to in the previous section, and the objective grating. The latter is mainly used to produce subsidiary images for photometry and astrometry and is discussed in sections 3.2 and 5.1.

Plane gratings are most commonly used in astronomical spectroscopes and are almost invariably incorporated into one or other of two designs discussed in the previous section, with the grating replacing the prism. These are the compact basic spectroscope (figure 4.2.1) which is sometimes called a Czerny–Turner system when it is based upon a grating, and the Littrow spectroscope (figure 4.2.2) which is called an Ebert spectroscope when based upon a grating and reflection optics.

Most of the designs of spectroscopes which use curved gratings are based upon the Rowland circle (figure 4.1.12). The Paschen–Runge mounting in fact is identical to that shown in figure 4.1.12. It is a common design for laboratory spectroscopes since wide spectral ranges can be accommodated, but its size and awkward shape make it less useful for astronomical purposes. A more compact design based upon the Rowland circle is called the Eagle spectroscope (figure 4.2.5). However, the vertical displacement of the slit and the spectrum (see the side view in figure 4.2.5) introduces some astigmatism. The Wadsworth design abandons the Rowland circle but still produces a stigmatic image through its use of a collimator (figure 4.2.6). The focal surface, however, becomes paraboloidal,

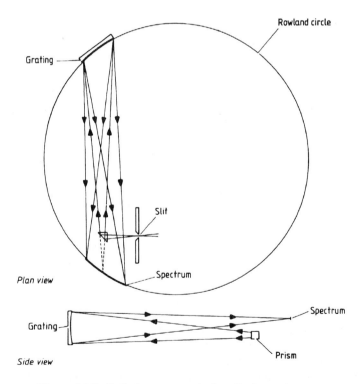

Figure 4.2.5. Optical arrangement of an Eagle spectroscope.

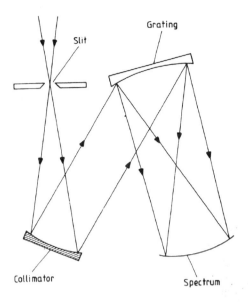

Figure 4.2.6. The Wadsworth spectroscope mounting.

and some spherical aberration and coma are introduced. Furthermore, the dispersion for a given grating, if it is mounted into a Wadsworth system, is only half of what it could be if the same grating were mounted into an Eagle system, since the spectrum is produced at the prime focus of the grating and not at its radius of curvature. With designs such as the Wadsworth and its variants, the imaging element is likely to be a Schmidt camera system (section 1.1) in order to obtain high quality images with a fast photographic system. Gratings can also be used as discussed earlier in various specialised applications such as nebular and prominence spectroscopes.

The major astronomical use of transmission gratings occurs at short wavelengths, and they then simply consist of opaque bars with totally clear spaces in between. There is no substrate since this would absorb the radiation. They can be allied to grazing incidence optical systems to produce low dispersion x-ray spectroscopes (section 1.3).

The design of a spectroscope is generally limited by the available size and quality of the grating, and these factors are of course governed by cost. The cost of a grating in turn is dependent upon its method of production. The best gratings are originals, which are produced in the following manner. A glass or other low expansion substrate is overcoated with a thin layer of aluminium. The grooves are then scored into the surface of the aluminium by lightly drawing a diamond across it. The diamond's point is precisely machined and shaped so that the required blaze is imparted to the rulings. An extremely high quality machine is required for controlling the diamond's movement, since not only must the grooves be straight and parallel, but their spacings must be uniform if Rowland and Lyman ghosts are to be avoided. The position of the diamond is therefore controlled by a precision screw thread, and is monitored interferometrically. We have seen that the resolution of a grating is dependent upon the number of grooves, while the dispersion is a function of the groove spacing (section 4.1). Thus ideally a grating should be as large as possible, and the grooves should be as close together as possible (at least until their separation is less than their operating wavelength). Unfortunately both of these parameters are limited by the wear on the diamond point. Thus it is possible to have large coarse gratings and small fine gratings, but not large fine gratings. The upper limits on size are about 0.5 m square, and on groove spacing, about 1500 lines mm^{-1}. A typical grating for an astronomical spectroscope might be 0.1 m across and have 500 lines mm^{-1}. Only rarely is it essential to have a grating produced by direct ruling. Most of the time a replica grating is adequate. Since many replicas can be produced from a single original their cost is only a small fraction of that of an original grating. Some loss of quality occurs during replication, but this is acceptable for most purposes. A replica improves on an original in one way however, and that is in its reflection efficiency. This is better than that of the original because the most highly burnished portions of the grooves are at the bottoms of the grooves on the original, but are transferred to the tops of the grooves on the replicas. The original has of course to be the inverse of the finally desired grating. The replicas are produced by covering the original with a thin coat of liquid plastic or epoxy resin, which is stripped off after it has set. The replica is then mounted onto a substrate, which may be appropriately curved if necessary, and then aluminised. More recently, large high-quality gratings have been produced holographically. An intense monochromatic laser beam is collimated and used to illuminate a photoresist covered surface. The reflections from the back of the blank interfere with the incoming radiation, and the photoresist is exposed along the nodal planes of the interference field. Etching away the surface then leaves a grating which can be aluminised and used directly, or it can have replicas formed from it as above. The wavelength of the illuminating radiation and its inclination can be altered to give almost any required groove spacing and blaze angle.

Techniques of spectroscopy

There are several problems and techniques which are peculiar to astronomical spectroscopy, and which are essential knowledge for the intending astrophysicist.

One of the greatest problems, and one which has been mentioned several times already, is that the image of the star may be broadened by atmospheric turbulence until its size is several times the slit width. Only a small percentage of the already pitifully small amount of light from a star therefore enters the spectroscope. The design of the spectroscope, and in particular the use of a large focal length of the collimator in comparison to that of the imaging element, can enable wider slits to be used. Even then, the slit is generally still too small to accept the whole stellar image, and the size of the spectroscope may have to become very large if reasonable resolutions and dispersions are to be obtained.

An alternative approach to the use of a large collimator focal length lies in the use of an image dissector. Several designs for these exist. The most simple concept is a bundle of optical fibres whose cross section matches the slit at one end and the star's image at the other. Thus apart from the reflection and absorption losses within the fibres, all the star's light is conducted into the spectroscope. Disadvantages of fibre optics are mainly the degradation of the focal ratio of the beam due to imperfections in the walls of the fibre so that not all the light is intercepted by the collimator and the multilayered structure of the normal commercially available units, which leads to other light losses because only the central core of fibre actually transmits the light. Thus specially designed fibre optic cables are usually required, and these are made 'in house' at the observatory needing them. They are usually much thicker than normal fibre optics and are formed from plastic or fused quartz. Even with these *ad hoc* modifications, the focal ratio may be degraded so that little of the potentially available gain is actually realised. A variant of the system may be used to obtain many simultaneous spectra of separate objects which are in the same field of view. The fibre optic strands are each made large enough to be able to contain the whole of an individual image. Each strand then has one of its ends positioned in the image plane of the telescope so that it intercepts one of the required images. The other ends are then aligned along the length of the spectroscope slit, so that up to 100 spectra may be photographed in a single exposure. One such system developed for use on the Anglo–Australian telescope, called Autofib, is reasonably typical. It has 64 optical fibres which feed the spectrograph and another 6 for guiding. All 70 fibres can be repositioned in under 10 minutes to an accuracy of 0.2″ using a computer-controlled electromagnetic positioning system. This approach has recently been developed further with the two-degree field (2dF) for the AAT. This uses correcting lenses to provide an unvignetted 2° field of view, over which 400 fibres can be positioned to feed a spectroscope. Schmidt cameras can also be used with advantage for multi-object spectroscopy because their wide fields of view provide more objects for observation, and their short focal ratios are well matched to transmission through optical fibres. Thus for example FLAIR (Fibre-Linked Array-Image Reformatter) on the UK Schmidt camera can obtain up to 140 spectra simultaneously over a 40 square-degree field of view.

An earlier type of image dissector is due to Bowen and is still in common use. It consists of a stack of overlapped mirrors (figure 4.2.7) which section the image and then rearrange the sections end to end to form a linear image suitable for matching to a spectroscope slit. The Bowen–Walraven image slicer uses multiple internal reflection. A prism with a chamfered side is used and has a thin plate attached to it. Because of the chamfered side, the plate only touches

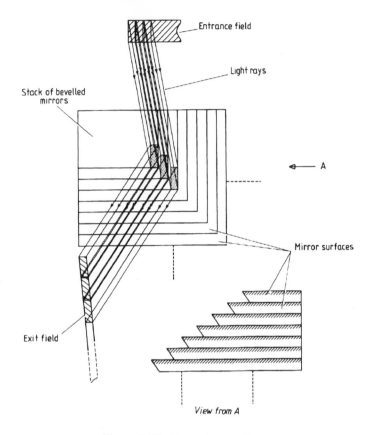

Figure 4.2.7. Bowen image slicer.

the prism along one edge. A light beam entering the plate is multiply internally reflected wherever the plate is not in contact with the prism, but is transmitted into the prism along the contact edge (figure 4.2.8).

Several other closely allied variants of these systems may be found from the literature.

Another problem in astronomical spectroscopy concerns the width of the spectrum. If the stellar image is held motionless at one point of the slit during the exposure, then the final spectrum may only be a few microns high. Not only is such a narrow spectrum very difficult to measure, but individual spectral features are recorded by only a few photographic grains and so the noise level is very high. Both of these difficulties can be overcome, although only at the expense of longer exposure times, by widening the spectrum artificially. There are several ways of doing this, such as introducing a cylindrical lens to make the images astigmatic, or by trailing the telescope during the exposure so that the image moves along the slit. Both these methods have drawbacks however,

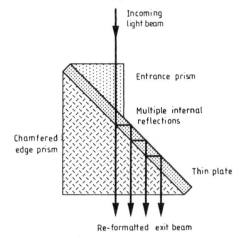

Figure 4.2.8. The Bowen–Walraven image slicer.

and an improved and generally preferred method is based upon the rocking of a plane piece of glass. The glass block's sides are accurately plane-parallel, and it is placed a short distance in front of the slit. If the block is at an angle to the optical axis of the telescope, then the image is displaced from its normal position but otherwise unaffected (figure 4.2.9). It is easy to arrange for this displacement to be along the line of the slit. Rocking the block backwards and forwards about an axis perpendicular to both the line of the slit and the telescope's optical axis then causes the star's image to track up and down the

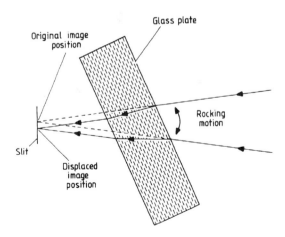

Figure 4.2.9. Image displacement by a plane-parallel glass plate.

slit. Providing that steep angles of incidence onto the block are avoided, there is little loss of light in the system. The length of the track of the star on the slit can be adjusted by varying the limits of the rocking motion, or by using blocks of differing thicknesses. Thus the final spectrum can be widened as desired, and in a uniform manner, provided that the exposure is much longer than the length of the period of the block's motion. Astronomical spectra are typically widened to about 0.1 to 1 mm.

Spectrographs operating in the near-infrared (1–5 μm) are of conventional designs, but need to be cooled to reduce the background noise level. Typically, an instrument such as the near-infrared camera and multi-object spectrometer (NICMOS) on board the Hubble space telescope has its main components cooled to 60 or 80 K, while the detectors are held at 4 K or less.

Imaging or three-dimensional spectroscopy is now becoming more widely used. This provides an image of an extended object, each point of which has its spectrum available (see also holography, p. 333). Some wide band, low dispersion techniques are available. A colour photograph or image, of course, is essentially a low resolution 3D spectrogram. A possible extension to existing colour imaging techniques which may have a high enough spectral resolution to be useful, is through the use of dye-doped polymers. These are thin films of polymers such as polyvinylbutyral, containing a dye such as chlorin which has a very narrow absorption band. Changing conditions within the substrate cause the wavelength of the absorption band to scan through a few nm, and potentially provide high efficiency direct imaging combined with spectral resolutions of perhaps 10^5 or more. Several instruments now in use employ a two-dimensional grid of short focal length micro-lenses to reduce the image from a telescope to an array of small dots. This, when fed into a suitably designed and oriented grism, will produce an array of non-overlapping short spectra. Each spectrum in the array having derived from a portion of the original image.

Most current three-dimensional spectroscopy, however, is of high spectral resolution and covers only a narrow bandwidth. A typical system would be based upon a Fabry–Perot etalon acting as a very narrow band filter. The etalon would be scanned in wavelength over an emission line from the object, while a series of images was obtained. The final data would thus be in the form of a 'cube' with the two-dimensional images arranged in wavelength order, and could be used, for example, to determine the velocity structure inside a gaseous nebula. Alternatively a series of long-slit spectra can be taken scanning across the image. Yet another approach, which has the merit of producing the image cube in a single observation, is to use an image slicer (figure 4.2.8) to feed a long-slit spectroscope. Provided that the sliced image is held in a fixed position on the entrance slit, each horizontal segment of the resulting spectrum corresponds to the spectrum of a single point in the original image.

Atmospheric dispersion is another difficulty which needs to be considered. Refraction in the Earth's atmosphere changes the observed position of a star from its true position (equation (5.1.17)). But the refractive index varies with wavelength. For example, at standard temperature and pressure, we have Cauchy's formula

$$\mu = 1.000287566 + \frac{1.3412 \times 10^{-18}}{\lambda^2} + \frac{3.777 \times 10^{-32}}{\lambda^4} \qquad (4.2.18)$$

so that the angle of refraction changes from one wavelength to another, and the star's image is drawn out into a very short vertical spectrum. In a normal basic telescope system (i.e. without the extra mirrors etc required, for example, by the Coudé system), the long wavelength end of the spectrum will be uppermost. To avoid spurious results, particularly when undertaking spectrophotometry, the atmospheric dispersion of the image must be arranged to lie along the slit, otherwise certain parts of the spectrum may be preferentially selected from the image. Alternatively, an atmospheric dispersion corrector (ADC) may be used. This is a low but variable dispersion direct vision spectroscope which is placed before the entrance slit of the main spectroscope and whose dispersion is equal and opposite to that of the atmosphere. The problem is usually only significant at large zenith distances, so that it is normal practice to limit spectroscopic observations to zenith angles of less than 45°. The increasing atmospheric absorption, and the tendency for telescope tracking to deteriorate at large zenith angles, also contribute to the wisdom of this practice.

In many cases it will be necessary to try and remove the degradation introduced into the observed spectrum because the spectroscope is not perfect. The many techniques and their ramifications for this process of deconvolution are discussed in section 2.1.

In order to determine radial velocities, it is necessary to be able to measure the actual wavelengths of the lines in the spectrum, and to compare these with their laboratory wavelengths. The difference, $\Delta\lambda$, then provides the radial velocity via the Doppler shift formula

$$v = \frac{\Delta\lambda}{\lambda}c = \frac{\lambda_{\text{Observed}} - \lambda_{\text{Laboratory}}}{\lambda_{\text{Laboratory}}}c \qquad (4.2.19)$$

where c is the velocity of light. The observed wavelengths of spectrum lines are most normally determined by comparison with the positions of emission lines in an artificial spectrum. This comparison spectrum is usually that of an iron or copper arc or comes from a low pressure gas emission lamp such as sodium or neon. The light from the comparison source is fed into the spectroscope and appears as two spectra on either side of the main spectrum (figure 4.2.10). The emission lines in the comparison spectra are at their rest wavelengths and these are known very precisely. A precision travelling microscope (for photographic spectra), or appropriate software (for CCD spectra) are used to measure the positions of both the comparison and the stellar (or other object) lines, and the observed positions of the latter found by interpolation. At badly light-polluted sites, the emission lines from terrestrial sources can appear in astronomical spectra and can then also be used as reference points.

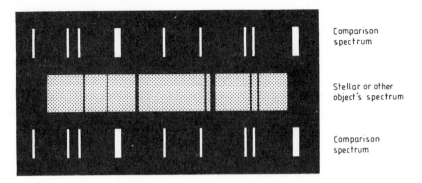

Figure 4.2.10. The wavelength comparison spectrum.

For very high precision determinations of radial velocity the cross-correlation spectroscope, originally devised by Griffin, is often used. This places a mask over the image of the spectrum, and re-images the radiation passing through the mask onto a point source detector. The mask is a negative version (either a photograph or artificially generated) of the spectrum. The mask is moved across the spectrum and when the two coincide exactly, there is a very sharp drop in the output from the detector. The position of the mask at the correlation point can then be used to determine the radial velocity with a precision, in the best cases, of a few metres per second. More conventional spectroscopes can provide velocities to a precision of a few metres per second for bright stars through the use of an absorption cell containing HF or iodine. This is placed in the light path, and superimposes artificial absorption lines with very precisely known wavelengths onto the stellar spectrum. With this system there is no error due to differing light paths as there may be with a conventional comparison spectrum.

A comparison is also needed for photographic and sometimes other types of spectra in order to reproduce the original intensities because the detector response is non-linear (see figure 2.2.5). The photometric calibration most usually consists of a second spectrum (or third if a wavelength comparison spectrum is also used) on the same photographic plate as the main spectrum. The calibration spectrum is of a number of artificial sources of known and differing brightnesses. Scanning these enables the characteristic curve of the photographic emulsion, or the response curve of another non-linear detector, to be found and so the photographic density to be converted back to the intensity of the original image.

The guiding of a telescope–spectroscope combination has already been mentioned. For bright objects the reflected image from slit jaws polished to a mirror finish can be viewed through an optical inspection system. Guiding becomes more difficult for faint objects or when an image dissector is in use. Either an accurate off-set guiding system on some nearby brighter star must be used, or some other system devised. One method that may be encountered quite frequently is based upon the exposure meter. Most major telescopes have such a device because unlike most other astronomical work, spectroscopy can often be continued through thin cloud, or an exposure can be restarted after the object has been briefly obscured. Thus an integrating exposure meter is an almost essential component of a spectroscope. If a read-out for the instantaneous signal to this is available, then guiding on faint objects may be simply accomplished by peaking the signal.

Future developments

The major foreseeable developments in spectroscopy seem likely to lie in the direction of improving the efficiency of existing systems rather than in any radically new systems or methods. The lack of efficiency of a typical spectroscope arises from two main sources. One is the loss of light within the overall system. To improve on this requires gratings of greater efficiency, reduced scattering and surface reflection, etc from the optical components, and so on. Since these factors are already quite good, and it is mostly the total number of components in the telescope–spectroscope combination which reduces the efficiency, improvements are likely to be slow and gradual. The other main cause of the lack of efficiency lies in the quantum efficiency of the detector. The IPCS and similar systems are already providing improvements of up to a factor of a hundred over the photographic spectrograph, and their application seems likely to spread rapidly, with cost as the main constraint. Charge-coupled devices (section 1.1) are now widely used in spectroscopy and are similarly of high quantum efficiency, but at longer wavelengths. Such systems are already of 50% quantum efficiency so that scope for further improvement is fairly limited. Thus overall the prospects for astronomical spectroscopy over the next decade seem to be more of the same with only marginal improvements over the best of the current systems being possible.

Exercises

4.2.1 Design a spectroscope (i.e. calculate its parameters) for use at a Coudé focus where the focal ratio is $f25$. A resolution of 10^{-3} nm is required over the spectral range 500 to 750 nm. A grating with 500 lines mm^{-1} is available, and the final recording of the spectrum is to be by photography.

4.2.2 Calculate the limiting B magnitude for the system designed in problem (4.2.1) when it is used on a 4 m telescope. The final spectrum is widened to 0.2 mm, and the longest practicable exposure is 8 hours.

4.2.3 Calculate the apex angle of the dense flint prism required to form a direct-vision spectroscope in combination with two 40° crown glass prisms, if the undeviated wavelength is to be 550 nm. (See section 1.1 for some of the data.)

Chapter 5

Other Techniques

5.1 ASTROMETRY

Background

Coordinate systems

The measurement of a star's position in the sky must be with respect to some coordinate system. There are several such systems in use by astronomers, but the commonest is that of right ascension and declination. This system, along with most of the others, is based upon the concept of the celestial sphere; which is a hypothetical sphere, centred upon the Earth and enclosing all objects observed by astronomers. The space position of an object is related to a position on the celestial sphere by a radial projection from the centre of the Earth. Henceforth in this section we talk about the position of an object as its position on the celestial sphere, and we ignore the differing radial distances that may be involved. We also extend the polar axis, the equatorial and orbital planes of the Earth until these too meet the celestial sphere (figure 5.1.1). These intersections are called the celestial north pole, the celestial equator, etc but usually there is no ambiguity if the 'celestial' qualification is omitted, so that they are normally referred to as the north pole, the equator and so on. The intersections (or nodes) of the ecliptic and the equator are the vernal and autumnal equinoxes. The former is also known as the first point of Aries from its position in the sky some two thousand years ago. The ecliptic is also the apparent path of the Sun across the sky during a year, and the vernal equinox is defined as the node at which the Sun passes from the southern to the northern hemisphere. This passage occurs within a day of 21 March each year. The position of a star or other object is thus given with respect to these reference points and planes.

The declination of an object, δ, is the angular distance north or south of the equator. The right ascension, α, is the angular distance around from the meridian (or great circle) which passes through the vernal equinox and the poles, measured in the same direction as the solar motion (figure 5.1.2). By

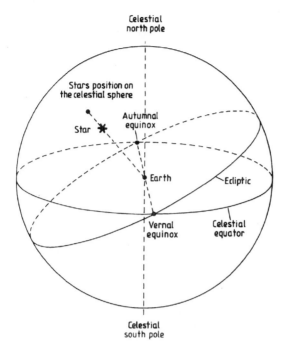

Figure 5.1.1. The celestial sphere.

convention, declination is measured from $-90°$ to $+90°$ in degrees, minutes and seconds of arc, and is positive to the north of the equator and negative to the south. Right ascension is measured from $0°$ to $360°$, in units of hours, minutes and seconds of time where

$$1 \text{ hour} = 15° \tag{5.1.1}$$
$$1 \text{ minute} = 15' \tag{5.1.2}$$
$$1 \text{ second} = 15''. \tag{5.1.3}$$

The Earth's axis moves in space with a period of about 25 750 years, a phenomenon known as precession. Hence the celestial equator and poles also move. The positions of the stars therefore slowly change with time. Catalogues of stars customarily give the date, or epoch, for which the stellar positions that they list are valid. To obtain the position at some other date, the effects of precession must be added to the catalogue positions

$$\delta_T = \delta_E + \left(\theta \sin \epsilon \cos \alpha_E\right) T \tag{5.1.4}$$
$$\alpha_T = \alpha_E + \left[\theta\left(\cos \epsilon + \sin \epsilon \sin \alpha_E \tan \delta_E\right)\right] T \tag{5.1.5}$$

where α_T and δ_T are the right ascension and declination of the object at an interval T after the epoch E, α_E, δ_E are the coordinates at the epoch and θ is

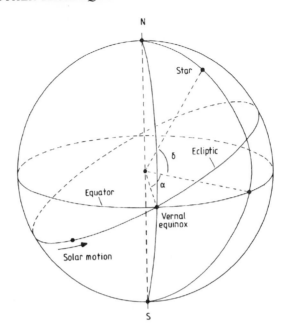

Figure 5.1.2. Right ascension and declination.

the precession constant

$$\theta = 50.40'' \text{ yr}^{-1}. \tag{5.1.6}$$

ε is the angle between the equator and the ecliptic—more commonly known as the obliquity of the ecliptic

$$\epsilon = 23°27'8''. \tag{5.1.7}$$

Commonly used epochs are the beginnings of the years 1900, 1950, 2000, etc. Other effects upon the position, such as nutation, proper motion, etc may also need to be taken into account in determining an up-to-date position for an object.

Alternative coordinate systems which are in use are celestial latitude and longitude, which are respectively the angular distance up or down from the ecliptic and the angular distance around the ecliptic from the vernal equinox, and galactic latitude and longitude, which are respectively the angular distance above or below the galactic plane, and the angular distance around the plane of the galaxy measured from the direction to the centre of the galaxy.

Position angle and separation

The separation of a visual double star is just the angular distance between its components. The position angle is the angle from the north, measured in the sense, north → east → south → west → north, from 0° to 360° (figure 5.1.3)

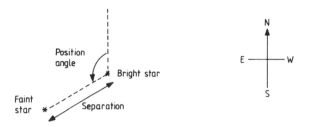

Figure 5.1.3. Position angle and separation of a visual double star (as seen on the sky directly).

of the fainter star with respect to the brighter star. The separation and position angle are related to the coordinates of the star by

$$\text{separation} = \left\{ \left[\left(\alpha_F - \alpha_B \right) \cos \delta_B \right]^2 + \left(\delta_F - \delta_B \right)^2 \right\}^{1/2} \qquad (5.1.8)$$

$$\text{position angle} = \tan^{-1} \left(\frac{\delta_F - \delta_B}{\left(\alpha_F - \alpha_B \right) \cos \delta_B} \right) \qquad (5.1.9)$$

where α_B and δ_B are the right ascension and declination of the brighter star and α_F and δ_F are the right ascension and declination of the fainter star with the right ascensions of both stars being converted to normal angular measures.

Micrometers

Micrometers are used in astrometry for two quite separate methods of observation. The first is in a transit telescope in order to obtain the absolute positions of stars, while the second is to determine the relative position of one star with respect to another.

Transit telescopes (figure 5.1.4), which are also sometimes called meridian circles, provide the basic measurements of stellar positions for the calibration of other astrometric methods. The principle of the instrument is simple, but great care is required in practice if it is to produce reliable results. The instrument is almost always a refractor, because of the greater stability of the optics, and it is pivoted only about a horizontal east–west axis. The telescope is thereby constrained to look only at points on the prime meridian. A star is observed as it crosses the centre of the field of view (i.e. when it transits the prime meridian), and the precise time of the passage is noted. The declination of the star is obtainable from the altitude of the telescope, while the right ascension is given by the local sidereal time at the observatory at the instant of transit,

$$\delta = A + \varphi - 90° \qquad (5.1.10)$$

$$\alpha = \text{LST} \qquad (5.1.11)$$

where A is the altitude of the telescope, φ is the latitude of the observatory and LST is the local sidereal time at the instant of transit. In order to achieve an

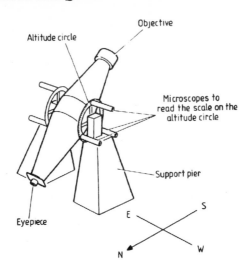

Figure 5.1.4. A transit telescope.

accuracy of a tenth of a second of arc in the absolute position of the star, a very large number of precautions and corrections are needed. A few of the more important ones are: temperature control; use of multiple or driven cross wires in the micrometer eyepiece; reversal of the telescope on its bearings; corrections for flexure, non parallel objective and eyepiece focal planes, rotational axis not precisely horizontal, rotational axis not precisely east–west, errors in the setting circles, incorrect position of the micrometer cross wire, personal setting errors, and so on. Then of course all the normal corrections for refraction, aberration, etc have also to be added.

Two more modern instruments which perform the same function as the transit telescope are the photographic zenith tube and the astrolabe. Both of these use a bath of mercury in order to determine the zenith position very precisely. The photographic zenith tube obtains photographs of stars which transit close to the zenith and provides an accurate determination of the time and latitude of an observatory, but is restricted in its observations of stars to those which culminate within a few tens of minutes of arc of the zenith. The astrolabe, however, observes near an altitude of 60° and so can give precise positions for a far wider range of objects. In its most developed form (figure 5.1.5) it is known as the Danjon or impersonal astrolabe since its measurements are independent of focusing errors. Two separate beams of light are fed into the objective by the 60° prism, one of the beams having been reflected from a bath of mercury. A Wollaston prism (section 5.2) is used to produce two focused beams parallel to the optic axis with the other two non-parallel emergent beams being blocked off. Two images of the star are then visible in the eyepiece. The Wollaston prism is moved along the optical axis to compensate for the sidereal motion of the

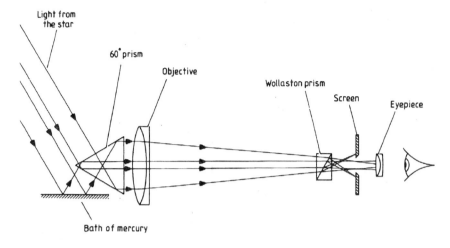

Figure 5.1.5. Optical paths in the impersonal astrolabe.

star. The measurement of the star's position is accomplished by moving the 60° prism along the optical axis until the two images merge into one. The position of the prism at a given time for this coincidence to occur can then be converted into the position of the star in the sky.

While the addition of one or more cross wires to an eyepiece can convert it into a micrometer when it is used in conjunction with a clock, it is the bifilar micrometer that is more normally brought to mind by the term micrometer. Although for most purposes the bifilar micrometer has been superseded by CCDs and photography, it is still the best method for measuring close visual double stars. It also finds application as a means of off-setting from a bright object to a faint or invisible one. Designs for the device vary in their details, but the most usual model comprises a fixed cross wire at the focus of an eyepiece, with a third thread parallel to one of the fixed ones. The third thread is displaced slightly from the others, although still within the eyepiece's field of sharp focus, and it may be moved perpendicularly to its length by a precision screw thread. The screw has a calibrated scale so that the position of the third thread can be determined to within one micron. The whole assembly is mounted on a rotating turret whose angular position may also be measured (figure 5.1.6).

To measure a double star with such a micrometer, the following procedure should be followed.

(*a*) Align one of the fixed cross wires in an east–west direction. This is accomplished by turning off the telescope drive and rotating the turret until the wire is parallel to the drifting motion of the star. Note the reading of the position angle scale.

(*b*) Centre the cross wires on the brighter star and rotate the turret until the

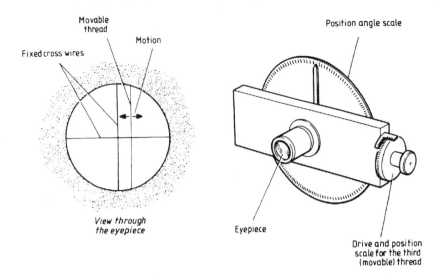

Figure 5.1.6. Bifilar micrometer.

cross wire which was aligned east–west now passes through both stars. Again note the reading of the position angle scale.

(c) Calculate the position angle (figure 5.1.3). The precise relationship between the two readings just obtained and the position angle will depend upon the optical path of the telescope. In cases of doubt a little experimentation with the telescope's motions will indicate the orientation of the field of view.

(d) Calibrate the micrometer screw reading. This is achieved by aligning the movable cross wire in a north–south direction (cf step *(a)*) and moving it until it is superimposed upon the fixed wire and noting the reading. It is then moved to, say, half way to the edge of the field of view, and the screw reading again noted. With the telescope drive turned off, the sidereal drift time, t, for a star to cross between the two parallel wires is measured. The angular distance between the parallel wires is then given by

$$\theta = 15t \cos \delta \qquad (5.1.12)$$

where θ is in seconds of arc and is the angular separation of the two parallel wires, t is in sidereal seconds of time and δ is the declination of the star. Alternatively if t is measured in solar seconds of time, we have

$$\theta = 15.04t \cos \delta. \qquad (5.1.13)$$

Dividing this angle by the difference between the two screw scale readings then gives the angular distance in seconds of arc which corresponds to one unit of the screw scale.

(e) The separation is then simply measured by setting the cross wires onto one star, aligning the movable wire so that it is perpendicular to the line joining the stars, and moving it until it bisects the other star's image. The calibration of the screw scale then allows the readings to be converted directly into the star's angular separation.

If the micrometer is used to determine the absolute position of one star from the known position of another, then the same procedure is used, and the unknown star's position, providing that the separation is small, is given by

$$\alpha = \alpha_0 + r \sin \theta \ \sec \delta_0 \qquad\qquad (5.1.14)$$

$$\delta = \delta_0 + r \cos \theta \qquad\qquad (5.1.15)$$

where α and δ are the required right ascension and declination, α_0 and δ_0 are the right ascension and declination of the comparison star, θ is the position angle and r is the separation. The practical use of the micrometer requires only a few additional steps. The most important of these is to make multiple observations at all stages of the procedure, either by the use of repeated observations or by using several parallel co-moving wires or both. An improved method of measuring the separation is to proceed as indicated through to stage *(e)*. The second screw reading, however, is then obtained by resetting the cross wires onto the second star and moving the movable wire until it bisects the image of the first star. The difference in the readings is then twice the separation. The advantage of this procedure is that if there is any systematic difference between the setting error of the cross wire and that of the movable wire, it will tend to cancel out. The cross wires are frequently formed from spiders' webs, and so must be illuminated in order to be visible. This may either be by direct illumination, when the wires are visible as bright lines against a dark field of view, or by illumination of the field of view, when they are visible as silhouettes. Other features which may be added include adjustments to centre the cross wires on the rotational axis of the turret, the ability to move or change the eyepiece, devices to reduce the contrast between stars of differing magnitudes, filters and filter holders, etc.

Another type of micrometer which has the advantage of being far less affected by scintillation is the double-image micrometer. Two images of each star are produced, and the device is adjusted so that one image of one star is superimposed upon one image of the other. The settings of the device can be calibrated, and then the separation and position angle follow directly. Commonly used methods of splitting the images are an objective cut into two along a diameter whose two halves may be tilted with respect to each other, and a Nicol prism (section 5.2). The advantage of systems of this type derives from the instantaneous determination of both separation and position angle, compared with the delay of several seconds or even minutes for obtaining the various readings from a bifilar micrometer, and from the reduced effect of scintillation since both images are likely to be affected by it in a similar manner.

Photography

The majority of astrometric work is undertaken from CCD images or photographs taken on long focus refractors. The position of an unknown star is obtained by comparison with the images of reference stars which also appear on the plate and whose positions are known. If absolute positions are to be determined, then the reference stars' positions must have been found via a transit instrument or astrolabe, etc. For relative positional work, such as that involved in the determination of parallax and proper motion, any very distant star may be used as a comparison star.

Refractors are favoured for this work because of their greater stability when compared with reflectors, because of their closed tubes, because their optics may be left undisturbed for many years and because flexure is generally less deleterious in them. A few specially designed reflectors are used successfully for astrometry, however. Against the advantages of the refractor their disadvantages must be set. These are primarily their residual chromatic aberration and coma, temperature effects, and slow changes in the optics such as the rotation of the elements of the objective in their mounting cells. Refractors in use for astrometry usually have focal ratios of $f15$ to $f20$, and range in diameter from about 0.3 to 1.0 metres. Their objectives are achromatic (section 1.1), and a combination of the CCD's or photographic emulsion's spectral sensitivity (sections 1.1 and 2.2) and filters is usually used so that the bandwidth is limited to a few tens of nanometres either side of their optimally corrected wavelength. Coma may need to be limited by stopping down the aperture, by introducing correcting lenses near the focal plane, or by compensating the measurements later, in the analysis stages, if the magnitudes and spectral types of the stars are well known. The instrument is guided during an exposure either by a second refractor mounted alongside it or by using an off-axis system to view stars through the main instrument. The guide telescope is often almost as large as the main one in order that the image scales are comparable and so accurate guiding is possible. Sometimes just the detector holder may be moved rather than the whole telescope since this will generally allow a finer control of the corrections to be exercised. Observations are usually obtained within a few degrees of the meridian in order to minimise the effects of atmospheric refraction and dispersion and to reduce changes in flexure. The position of the focus may change with temperature and needs to be checked frequently throughout the night. Other corrections for such things as plate tilt collimation errors, astigmatism, distortion, emulsion movement, etc need to be known and their effects allowed for when reducing the raw data.

It is advantageous to have the star of interest and its reference stars of similar brightnesses. This may be effected by the use of variable density filters, by the use of a rotating chopper in front of the detector, or by the use of an objective grating (section 4.2). The latter device is arranged so that pairs of images, which in fact are the first-order or higher order spectra of the star, appear on either side of it. By adjusting the spacing of the grating the brightness of these secondary images for the brighter stars may be arranged to be comparable with the brightness of the primary images of the fainter stars. The position of the brighter star is then taken as the average of the two positions of its secondary images. The apparent magnitude, m, of such a secondary image resulting from the nth order spectrum is given by

$$m = m_0 + 5\log(N\eta\mathrm{cosec}\,\eta) \qquad (5.1.16)$$

where m_0 is the apparent magnitude of the star, η is given by

$$\eta = \pi nd/D \qquad (5.1.17)$$

where D is the separation of the grating bars, d is the width of the gap between the grating bars and N is the total number of slits across the objective.

Photography or CCD imaging may usefully be used to observe double stars when the separation is greater than about three seconds of arc. Many exposures are made with a slight shift of the telescope between each exposure. Averaging the measurements can give the separation to better than a hundredth of a second of arc and the position angle to within a few minutes of arc. For double stars with large magnitude differences between their components, an objective grating can be used as already mentioned, or the shape of the aperture can be changed so that the fainter star lies between two diffraction spikes and has an improved noise level. The latter technique is a variation of the technique of apodisation mentioned in section 2.5.

Other techniques may be useful in particular cases, especially for very close double stars. Interferometry, speckle interferometry and lunar occultation are the main such techniques and are considered in more detail in sections 2.5, 2.6 and 2.7 respectively. Reflectors and Schmidt cameras may be used when the positions of very faint or very large numbers of stars are needed. The relatively recent development of automatic measuring and processing machines has revolutionised statistical astrometry, which is concerned with the motions of large numbers of stars.

The next major development in the subject will be satellite-borne astrographic telescopes. The first of these, called Hipparcos (High-Precision Parallax Collecting Satellite), was launched by ESA in 1990. Unfortunately the apogee boost motor failed and instead of going into a geostationary orbit, it eventually ended up in a $10\frac{1}{2}$ hour orbit with a perigee of 6800 km and apogee of 42 000 km. However, by using more ground stations and extending the mission duration to 3 years, the original aim of measuring the positions of 120 000 stars to an accuracy of 0.002$''$ was achieved. The Hipparcos telescope was fed by two flat mirrors at an angle of 29° to each other and determined the relative phase difference between two stars viewed via the two mirrors, in order to obtain their angular separation. The satellite rotated once every two hours and measured all the stars down to 9.0^{m} about eighty times. Each measurement was from a slightly different angle and the final positional catalogue was obtained by the processing of some 10^{12} bits of such information. A proposal for the future is for a continuously rotating spacecraft bearing several interferometers and capable of measuring positions to an accuracy of $5 \times 10^{-6}\,''$.

Electronic images

Many if not most images are now obtained via electronic detectors, usually CCDs (section 1.1). These devices have the enormous advantage for astrometry that the stellar or other images can be identified with individual pixels whose positions are automatically defined by the structure of the detector. The measurement of the relative positions of objects on the image is therefore vastly simplified. Software to fit a suitable point spread function (section 2.1) to the image enables the position of its centroid to be defined to a fraction of the size of the individual pixels.

The disadvantage of electronic images is their small scale. Most CCD chips are only a few centimetres in size compared with tens of centimetres for astrographic photographic plates. They therefore cover only a small region of the sky and it can become difficult to find suitable comparison stars for the measurements. The greater dynamic range of CCDs compared with photographic emulsion and the use of anti-blooming techniques (section 1.1), however, enable the positions of many more stars to be usefully measured. In other respects, the processing and reduction of electronic images for astrometric purposes is the same as that for a photographic image.

Measurement and reduction

Once an image of a star field has been obtained, the positions of the stars' images must be measured to a high degree of accuracy and the measurements converted into right ascension and declination or separation and position angle. The measuring is undertaken in a manner similar to that used to measure spectra, except that two-dimensional positions are required. Thus for photographs the specimen table can be moved in two axes and its position determined with an accuracy of about one micron in both. The measurement procedure follows that outlined for spectra except that the plate may be inverted as well as reversed so that setting errors tend to cancel out. Errors in the positions obtained from a single plate are generally in the region of $0.02''$ to $0.03''$ at best. The measuring procedure is a tedious process and, as in the case of spectroscopic measurement, a number of automatic measuring machines have been developed over the last decade. In some ways the problem of automating the measurement of these plates is easier than in the spectroscopic case since almost all the images are the same shape and are isolated from each other, hence completely automatic measuring machines have been in use for several years. The machines are usually computer controlled and also use the computer to undertake some or all of the data reduction. Their designs vary quite widely, but most have the photograph mounted on a table which moves slowly in one axis, while it is scanned rapidly in the other. The scanning spot can be from a laser or from a television system or a variety of other sources. The position of an image is determined by fitting a one- or two-dimensional Gaussian profile to the output. Currently the state of the art is represented by the UK's Automatic Plate Measuring Machine (APM) which has a linear accuracy of better than half a micron and a photometric accuracy better than 0.001D (0.2%). As well as measuring position, some of the machines also measure brightness and can sort the images into types of object (stars, galaxies, asteroid trails, etc). Currently the machines are limited by the speed of their computers. At the time of writing, the fastest of the measuring machines can completely process about two large Schmidt plates per day—a possible total of one million images needing to be measured in that time. CCDs give positions more directly (see above).

The process of converting the measurements into the required information is known as reduction. For accurate astrometry it is a lengthy process. A number of corrections have already been mentioned, and in addition to those, there are the errors of the measuring machine itself to be corrected, and the distortion introduced by the photograph to be taken into account. The latter is caused by the projection of the curved sky onto the flat photographic plate, or in the case of Schmidt cameras its projection onto the curved focal plane of the camera, followed by the flattening of the plate after it is taken out of the plate holder. For the simpler case of the astrograph, the projection is shown in figure 5.1.7, and the relationship between the coordinates on the flat plate and in the curved sky is given by

$$\rho = \tan^{-1}(y/F) \qquad (5.1.18)$$

$$\tau = \tan^{-1}(x/F) \qquad (5.1.19)$$

where x and y are the coordinates on the plate, τ and ρ are the equivalent coordinates on the image of the celestial sphere and F is the focal length of the telescope.

One instrument, which, while it is not strictly of itself a part of astrometry, is very frequently used to identify the objects which are to be measured, is called the blink comparator. In this machine the worker looks alternately at two images of the same star field obtained some time apart from each other, the frequency of the interchange being about two hertz. Stars, asteroids or comets, etc which have moved in the interval between the photographs then call attention to themselves by appearing to jump backwards and forwards while the remainder of the stars are stationary. The device can also pick out variable stars for further study by photometry (Chapter 3) since although these remain stationary, they appear to blink on and off (hence the name of the instrument).

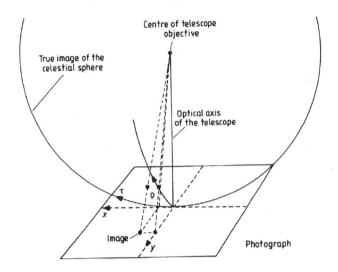

Figure 5.1.7. Projection onto a photographic plate.

Exercises

5.1.1   If the axis of a transit telescope is accurately horizontal, but is displaced north or south of the east–west line by E seconds of arc, then show that for stars on the equator

$$\Delta\alpha^2 + \Delta\delta^2 \simeq E^2 \sin^2 \varphi$$

where $\Delta\alpha$ and $\Delta\delta$ are the errors in the determinations of right ascension and declination in seconds of arc and φ is the latitude of the observatory.

5.1.2   Calculate the widths of the slits required in an objective grating for use on a one metre telescope if the third-order image of Sirius A is to be of the same brightness as the primary image of Sirius B. The total number of slits in the grating is 99. The apparent magnitude of Sirius A is -1.5, while that of Sirius B is $+8.5$.

5.2   POLARIMETRY

Background

Although the discovery of polarised light from astronomical sources goes back to the beginning of the last century, when Arago detected its presence in moonlight, the extensive development of its study is a relatively recent phenomenon. This is largely due to the technical difficulties which are involved, and to the lack of any expectation of polarised light from stars by astronomers. Many phenomena,

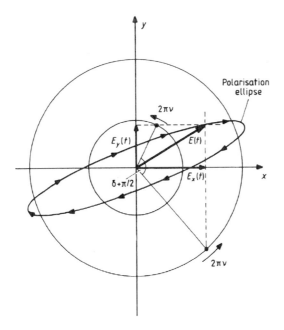

Figure 5.2.1. The x and y components of the elliptically polarised component of partially elliptically polarised light.

however, can contribute to the polarisation of radiation, and so, conversely, its observation can potentially provide information upon an equally wide range of basic causes.

Stokes' parameters

Polarisation of radiation is simply the non-random angular distribution of the electric vectors of the photons in a beam of radiation. Customarily two types are distinguished—linear and circular polarisations. In the former the electric vectors are all parallel and their direction is constant, while in the latter, the angle of the electric vector rotates with time at the frequency of the radiation. These are not really different types of phenomena, however, and all types of radiation may be considered to be different aspects of partially elliptically polarised radiation. This has two components, one of which is unpolarised, the other being elliptically polarised. Elliptically polarised light is similar to circularly polarised light in that the electric vector rotates at the frequency of the radiation, but in addition the magnitude varies at twice that frequency, so that plotted on a polar diagram the electric vector would trace out an ellipse (see figure 5.2.1 for example). The properties of partially elliptically polarised light are completely described by four parameters which are called the Stokes' parameters. These fix the intensity of the unpolarised light, the degree of ellipticity, the direction of the major axis of the ellipse, and the sense (left- or right-handed rotation) of the elliptically polarised light. If the radiation is imagined to be propagating along the z axis of a three-dimensional rectangular coordinate system, then the elliptically polarised component at a point along the z axis may have its electric vector resolved into components along the x and y axes (figure 5.2.1), these being given by

$$E_x(t) = e_1 \cos(2\pi v t) \tag{5.2.1}$$

$$E_y(t) = e_2 \cos(2\pi v t + \delta) \tag{5.2.2}$$

where v is the frequency of the radiation, δ is the phase difference between the x and y components and e_1, and e_2 are the amplitudes of the x and y components. It is then a tedious but straightforward matter to show that

$$a = \left(\frac{e_1^2 + e_2^2}{1 + \tan^2[\frac{1}{2} \sin^{-1}\{[2e_1e_2/(e_1^2 + e_2^2)] \sin \delta\}]} \right)^{1/2} \tag{5.2.3}$$

$$b = a \tan[\frac{1}{3} \sin^{-1}\{[2e_1e_2/(e_1^2 + e_2^2)] \sin \delta\}] \tag{5.2.4}$$

$$a^2 + b^2 = e_1^2 + e_2^2 \tag{5.2.5}$$

$$\psi = \frac{1}{2} \tan^{-1}\{[2e_1e_2/(e_1^2 - e_2^2)] \cos \delta\} \tag{5.2.6}$$

where a and b are the semi-major and semi-minor axes of the polarisation ellipse and ψ is the angle between the x axis and the major axis of the polarisation ellipse. The Stokes' parameters are then defined by

$$Q = e_1^2 - e_2^2 = \frac{a^2 - b^2}{a^2 + b^2} \cos(2\psi) I_p \tag{5.2.7}$$

$$U = 2e_1e_2 \cos \delta = \frac{a^2 - b^2}{a^2 + b^2} \sin(2\psi) I_p \tag{5.2.8}$$

$$V = 2e_1e_2 \sin \delta = \frac{2ab}{a^2 + b^2} I_p \tag{5.2.9}$$

where I_p is the intensity of the polarised component of the light. From equations (5.2.7), (5.2.8) and (5.2.9) we have

$$I_p = \left(Q^2 + U^2 + V^2 \right)^{1/2} \tag{5.2.10}$$

The fourth Stokes' parameter, I, is just the total intensity of the partially polarised light

$$I = I_u + I_p \tag{5.2.11}$$

where I_u is the intensity of the unpolarised component of the radiation. (*Note:* The notation and definitions of the Stokes' parameters can vary. While that given here is probably the commonest usage, a check should always be carried out in individual cases to ensure that a different usage is not being employed.)

The degree of polarisation, π, of the radiation is given by

$$\pi = \frac{(Q^2 + U^2 + V^2)^{1/2}}{I} = \frac{I_\mathrm{p}}{I} \qquad (5.2.12)$$

while the degree of linear polarisation, π_L, and the degree of ellipticity, π_e, are

$$\pi_\mathrm{L} = \frac{(Q^2 + U^2)^{1/2}}{I} \qquad (5.2.13)$$

$$\pi_\mathrm{e} = V/I. \qquad (5.2.14)$$

When $V = 0$ (i.e. the phase difference, δ, is 0 or π radians) we have linearly polarised radiation. The degree of polarisation is then equal to the degree of linear polarisation, and is the quantity which is commonly determined experimentally

$$\pi = \pi_\mathrm{L} = \frac{I_\mathrm{max} - I_\mathrm{min}}{I_\mathrm{max} + I_\mathrm{min}} \qquad (5.2.15)$$

where I_max and I_min are the maximum and minimum intensities which are observed through a polariser as it is rotated. The value of π_e is positive for right-handed, and negative for left-handed radiation.

When several incoherent beams of radiation are mixed, their various Stokes' parameters combine individually by simple addition. A given monochromatic partially elliptically polarised beam of radiation may therefore have been formed in many different ways, and these are indistinguishable by the intensity and polarisation measurements alone. It is often customary therefore to regard partially elliptically polarised light as formed from two separate components, one of which is unpolarised and the other completely elliptically polarised. The Stokes' parameters of these components are then:

	I	Q	U	V
Unpolarised component	I_u	0	0	0
Elliptically polarised component	I_p	Q	U	V

and the normalised Stokes' parameters for more specific mixtures are:

Type of radiation	Stokes' parameters			
	I/I	Q/I	U/I	V/I
Right-hand circularly polarised (clockwise)	1	0	0	1
Left-hand circularly polarised (anticlockwise)	1	0	0	−1
Linearly polarised at an angle ψ to the x axis	1	$\cos 2\psi$	$\sin 2\psi$	0

The Stokes' parameters are related to more familiar astronomical quantities by

$$\theta = \tfrac{1}{2}\tan^{-1}(U/Q) = \psi \tag{5.2.16}$$

$$e = \left\{ 1 - \tan^2 \left[\frac{1}{2}\sin^{-1}\left(\frac{V}{(Q^2 + U^2 + V^2)^{1/2}} \right) \right] \right\}^{1/2} \tag{5.2.17}$$

where θ is the position angle (section 5.1) of the semi-major axis of the polarisation ellipse (when the x axis is aligned north–south) and e is the eccentricity of the polarisation ellipse, and is 1 for linearly polarised radiation and 0 for circularly polarised radiation.

Optical components for polarimetry

Polarimeters usually contain a number of components which are optically active in the sense that they alter the state of polarisation of the radiation. They may be grouped under three headings: polarisers, converters and depolarisers. The first produces linearly polarised light, the second converts elliptically polarised light into linearly polarised light, or vice versa, while the last eliminates polarisation. Most of the devices rely upon birefringence for their effects, so we must initially discuss some of its properties before looking at the devices themselves.

Birefringence

The difference between a birefringent material and a more normal optical one may best be understood in terms of the behaviour of the Huygens' wavelets. In a normal material the refracted ray can be constructed from the incident ray by taking the envelope of the Huygens' wavelets, which spread out from the incident surface with a uniform velocity. In a birefringent material, however, the velocities of the wavelets depend upon the polarisation of the radiation, and for at least one component the velocity will also depend on the orientation of the ray with respect to the structure of the material. In some materials it is possible to find a direction for linearly polarised radiation for which the wavelets expand spherically from the point of incidence as in 'normal' optical materials. This ray is then termed the ordinary ray and its behaviour may be described by normal geometrical optics. The ray which is polarised orthogonally to the ordinary ray is then termed the extraordinary ray, and wavelets from its point of incidence spread out elliptically (figure 5.2.2). The velocity of the extraordinary ray thus varies with direction. We may construct the two refracted rays in a birefringent material by taking the envelopes of their wavelets as before (figure 5.2.3). The direction along which the velocities of the ordinary and extraordinary rays are equal is called the optic axis of the material. When the velocity of the extraordinary ray is in general larger than that of the ordinary ray (as illustrated in figures 5.2.2 and 5.2.3), then the birefingence is negative. It is positive when

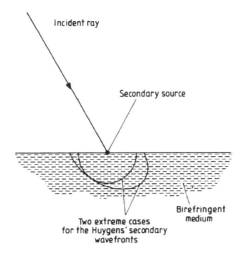

Incident ray

Secondary source

Two extreme cases
for the Huygens' secondary
wavefronts

Birefringent
medium

Figure 5.2.2. Huygens' secondary wavelets in a birefringent medium.

the situation is reversed. The degree of the birefringence may be obtained from the principal extraordinary refractive index, μ_E. This is the refractive index corresponding to the maximum velocity of the extraordinary ray for negative materials, and the minimum velocity for positive materials. It will be obtained from rays travelling perpendicularly to the optic axis of the material. The degree of birefringence, which is often denoted by the symbol J, is then simply the difference between the principal extraordinary refractive index and the refractive index for the ordinary ray, μ_0

$$J = \mu_E - \mu_0 \qquad (5.2.18)$$

Most crystals exhibit natural birefringence, and it can be introduced into many more and into amorphous substances such as glass by the presence of strain in the material. One of the most commonly encountered birefringent materials is calcite. The cleavage fragments form rhombohedrons and the optic axis then joins opposite blunt corners if all the edges are of equal length (figure 5.2.4). The refractive index of the ordinary ray is 1.658, while the principal extraordinary refractive index is 1.486, giving calcite the very high degree of birefringence of -0.172.

Crystals such as calcite are uniaxial and have only one optic axis. Uniaxial crystals belong to the tetragonal or hexagonal crystallographic systems. Crystals which belong to the cubic system are not usually birefringent, while crystals in the remaining systems—orthorhombic, monoclinic, or triclinic—generally produce biaxial crystals. In the latter case, normal geometrical optics breaks down completely and all rays are extraordinary.

Some crystals such as quartz which are birefringent ($J = +0.009$), are in

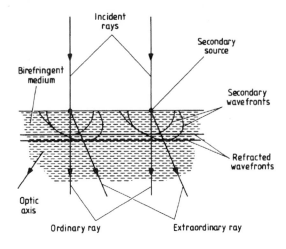

Figure 5.2.3. Formation of ordinary and extraordinary rays in a birefringent medium.

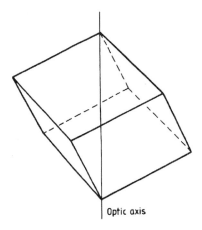

Figure 5.2.4. An equilateral calcite cleavage rhomb and its optic axis.

addition optically active. This is the property whereby the plane of polarisation of a beam of radiation is rotated as it passes through the material. Many substances other than crystals, including most solutions of organic chemicals, can be optically active. Looking down a beam of light, against the motion of the photons, a substance is called dextro-rotatory or right handed if the rotation of the plane of vibration is clockwise. The other case is called laevo-rotatory or left handed. (Unfortunately and confusingly the opposite convention is also in occasional use.)

The description of the behaviour of an optically active substance in terms of

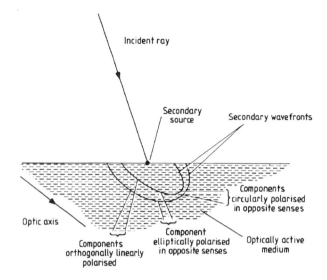

Figure 5.2.5. Huygens' secondary wavelets in an optically active medium.

the Huygens' wavelets is rather more difficult than was the case for birefringence. Incident light is split into two components as previously, but the velocities of both components vary with angle, and in no direction do they become equal. The optic axis therefore has to be taken as the direction along which the difference in velocities is minimised. Additionally the nature of the components changes with angle as well. Along the optic axis they are circularly polarised in opposite senses, while perpendicular to the optic axis they are orthogonally linearly polarised. Between these two extremes the two components are elliptically polarised to varying degrees and in opposite senses (figure 5.2.5).

Polarisers

These are devices which only allow the passage of light which is linearly polarised in some specified direction. There are several varieties which are based upon birefringence, of which the Nicol prism is the best known. This consists of two calcite prisms cemented together by Canada balsam. Since the refractive index of Canada balsam is 1.55, it is possible for the extraordinary ray to be transmitted, while the ordinary ray is totally internally reflected (figure 5.2.6). The Nicol polariser has the drawbacks of displacing the light beam and of introducing some elliptical polarisation into the emergent beam. Its inclined faces also introduce additional light losses by reflection. Various other designs of polarisers have therefore been developed, some of which in fact produce mutually orthogonally polarised beams with an angular separation. Examples of several such designs are shown in figure 5.2.7.

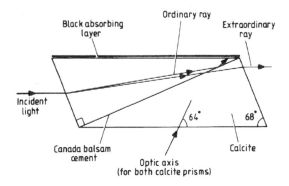

Figure 5.2.6. The Nicol prism.

The ubiquitous polarising sunglasses are based upon another type of polariser. They employ dichroic crystals which have nearly 100% absorption for one plane of polarisation and less than 100% for the other. Generally the dichroism varies with wavelength so that these polarisers are not achromatic. Usually, however, they are sufficiently uniform in their spectral behaviour to be usable over quite wide wavebands. The polarisers are produced commercially in sheet form and contain many small aligned crystals rather than one large one. The two commonly used compounds are polyvinyl alcohol impregnated with iodine and polyvinyl alcohol catalysed to polyvinylene by hydrogen chloride. The alignment is achieved by stretching the film. The use of microscopic crystals and the existence of the large commercial market means that dichroic polarisers are far cheaper than birefringent polarisers, and so they may be used even when their performance is poorer than that of the birefringent polarisers.

Polarisation by reflection can be used to produce a polariser. A glass plate inclined at the Brewster angle will reflect a totally polarised beam. However, only a small percentage (about 7.5% for crown glass) of the incident energy is reflected. Reflection from the second surface will reinforce the first reflection, however, and then several plates may be stacked together to provide further reflections (figure 5.2.8). The transmitted beam is only partially polarised. However, as the number of plates is increased, the total intensity of the reflected beams (ignoring absorption) will approach half of the incident intensity (figure 5.2.9). Hence the transmitted beam will approach complete polarisation. In practice therefore, the reflection polariser is used in transmission, since the problem of recombining the multiple reflected beams is thereby avoided and the beam suffers no angular deviation.

Outside the optical, near-infrared and near-ultraviolet regions, the polarisers tend to be somewhat different in nature. In the radio region, a linear dipole is naturally only sensitive to radiation polarised along its length. Helical antennae (section 1.2) can likewise be used for the detection of elliptically polarised

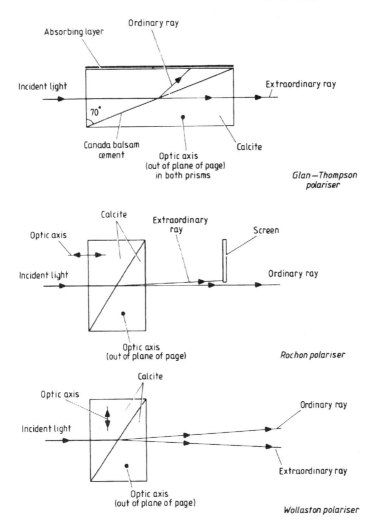

Figure 5.2.7. Examples of birefringence polarisers.

radiation. In the microwave region and the medium to far infrared, wire grid polarisers are suitable. These, as their name suggests, are grids of electrically conducting wires. Their spacing is about five times their thickness, and they transmit the component of the radiation which is polarised along the length of the wires. They work efficiently for wavelengths longer than their spacing. In the x-ray region, Bragg reflection is polarised (section 1.3) and so a rotating Bragg spectrometer can also act as a linear polarisation detector.

The behaviour of a polariser may be described mathematically by its effect upon the Stokes' parameters of the radiation. This is most easily accomplished

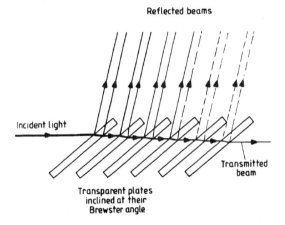

Figure 5.2.8. A reflection polariser.

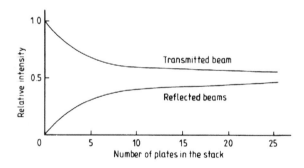

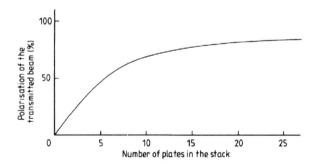

Figure 5.2.9. Properties of reflection polarisers, assuming negligible absorption and a refractive index of 1.5.

by writing the parameters as a column vector. The effect of the polariser, and also of the other optical components which we discuss later in this section, may then be represented by a matrix multiplying the vector on the left. The technique is sometimes given the name of the Mueller calculus. There is also an alternative, and to some extent complementary formulation, termed the Jones calculus. The reader is referred to items in the bibliography for further details of the latter system. In the Mueller calculus the effect of passing the beam through several optical components is simply found by successive matrix multiplications of the Stokes' vector. The first optical component's matrix is closest to the original Stokes' vector, and the matrices of subsequent components are positioned successively further to the left as the beam passes through the optical system. For a perfect polariser whose transmission axis is at an angle, θ, to the reference direction, we have

$$
\begin{bmatrix} I' \\ Q' \\ U' \\ V' \end{bmatrix} = \begin{bmatrix} \frac{1}{2} & \frac{1}{2}\cos 2\theta & \frac{1}{2}\sin 2\theta & 0 \\ \frac{1}{2}\cos 2\theta & \frac{1}{2}\cos^2 2\theta & \frac{1}{2}\cos 2\theta \sin 2\theta & 0 \\ \frac{1}{2}\sin 2\theta & \frac{1}{2}\cos 2\theta \sin^2 2\theta & \frac{1}{2}\sin^2 2\theta & 0 \\ 0 & 0 & 0 & 0 \end{bmatrix} \begin{bmatrix} I \\ Q \\ U \\ V \end{bmatrix} \tag{5.2.19}
$$

where the primed Stokes' parameters are for the beam after passage through the polariser and the unprimed ones are for it before such passage.

Converters

These are devices which alter the type of polarisation and/or its orientation. They are also known as retarders or phase plates. We have seen earlier (equations (5.2.1) and (5.2.2) and figure 5.2.1) that elliptically polarised light may be resolved into two orthogonal linear components with a phase difference. Altering that phase difference will alter the degree of ellipticity (equations 5.2.3 and 5.2.4) and inclination (equation 5.2.6) of the ellipse. We have also seen that the velocities of mutually orthogonal linearly polarised beams of radiation will in general differ from each other when the beams pass through a birefringent material. From inspection of figure 5.2.6 it will be seen that if the optic axis is rotated until it is perpendicular to the incident radiation, then the ordinary and extraordinary rays will travel in the same direction. Thus they will pass together through a layer of a birefringent material which is oriented in this way, and will recombine upon emergence, but with an altered phase delay, due to their differing velocities. The phase delay, δ', is given to a first-order approximation by

$$
\delta' = \frac{2\pi d}{\lambda} J \tag{5.2.20}
$$

where d is the thickness of the material and J is the birefringence of the material. Now let us define the x axis of figure 5.2.1 to be the polarisation direction of the extraordinary ray. We then have from equation (5.2.6) the intrinsic phase difference, δ, between the components of the incident radiation

$$\delta = \cos^{-1}\left(\frac{e_1^2 - e_2^2}{2e_1 e_2}\tan 2\psi\right). \tag{5.2.21}$$

The ellipse for the emergent radiation then has a minor axis given by

$$b' = a'\tan\left[\tfrac{1}{2}\sin^{-1}\left\{\left[2e_1 e_2/(e_1^2 + e_2^2)\right]\sin(\delta + \delta')\right\}\right] \tag{5.2.22}$$

where the primed quantities are for the emergent beam. So

$$b' = 0 \qquad \text{for} \qquad \delta + \delta' = 0 \tag{5.2.23}$$

$$b' = a' \qquad \text{for} \qquad \delta + \delta' = \sin^{-1}\left[(e_1^2 + e_2^2)/2e_1 e_2\right] \tag{5.2.24}$$

and also

$$\psi' = \tfrac{1}{2}\tan^{-1}\left[2e_1 e_2/(e_1^2 - e_2^2)\cos(\delta + \delta')\right]. \tag{5.2.25}$$

Thus

$$\psi' = -\psi \qquad \text{for} \qquad \delta' = 180° \tag{5.2.26}$$

and

$$a' = a \qquad \text{and} \qquad b' = b. \tag{5.2.27}$$

Thus we see that, in general, elliptically polarised radiation may have its degree of ellipticity altered and its inclination changed by passage through a converter. In particular cases it may be converted into linearly polarised or circularly polarised radiation, or its orientation may be reflected about the fast axis of the converter.

In real devices, the value of δ' is chosen to be 90° or 180° and the resulting converters are called quarter-wave plates or half-wave plates respectively, since one beam is delayed with respect to the other by a quarter or a half of a wavelength. The quarter-wave plate is then used to convert elliptically or circularly polarised light into linearly polarised light or vice versa, while the half-wave plate is used to rotate the plane of linearly polarised light.

Many substances can be used to make converters, but mica is probably the commonest because of the ease with which it may be split along its cleavage planes. Plates of the right thickness (about 40 μm) are therefore simple to obtain. Quartz cut parallel to its optic axis can also be used, while for ultraviolet work magnesium fluoride is suitable. Amorphous substances may be stretched or compressed to introduce stress birefringence. It is then possible to change the phase delay in the material by changing the amount of stress. Extremely high acceptance angles and chopping rates can be achieved with low power consumption if the material is driven by a small acoustic transducer at one of its natural frequencies. This is then commonly called a photoelastic modulator. An electric field can also induce birefringence. Along the direction of the field, the phenomenon is called the Pockels effect, while perpendicular to the field

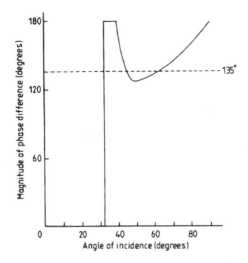

Figure 5.2.10. Differential phase delay upon internal reflection in a medium with a refractive index of 1.6.

it is called the Kerr effect. Suitable materials abound and may be used to produce quarter- or half-wave plates. Generally these devices are used when rapid switching of the state of birefringence is needed, as for example in Babcock's solar magnetometer (section 5.3). In glass the effects take up to a minute to appear and disappear, but in other substances the delay can be far shorter. The Kerr effect in nitrobenzene, for example, permits switching at up to one megahertz, and the Pockels effect can be used similarly with ammonium dihydrogen phosphate or potassium dihydrogen phosphate.

All the above converters will normally be chromatic and usable over only a restricted wavelength range. Converters which are more nearly achromatic can be produced which are based upon the phase changes which occur in total internal reflection. The phase difference between components with electric vectors parallel and perpendicular to the plane of incidence is shown in figure 5.2.10. For two of the angles of incidence (approximately 45° and 60° in the example shown) the phase delay is 135°. Two such reflections therefore produce a total delay of 270°, or, as one may also view the situation, an advance of the second component with respect to the first by 90° . The minimum value of the phase difference shown in figure 5.2.10 is equal to 135° when the refractive index is 1.497. There is then only one suitable angle of incidence and this is 51°47′. For optimum results, the optical design should approach as close to this ideal as possible. A quarter-wave retarder which is nearly achromatic can thus be formed using two total internal reflections at appropriate angles. The precise angles usually have to be determined by trial and error since additional phase changes are produced when the beam interacts with the entrance and exit faces

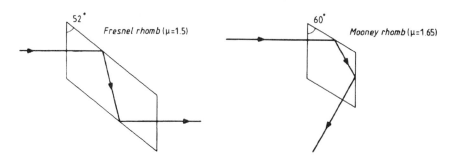

Figure 5.2.11. Quarter-wave retarders using total internal reflection.

of the component. Two practical designs—the Fresnel rhomb and the Mooney rhomb—are illustrated in figure 5.2.11. Half-wave retarders can be formed by using two such quarter-wave retarders in succession.

Pseudo quarter- and half-wave plates that are usable over several hundred nanometres in the visual can be made by combining three retarders. The Pancharatnam design employs two retarders with the same delay and orientation, together with a third sandwiched between the first two. The inner retarder is rotated with respect to the other two, and possibly has a differing delay. A composite 'half-wave plate' with actual delays in the range $180° \pm 2°$ over wavelengths from 300 to 1200 nm can, for example, be formed from three $180°$ retarders with the centre one oriented at about $60°$ to the two outer ones. Quartz and magnesium fluoride are commonly used materials for the production of such super-achromatic waveplates.

The Mueller matrices for converters are

$$\begin{bmatrix} 1 & 0 & 0 & 0 \\ 0 & \cos^2 2\psi & \cos 2\psi \sin 2\psi & -\sin 2\psi \\ 0 & \cos 2\psi \sin 2\psi & \sin^2 2\psi & \cos 2\psi \\ 0 & \sin 2\psi & -\cos 2\psi & 0 \end{bmatrix} \tag{5.2.28}$$

for a quarter-wave plate, and

$$\begin{bmatrix} 1 & 0 & 0 & 0 \\ 0 & \cos^2 2\psi - \sin^2 2\psi & 2\cos 2\psi \sin 2\psi & 0 \\ 0 & 2\cos 2\psi \sin 2\psi & \sin^2 2\psi - \cos^2 2\psi & 0 \\ 0 & 0 & 0 & -1 \end{bmatrix} \tag{5.2.29}$$

for a half-wave plate, while in the general case we have

$$\begin{bmatrix} 1 & 0 & 0 & 0 \\ 0 & \cos^2 2\psi + \sin^2 2\psi \cos \delta & (1 - \cos \delta)\cos 2\psi \sin 2\psi & -\sin 2\psi \sin \delta \\ 0 & (1 - \cos \delta)\cos 2\psi \sin 2\psi & \sin^2 2\psi + \cos^2 2\psi \cos \delta & \cos 2\psi \sin \delta \\ 0 & \sin 2\psi \sin \delta & -\cos 2\psi \cos \delta & \cos \delta \end{bmatrix} . \tag{5.2.30}$$

Depolarisers

The ideal depolariser would accept any form of polarised radiation and produce unpolarised radiation. No such device exists, but pseudo-depolarisers can be made. These convert the polarised radiation into radiation which is unpolarised when it is averaged over wavelength, time or area.

A monochromatic depolariser can be formed from a rotating quarter-wave plate which is in line with a half-wave plate rotating at twice its rate. The emerging beam at any given instant will have some form of elliptical polarisation, but this will change rapidly with time, and the output will average to zero polarisation over several rotations of the plates.

The Lyot depolariser averages over wavelength. It consists of two retarders with phase differences very much greater than 360°. The second plate has twice the thickness of the first and its optic axis is rotated by 45° with respect to that of the first. The emergent beam will be polarised at any given wavelength, but the polarisation will vary very rapidly with wavelength. In the optical region, averaging over a waveband a few tens of nanometres wide is then sufficient to reduce the net polarisation to one per cent of its initial value.

If a retarder is left with a rough surface and immersed in a liquid whose refractive index is the average refractive index of the retarder, then a beam of light will be undisturbed by the roughness of the surface because the hollows will be filled in by the liquid. However, the retarder will vary, on the scale of its roughness, in its effect. Thus the polarisation of the emerging beam will vary on the same scale, and a suitable choice for the parameters of the system can lead to an average polarisation of zero over the whole beam.

The Mueller matrix of an ideal depolariser is simply

$$\begin{bmatrix} 1 & 0 & 0 & 0 \\ 0 & 0 & 0 & 0 \\ 0 & 0 & 0 & 0 \\ 0 & 0 & 0 & 0 \end{bmatrix}. \tag{5.2.31}$$

Polarimeters

A polarimeter is an instrument which measures the state of polarisation, or some aspect of the state of polarisation, of a beam of radiation. Ideally the values of all four Stokes' parameters should be determinable, together with their variations with time, space and wavelength. In practice this is rarely possible, at least for astronomical sources. Most of the time only the degree of linear polarisation and its direction are found. Astronomical polarimeters normally use either photographic or photoelectric detection. The latter type are usually variants of some of the photometer designs discussed in section 3.2. We deal with the former type of polarimeter first.

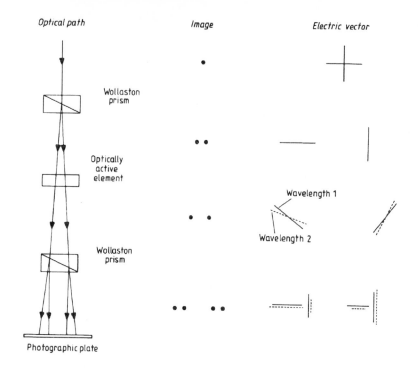

Figure 5.2.12. Öhman's multicolour polarimeter.

Photographic polarimeters

At its simplest, photographic polarimetry is just photographic photometry (section 3.2) with the addition of a polariser into the light path. Polaroid film is most frequently used as the polariser. Several separate exposures are made through it with the polariser rotated relative to the image plane by some fixed angle each time. The errors of the photographic photometric method are usually 0.01 to 0.05 magnitudes or more. This technique is therefore of little use for most stellar polarimetry, since the degree of polarisation to be measured is normally 1% or less. It is especially useful, however, for highly polarised extended objects such as the Crab nebula.

The photographic method can be improved considerably by the use of a double-beam polariser such as the Wollaston prism. Two images are produced for each star in light which is mutually orthogonally linearly polarised. Again, several exposures with a rotation of the polariser are needed. Multicolour observations can be made by using two Wollaston prisms with an interposed optically active element (figure 5.2.12). The degree of rotation introduced by the latter is generally wavelength-dependent. Hence four images result in which one pair have a similar colour balance and mutually orthogonal polarisations, while the second pair are also orthogonally polarised but have a different colour balance from the first pair.

Treanor has devised a photographic method which requires only a single exposure. A polariser and a slightly inclined plate of glass are rotated in front of the emulsion. The inclined plate displaces the instantaneous image by a small amount, and its rotation spreads this out into a ring. The polariser then modulates the intensity around the ring if the original source is polarised. By adding a neutral density filter to cover a segment of the rotor, a photographic intensity calibration can be included into the system as well.

Circular polarisation can be studied photographically using large flat plastic quarter-wave plates. These are placed before the polariser and convert the circular (or elliptical) radiation into linearly polarised radiation. The detection of this then proceeds along the lines just discussed. The detection threshold is currently about 10% polarisation for significantly accurate measurements.

Photoelectric polarimeters

Most of these devices bear a distinct resemblance to photoelectric photometers (section 3.2), and indeed, many polarimeters can also function as photometers. The major differences, apart from the components which are needed to measure the polarisation, arise from the necessity of reducing or eliminating the instrumentally-induced polarisation. The instrumental polarisation originates primarily in inclined planar reflections or in the detector. Thus Newtonian and Coudé telescopes cannot be used for polarimetry because of their inclined subsidiary mirrors. Inclined mirrors must also be avoided within the polarimeter. Cassegrain and other similar designs should not in principle introduce any net polarisation because of their symmetry about the optical axis. However, problems may arise even with these telescopes if mechanical flexure or poor adjustment move the mirrors and/or the polarimeter from the line of symmetry.

Many detectors, especially photomultipliers (section 1.1), are sensitive to polarisation, and furthermore this sensitivity may vary over the surface of the detector. If the detector is not shielded from the Earth's magnetic field, the sensitivity may vary with the position of the telescope as well. In addition, although an ideal Cassegrain system will produce no net polarisation, this sensitivity to polarisation arises because rays on opposite sides of the optical axis are polarised in opposite but equal ways. The detector non-uniformity may then result in incomplete cancellation of the telescope polarisation. Thus a Fabry lens and a depolariser immediately before the detector are almost always essential components of a polarimeter.

Alternatively the instrumental polarisation can be reversed using a Bowen compensator. This comprises two retarders with delays of about $\lambda/8$ which may be rotated with respect to the light beam and to each other until their polarisation effects are equal and opposite to those of the instrument. Small amounts of instrumental polarisation can be corrected by the insertion of a tilted

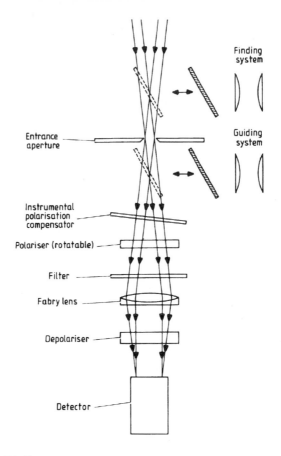

Figure 5.2.13. The basic polarimeter: schematic design and optical paths.

glass plate into the light beam to introduce opposing polarisation. In practice an unpolarised source is observed, and the plate tilted and rotated until the observed polarisation is minimised or eliminated.

There are many detailed designs of polarimeter, but we may group them into single- and double-channel devices, with the former further subdivided into discrete and continuous systems. The single-channel discrete polarimeter is the basis of most of the other designs, and its main components are shown in figure 5.2.13, although the order is not critical for some of them. Since stellar polarisations are usually of the order of 1% or less, it is also usually necessary to incorporate a chopper and use a phase-sensitive detector (section 3.2) plus integration in order to obtain sufficiently accurate measurements. Switching between the sky and background is also vital since the background radiation is very likely to be polarised. The polariser is rotated either in steps or continuously, but at a rate which is slow compared with the chopping and integration times. The output for a linearly polarised source is then modulated with the position of the polariser.

The continuous single-beam systems use a rapidly rotating polariser. This gives an oscillating output for a linearly polarised source, and this may be fed directly into a computer or into a phase-sensitive detector which is driven at the same frequency. Alternatively a quarter-wave plate and a photoelastic modulator can be used. The quarter-wave plate produces circularly polarised light, and this is then converted back into linearly polarised light by the photoelastic modulator but with its plane of polarisation rotated by 90° between alternate peaks of its cycle. A polariser which is stationary will then introduce an intensity variation with time, which may be detected and analysed as before. The output of the phase-sensitive detector in systems such as these is directly proportional to the degree of polarisation, with unpolarised sources giving zero output. They may also be used as photometers to determine the total intensity of the source, by inserting an additional fixed polariser before the entrance aperture, to give in effect a source with 100% polarisation and half the intensity of the actual source (provided that the intrinsic polarisation is small). An early polarimeter system which is related to these is due to Dolfus. It uses a quarter-wave plate to convert the linearly polarised component of the radiation into circularly polarised light. The beam is then chopped by a rotating disc which contains alternate sectors made from zero- and half-wave plates. The circular polarisation then alternates in its direction but remains circularly polarised. Another quarter-wave plate then converts the circularly polarised light back to linearly polarised light, but the plane of its polarisation switches through 90° at the chopping frequency. Finally a polariser is used to eliminate one of the linearly polarised components, so that the detector produces a square wave output.

The continuous single-beam polarimeter determines both linear polarisation components quasi-simultaneously, but it still requires a separate observation to acquire the total intensity of the source. In double-beam devices, the total intensity is found with each observation, but several observations are required to determine the polarisation. These instruments are essentially similar to that shown in figure 5.2.13, but a double-beam polariser such as a Wollaston prism is used in place of the single-beam polariser shown there. Two detectors then observe both beams simultaneously. The polariser is again rotated either in steps or continuously but slowly in order to measure the polarisation. Alternatively a half-wave plate can be rotated above a fixed Wollaston prism. The effects of scintillation and other atmospheric changes are much reduced by the use of instruments of these types.

Spectropolarimetry, which provides information on the variation of polarization with wavelength (see also Öhman's multicolour polarimeter, above) can be realised by several methods. Almost any spectroscope (section 4.2) suited for long slit spectroscopy can be adapted to spectropolarimetry. A quarter- or half-wave plate and a block of calcite are placed before the slit. The calcite is oriented so that the ordinary and extraordinary images lie along the length of the slit. Successive exposures are then obtained with the wave plate rotated between each exposure. For linear spectropolarimetry four positions separated by 22.5° are needed for the half-wave plate, while for circular spectropolarimetry just two orthogonal positions are required for the quarter-wave plate.

A few designs have been produced, particularly for infrared work, in which a dichroic mirror is used. These are mirrors which reflect at one wavelength and are transparent at another. They are not related to the other form of dichroism which was discussed earlier in this section. A thinly plated gold mirror reflects infrared radiation but transmits visible radiation for example. Continuous guiding on the object being studied is therefore possible, providing that it is emitting visible radiation, through the use of such a mirror in the polarimeter. Unfortunately the reflection introduces instrumental polarisation, but this can be reduced by incorporating a second 90° reflection whose plane is orthogonal to that of the first into the system. The advent of infrared detector arrays (section 1.1) enables many of the optical designs (above) to be used at longer wavelengths. In the ultraviolet little work has been undertaken to date. The solar maximum mission satellite, however, did include an ultraviolet spectrometer into which a magnesium fluoride retarder could be inserted, so that some measurements of the polarisation in strong spectrum lines could be made.

Data reduction and analysis

The output of a polarimeter is usually in the form of a series of intensity measurements for varying angles of the polariser. These must be corrected for instrumental polarisation if this has not already been removed or has only been incompletely removed by the use of an inclined plate. The atmospheric contribution to the polarisation must then be removed by comparison of the observations of the star and its background. A list of standard comparison stars is given in Appendix VI from which these corrections may be found empirically. Other photometric corrections, such as for atmospheric extinction (section 3.2) must also be made, especially if the observations at differing orientations of the polariser are made at significantly different times so that the position of the source in the sky has changed.

The normalised Stokes' parameters are then obtained from

$$\frac{Q}{I} = \frac{I(0) - I(90)}{I(0) + I(90)} \tag{5.2.32}$$

and

$$\frac{U}{I} = \frac{I(45) - I(135)}{I(45) + I(135)} \tag{5.2.33}$$

where $I(\theta)$ is the intensity at a position angle θ for the polariser. Since for most astronomical sources the elliptically polarised component of the radiation is 1%

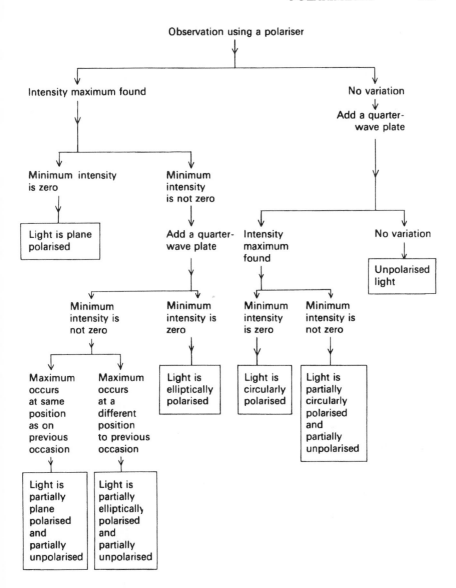

Figure 5.2.14. Flowchart of polarimetric analysis.

or less of the linearly polarised component, the degree of linear polarisation, π_L, can then be found via equation (5.2.13), or more simply from equation (5.2.15) if sufficient observations exist to determine I_{max} and I_{min} with adequate precision. The position angle of the polarisation may be found from equation (5.2.16), or again directly from the observations if these are sufficiently closely spaced. The fourth Stokes' parameter, V, is rarely determined in astronomical polarimetry. However, the addition of a quarter-wave plate to the optical train of the polarimeter is sufficient to enable it to be found if required (section 5.4 for example). The reason for the lack of interest is, as already remarked, the very low levels of circular polarisation which are usually found. The first Stokes' parameter, I, is simply the intensity of the source, and so if it is not measured directly as a part of the experimental procedure, can simply be obtained from

$$I = I(\theta) + I(\theta + 90°).$$ (5.2.34)

A flowchart to illustrate the steps in an observation and analysis sequence to determine all four Stokes' parameters is shown in figure 5.2.14.

Exercises

5.2.1   Obtain equation (5.2.6) from equations (5.2.1) and (5.2.2).

5.2.2   Show, using the Mueller calculus, that the effect of two ideal half-wave plates upon a beam of radiation of any form of polarisation is zero.

5.3   SOLAR STUDIES

Introduction

The Sun is a reasonably typical star and as such it can be studied by many of the techniques discussed elsewhere in this book. In one major way, however, the Sun is quite different from all other stars, and that is that it may easily be studied using an angular resolution smaller than its diameter. This, combined with the vastly greater amount of energy which is available compared with that from any other star, has led to the development of many specialised instruments and observing techniques. Many of these have already been mentioned or discussed in some detail in other sections. The main such techniques include: solar cosmic ray and neutron detection (section 1.4); neutrino detection (section 1.5); radio observations (sections 1.2 and 2.5); x-ray and gamma ray observations (section 1.3); radar observations (section 2.8); magnetometry (section 5.4); prominence spectroscopes (section 4.2). The other remaining important specialist instrumentation and techniques for solar work are covered in this section.

WARNING

Studying the Sun can be dangerous. Permanent eye damage or even blindness can easily occur through looking directly at the Sun, even with the naked eye. Smoked glass, exposed film, etc are *not* safe; only filters made and sold specifically for viewing the Sun should ever be used.

Herschel wedges (see below) may be suitable for use with *small* telescopes, but care must be taken to ensure that the rejected solar radiation does not cause burns or start fires.

Eyepiece projection is safe from the point of view of personal injury, but can damage the eyepiece and/or the telescope, especially with the shorter focal ratio instruments.

This warning should be carefully noted by all intending solar observers, and passed onto inexperienced persons with whom the observer may be working. In particular, the dangers of solar observing should be made quite clear to any groups of laymen touring the observatory, because the temptation to try looking at the Sun for themselves is very great, especially when there is an eclipse.

Solar telescopes

Almost any telescope may be used to image the Sun. While specialised telescopes are often built for professional work (see later discussion), small conventional telescopes are frequently used by amateurs or for teaching or training purposes. Since it is dangerous to look at the Sun through an unadapted telescope (see the warning above), special methods must be used. For small telescopes, the three main such methods are the use of a Herschel wedge, eyepiece projection, or a full aperture filter. The Herschel wedge is shown in figure 5.3.1. It is a thin prism with unsilvered faces. The first face is inclined at 45° to the optical axis and thus reflects about 5% of the solar radiation into an eyepiece in the normal star diagonal position. The second face also reflects about 5% of the radiation, but its inclination to the optical axis is not 45°, and so it is easy to arrange for this radiation to be intercepted by a baffle before it reaches the eyepiece. The remaining radiation passes through the prism and can be absorbed in a heat trap, or more commonly (and dangerously) just allowed to emerge as an intense beam of radiation at the rear end of the telescope. The radiation entering the eyepiece may comfortably be observed, provided that an infrared rejection filter is incorporated into the system, for telescopes up to about 0.1 m in diameter. Larger telescopes will need to be stopped down to an equivalent area if they are to be used with a Herschel wedge.

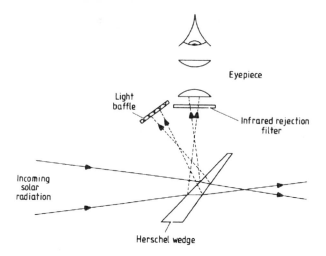

Figure 5.3.1. Optical paths in a Herschel wedge.

The principle of eyepiece projection is shown in figure 5.3.2, and it is the same as the eyepiece projection method of photography (section 2.2). The eyepiece is moved outwards from its normal position for visual use, and a real image of the Sun is obtained which may easily be projected onto a sheet of white paper or cardboard, etc for viewing purposes. From equation (2.2.10), we have the effective focal length (EFL) of the telescope

$$\text{EFL} = F(d - e)/e \qquad (5.3.1)$$

where F is the focal length of the objective, e is the focal length of the eyepiece, d is the projection distance (see figure 5.3.2). The final image size, D, for the Sun is then simply

$$D = 0.0093F(d - e)/e. \qquad (5.3.2)$$

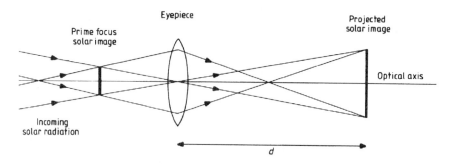

Figure 5.3.2. Projection of the solar image.

There are two drawbacks to this method. The first is that a real image of the Sun is formed at the prime focus of the objective. If this is larger than the eyepiece acceptance area, or if this image of the Sun is intercepted by a part of the telescope's structure during acquisition, then structural damage may occur through the intense heat in the beam. The second drawback is that the eyepiece lenses will absorb a small fraction of the solar energy, and this may stress the glass to the point at which it may fracture. Even if this does not occur, the heating will distort the lenses and reduce the quality of the final image. The technique is thus normally limited to telescopes with apertures of 0.15 m or less. Larger instruments will again need to be stopped down to about this area if they are to be used.

Full aperture filters are produced commercially for use on small telescopes. They consist of a very thin Mylar film which has been aluminised. They cover the whole aperture of the telescope as their name suggests, and may be made sufficiently opaque for use on telescopes up to 0.5 m or so in diameter. The film is so thin that very little degradation of the image occurs; however, this also makes the film liable to damage, and its useful life when in regular use is therefore quite short.

Larger instruments built specifically for solar observing come in several varieties. They are designed to try and overcome the major problem of the solar observer, which is turbulence within and outside the instrument, while taking advantage of the observational opportunity afforded by the plentiful supply of radiation. The most intractable of the problems is that of external turbulence. This arises from the differential heating of the observatory, its surroundings, and the atmosphere, which leads to the generation of convection currents. The effect reduces with height above the ground, and so one solution is to use a tower telescope. This involves a long focal length telescope which is fixed vertically with its objective supported on a tower at a height of several tens of metres. The solar radiation is fed into it using a coelostat (section 1.1). The other main approach to the problem (the two approaches are not necessarily mutually exclusive) is to try and reduce the extent of the turbulence. The planting of low growing tree or bush cover of the ground is one method of doing this, another is to surround the observatory by water. Within the telescope, the solar energy can also create convection currents. Sometimes these can be minimised by careful design of the instrument, or the telescope may be sealed and filled with helium whose high conductivity helps to keep the components cool and minimise convection. But the ultimate solution to them is to evacuate the optical system, so that all the light paths after the objective, or a window in front of the objective, are in a vacuum. Optically the telescopes are fairly conventional (section 1.1) except that very long focal lengths can be used to give large image scales since there is plenty of available light. It is generally undesirable to fold the light paths since the extra reflections will introduce extra scattering and distortion into the image. Thus the instruments are very cumbersome, and most are fixed in position and use a coelostat to acquire the solar radiation.

Conventional large optical telescopes can be used for solar infrared observations if they are fitted with a suitable mask. The mask covers the whole of the aperture of the telescope and protects the instrument from the visible solar radiation. It also contains one or more apertures filled with infrared transmission filters. The resulting combination provides high resolution infrared images without running the normal risk of thermal damage when a large telescope is pointed towards the Sun.

Smaller instruments may conveniently be mounted onto a solar spar. This is an equatorial mounting which is driven so that it tracks the Sun, and which has a general-purpose mounting bracket in the place normally reserved for a telescope. Special equipment may then be attached as desired. Often several separate instruments will be mounted on the spar together and may be in use simultaneously.

Spectrohelioscope

This is a monochromator which is adapted to provide an image of the whole or a substantial part of the Sun. It operates as a normal spectroscope, but

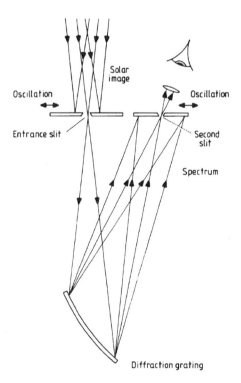

Figure 5.3.3. Principle of the spectrohelioscope.

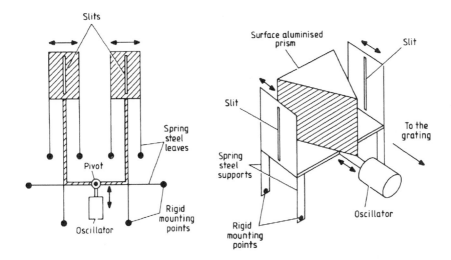

Figure 5.3.4. Mechanical arrangements for comoving the two slits of a spectrohelioscope.

with a second slit at the focal plane of the spectrum, whose position is aligned with the position of the desired wavelength in the spectrum (figure 5.3.3). The spectroscope's entrance slit is then oscillated so that it scans across the solar image. As the entrance slit moves, so also will the position of the spectrum. Hence the second slit must move in sympathy with the first if it is to continue to isolate the required wavelength. Some mechanical arrangements for ensuring that the slits remain in their correct mutual alignments are shown in figure 5.3.4. In the first of these the slit's movements are in antiphase with each other, and this is probably the commonest arrangement. In the second system the slits are in phase and the mechanical arrangements are far simpler. As an alternative to moving the slits, rotating prisms can be used to scan the beam over the grating and the image plane. If the frequency of the oscillation is higher than 15 to 20 Hz, then the monochromatic image may be viewed directly by eye. Alternatively, a photograph may be taken and a spectroheliogram produced.

Usually the wavelength which is isolated in a spectrohelioscope is chosen to be that of a strong absorption line, so that particular chromospheric features such as prominences, plages, mottling, etc are highlighted. In the visual region birefringent filters (see below) can be used in place of spectrohelioscopes, but for ultraviolet imaging and scanning, satellite-based spectrohelioscopes are used exclusively.

Narrow band filters

The spectrohelioscope produces an image of the Sun over a very narrow range of wavelengths. However, a similar result can be obtained by the use of a very

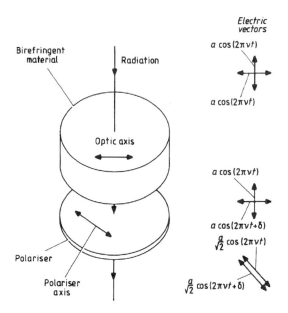

Figure 5.3.5. Basic unit of a birefringent monochromator.

narrow band filter in the optical path of a normal telescope. Since the bandwidth of the filter must lie in the region of 0.01 to 0.5 nm if the desired solar features are to be visible, normal dye filters or interference filters (section 4.1) are not suitable. Instead a filter based upon a birefringent material (section 5.2) has been developed by Lyot. It has various names—quartz monochromator, Lyot monochromator, or birefringent filter being amongst the commonest.

Its operational principle is based upon a slab of quartz or other birefringent material which has been cut parallel to its optic axis. As we saw in section 5.2, the extraordinary ray will then travel more slowly than the ordinary ray, and so the two rays will emerge from the slab with a phase difference (figure 5.3.5). The rays then pass through a sheet of polaroid film whose axis is midway between the directions of polarisation of the ordinary and extraordinary rays. Only the components of each ray which lie along the polaroid's axis are transmitted by it. Thus the rays emerging from the polaroid film have parallel polarisation directions and a constant phase difference, and so they can mutually interfere. If the original electric vectors of the radiation, E_o and E_e, along the directions of the ordinary and extraordinary rays' polarisations respectively, are given at some point by

$$E_o = E_e = a\cos(2\pi \nu t) \tag{5.3.3}$$

then after passage through the quartz, we will have at some point

$$E_o = a\cos(2\pi \nu t) \tag{5.3.4}$$

and

$$E_e = a \cos(2\pi \nu t + \delta) \tag{5.3.5}$$

where δ is the phase difference between the two rays. After passage through the polaroid, we then have

$$E_{45,o} = \frac{a}{\sqrt{2}} \cos(2\pi \nu t) \tag{5.3.6}$$

$$E_{45,e} = \frac{a}{\sqrt{2}} \cos(2\pi \nu t + \delta) \tag{5.3.7}$$

where $E_{45,x}$ is the component of the electric vector along the polaroid's axis for the xth component. Thus the total electric vector of the emerging radiation, E_{45}, will be given by

$$E_{45} = \frac{a}{\sqrt{2}}[\cos(2\pi \nu t) + \cos(2\pi \nu t + \delta)] \tag{5.3.8}$$

and so the emergent intensity of the radiation, I_{45}, is

$$I_{45} = 2a^2 \quad \text{for} \quad \delta = 2n\pi \tag{5.3.9}$$

$$I_{45} = 0 \quad \text{for} \quad \delta = (2n + 1)\pi \tag{5.3.10}$$

where n is an integer. Now

$$\delta = 2\pi c \Delta t / \lambda \tag{5.3.11}$$

where Δt is the time delay introduced between the ordinary and extraordinary rays by the material. If v_o and v_e are the velocities of the ordinary and extraordinary rays in the material then

$$\Delta t = \frac{T}{v_o} - \frac{T}{v_e} \tag{5.3.12}$$

$$= T\left(\frac{v_e - v_o}{v_e v_o}\right) \tag{5.3.13}$$

$$= \frac{TJ}{c} \tag{5.3.14}$$

where J is the birefringence of the material (section 5.2) and T is the thickness of the material. Thus

$$\delta = \frac{2\pi c(TJ/c)}{\lambda} \tag{5.3.15}$$

$$= \frac{2\pi TJ}{\lambda}. \tag{5.3.16}$$

The emergent ray therefore reaches a maximum intensity at wavelengths, λ_{max}, given by

$$\lambda_{max} = \frac{TJ}{n} \tag{5.3.17}$$

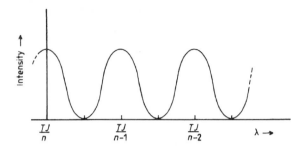

Figure 5.3.6. Spectrum of the emerging radiation from a basic unit of a birefringent monochromator.

and is zero at wavelengths, λ_{min}, given by

$$\lambda_{min} = \frac{2TJ}{(2n+1)} \tag{5.3.18}$$

(see figure 5.3.6).

Now if we require the eventual filter to have a whole bandwidth of $\Delta\lambda$ centred on λ_c, then the parameters of the basic unit must be such that one of the maxima in figure 5.3.6 coincides with λ_c, and the width of one of the fringes is $\Delta\lambda$; that is

$$\lambda_c = \frac{TJ}{n_c} \tag{5.3.19}$$

$$\Delta\lambda = TJ\left[\left(n_c - \tfrac{1}{2}\right)^{-1} - \left(n_c + \tfrac{1}{2}\right)^{-1}\right]. \tag{5.3.20}$$

Since selection of the material fixes J, and n_c is deviously related to T, the only truly free parameter of the system is T, and thus for a given filter we have

$$T = \frac{\lambda_c}{2J}\left[\frac{\lambda_c}{\Delta\lambda} + \left(\frac{\lambda_c^2}{\Delta\lambda^2} + 1\right)^{1/2}\right] \tag{5.3.21}$$

and

$$n_c = JT/\lambda_c. \tag{5.3.22}$$

Normally, however,

$$\lambda_c \gg \Delta\lambda \tag{5.3.23}$$

and quartz is the birefringent material, for which in the visible

$$J = +0.0092 \tag{5.3.24}$$

so that

$$T \simeq 109\lambda_c^2/\Delta\lambda. \tag{5.3.25}$$

Thus for a quartz filter to isolate the Hα line at 656.2 nm with a bandwidth of 0.1 nm, the thickness of the quartz plate should be about 470 mm.

Now from just one such basic unit, the emergent spectrum will contain many closely spaced fringes as shown in figure 5.3.5. But if we now combine it with a second basic unit oriented at 45° to the first, and whose central frequency is still λ_c, but whose bandwidth is $2\Delta\lambda$, then the final output will be suppressed at alternate maxima (figure 5.3.7). From equation (5.3.25) we see that the thickness required for the second unit, for it to behave in this way, is just half the thickness of the first unit. Further basic units may be added, with thicknesses 1/4, 1/8, 1/16, etc that of the first, whose transmissions continue to be centred upon λ_c, but whose bandwidths are 4, 8, 16, etc times that of the original unit. These continue to suppress additional unwanted maxima. With six such units, the final spectrum has only a few narrow maxima which are separated from λ by multiples of $32\Delta\lambda$ (figure 5.3.8). At this stage the last remaining unwanted maxima are sufficiently well separated from λ_c for them to

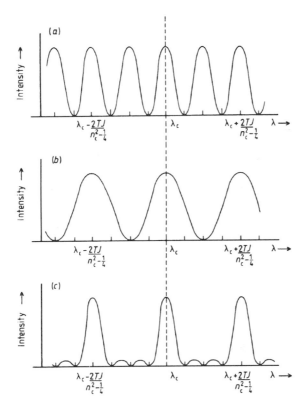

Figure 5.3.7. Transmission curves of birefringent filter basic units. (a) Basic unit of thickness T; (b) basic unit of thickness, $\frac{1}{2}T$; (c) combination of the units in (a) and (b).

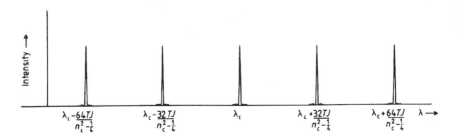

Figure 5.3.8. Transmission curve of a birefringent filter comprising six basic units with a maximum thickness T.

be eliminated by conventional dye or interference filters, so that just the desired transmission curve remains. One further refinement to the filter is required and that is to ensure that the initial intensities of the ordinary and extraordinary rays are equal, since this was assumed in obtaining equation (5.3.3). This is easily accomplished, however, by placing an additional sheet of polaroid before the first unit, which has its transmission axis at 45° to the optic axis of that first unit. The complete unit is shown in figure 5.3.9. Neglecting absorption and other losses, the peak transmitted intensity is half the incident intensity because of the use of the first polaroid. The properties of quartz are temperature dependent and so the whole unit must be enclosed and its temperature controlled to a fraction of a degree when it is in use. The temperature dependence allows the filter's central wavelength to be tuned slightly by changing the operating temperature. Much larger changes in wavelength may be encompassed by splitting each retarder into two opposed wedges. Their combined thicknesses can then be varied by displacing one set of wedges with respect to the other in a direction perpendicular

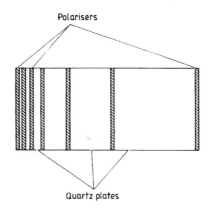

Figure 5.3.9. Six-element quartz birefringent filter.

to the axis of the filter. The wavelength may also be decreased by slightly tilting the filter to the direction of the incoming radiation, so increasing the effective thicknesses of the quartz plates. For the same reason, the beam of radiation must be collimated for its passage through the filter, or the bandwidth will be increased.

A closely related filter is due to Solč. It uses only two polarisers placed before and after the retarders. All the retarders are of the same thickness, and they have their optic axes alternately oriented with respect to the axis of the first polariser in clockwise and anticlockwise directions at some specified angle. The output consists of a series of transmission fringes whose spacing in wavelength terms increases with wavelength. Two such units of differing thicknesses can then be combined so that only at the desired wavelength do the transmission peaks coincide.

Another device for observing the Sun over a very narrow wavelength range is the Magneto-Optical Filter (MOF). This, however, can only operate at the wavelengths of strong absorption lines due to gases. It comprises two polarisers on either side of a gas cell. The polarisers are oriented orthogonally to each other if they are linear or are left- and right-handed if circular in nature. In either case, no light is transmitted through the system. A magnetic field is then applied to the gas cell, inducing the Zeeman effect. The Zeeman components of the lines produced by the gas are then linearly and/or circularly polarised (section 5.4) and so permit the partial transmission of light at their wavelengths through the whole system. The gases currently used in MOFs are vapours of sodium or potassium.

Relatively inexpensive Hα filters can be made as solid Fabry–Perot etalons (section 4.1). A thin fused-silica spacer between two optically flat dielectric mirrors is used together with a blocking filter. Peak transmissions of several tens of percent and bandwidths of better than a tenth of a nanometre can be achieved in this way.

Coronagraph

This instrument enables observations of the corona to be made at times other than during solar eclipses. It does this by producing an artificial eclipse. The principle is very simple; an occulting disc at the prime focus of a telescope obscures the photospheric image while allowing that of the corona to pass by. The practice, however, is considerably more complex, since scattered and/or diffracted light, etc in the instrument and the atmosphere can still be several orders of magnitude brighter than the corona. Extreme precautions have therefore to be taken in the design and operation of the instrument in order to minimise this extraneous light.

The most critical of these precautions lies in the structure of the objective. A single simple lens objective is used in order to minimise the number of surfaces involved and it is formed from a glass blank which is as free from bubbles, striae and other imperfections as possible. The surfaces of the lens are polished

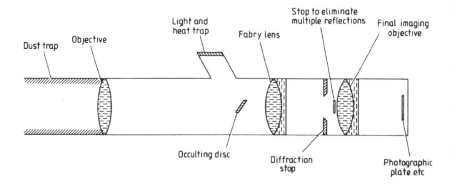

Figure 5.3.10. Schematic optical arrangement of a coronagraph.

with extreme care in order to eliminate all scratches and other surface markings. In use, they are kept dust-free by being tightly sealed when not in operation, and by the addition of a very long tube, which is lined with grease, in front of the objective to act as a dust trap.

The occulting disc is a polished metal cone or an inclined mirror so that the photospheric radiation may be safely reflected away to a separate light and heat trap. Diffraction from the edges of the objective is eliminated by imaging the objective with a Fabry lens after the occulting disc, and by using a stop which is slightly smaller than the image of the objective to remove the edge effects. Alternatively the objective can be apodised (section 2.5); its transparency decreases from its centre to its edge in a Gaussian fashion. This leads to full suppression of the diffraction halo although with some loss in resolution. A second occulting disc before the final objective removes the effects of multiple reflections within the first objective. The final image of the corona is produced by this second objective, and this is placed after the diffraction stop. The full system is shown in figure 5.3.10.

The atmospheric scattering can only be reduced by a suitable choice of observing site. Thus the early coronagraphs are to be found at high altitude observatories, while more recently they have been mounted on spacecraft in order to eliminate the problem of the Earth's atmosphere entirely.

The use of a simple lens results in a chromatic prime focus image. A filter must therefore normally be added to the system. This is desirable in any case since the coronal spectrum is largely composed of emission lines superimposed upon a diluted solar photospheric spectrum. Selection of a narrow band filter which is centred upon a strong coronal emission line therefore improves considerably the contrast of the final image. White light or wideband imaging of the corona is only possible using Earth-based instruments on rare occasions, and it can only be attempted under absolutely optimum observing conditions. Satellite-borne instruments may be so used more routinely.

Improvements to the basic coronagraph may be justified for balloon or satellite-borne instruments, since the sky background is then less than the scattered light in a more normal device, and have taken two different forms. A reflecting objective can be used. This is formed from an uncoated off-axis parabola. Most of the light passes through the objective and is absorbed. Bubbles and striations in the glass are of much less importance since they cause scattering primarily in the forward direction. The mirror is uncoated since metallic films are sufficiently irregular to a cause a considerable amount of scattering. In other respects, the coronagraph then follows the layout of figure 5.3.10. The second approach to the improvement of coronagraphs is quite different and consists simply of producing the artificial eclipse outside the instrument rather than at its prime focus. An occulting disc is placed well in front of the first objective of an otherwise fairly conventional coronagraph. The disc must be large enough to ensure that the first objective lies entirely within its umbral shadow. The inner parts of the corona will therefore be badly affected by vignetting. However this is of little importance since these are the brightest portions of the corona, and it may even be advantageous since it will reduce the dynamic range which must be covered by the detector. A simple disc produces an image with a bright central spot due to diffraction, but this can be eliminated by using a disc whose edge is formed into a zig-zag pattern of sharp teeth, or by the use of multiple occulting discs. By such means the instrumentally scattered light can be reduced to 10^{-4} of that of a basic coronagraph.

The final image in a coronagraph may be imaged directly, but more commonly is fed to a spectroscope, photometer or other ancillary instrument. From the Earth the corona may normally only be detected out to about one solar radius, but satellite-based coronagraphs have been used successfully out to six solar radii or more.

Devices similar to the coronagraph are also sometimes carried on spacecraft so that they may observe the atmospheres of planets, while shielding the detector from the radiation from the planet's surface. The contrast is generally far smaller than in the solar case so that such planetary 'coronagraphs' can be far simpler in design. Terrestrial and spacecraft-borne stellar coronagraphs are also under construction at the time of writing. These are designed to allow the detection of faint objects close to much brighter objects, such as stellar companions and accretion discs, by eliminating the bright stellar image in a similar manner to the removal of the solar image in a solar coronagraph. With the addition of suitable apodisation stops (section 4.1) and adaptive optics (section 1.1) it may even be possible to detect directly non-solar system planets.

Solar oscillations

Whole-body vibrations of the Sun reveal much about its inner structure. They are small scale effects and their study requires very precise measurements of the velocities of parts of the solar surface. The resonance scattering spectrometer originated by the solar group at the University of Birmingham is capable of detecting motions of a few tens of millimetres per second. It operates by passing the 770 nm potassium line from the Sun through a container of heated potassium vapour. A magnetic field is directed through the vapour and the solar radiation is circularly polarised. Depending upon the direction of the circular polarisation either the light from the long wave wing (I_L) or the short wave wing (I_S) of the line is resonantly scattered. By switching the direction of the circular polarisation rapidly, the two intensities may be measured nearly simultaneously. The line of sight velocity is then given by

$$v = k \frac{I_S - I_L}{I_S + I_L} \tag{5.3.26}$$

where k is a constant with a value around 3 km s^{-1}.

For the Solar and Heliospheric Observatory (SOHO) spacecraft, a similar device based upon sodium D line resonance scattering can detect motions to better than 1 mm s^{-1}.

Other solar observing methods

Slitless spectroscopes (section 4.2) are of considerable importance in observing regions of the Sun whose spectra consist primarily of emission lines. They are simply spectroscopes in which the entrance aperture is large enough to accept a significant proportion of the solar image. The resulting spectrum is therefore a series of monochromatic images of this part of the Sun in the light of each of the emission lines. They have the advantage of greatly improved speed over a slit spectroscope combined with the obvious ability to obtain spectra simultaneously from different parts of the Sun. Their most well known application is during solar eclipses, when the 'flash spectrum' of the chromosphere may be obtained in the few seconds immediately after the start of totality or just before its end. More recently, they have found widespread use as spacecraft-borne instrumentation for observing solar flares in the ultraviolet part of the spectrum.

One specialised satellite-borne instrument based upon slitless spectroscopes is the Lyα camera. The Lyman α line is a strong emission line. If the image of the whole disc of the Sun from an ultraviolet telescope enters a slitless spectroscope, then the resulting Lyα image may be isolated from the rest of the spectrum by a simple diaphragm, with very little contamination from other wavelengths. A second identical slitless spectroscope whose entrance aperture is this diaphragm, and whose dispersion is perpendicular to that of the first spectroscope, will then provide a stigmatic spectroheliogram at a wavelength of 121.6 nm.

Another specialised instrument, also frequently a part of spacecraft instrumentation, is the pyroheliometer. This measures the energy emitted by the Sun over a very wide range of wavelengths. At its simplest it consists of a pseudo black-body cavity which traps the radiation. The temperature of the cavity is then monitored and provides a measure of the radiation density.

In the radio region, solar observations tend to be undertaken by fairly conventional equipment although there are now a few dedicated solar radio telescopes. One exception to this, however, is the use of multiplexed receivers to provide a quasi-instantaneous radio spectrum of the Sun. This is of particular value for the study of solar radio bursts, since their emitted bandwidth may be quite narrow, and their wavelength drifts rapidly with time. These radio spectroscopes and the acousto-optical radio spectroscope were discussed more fully in section 1.2. The data from them is usually presented as a frequency/time plot, from whence the characteristic behaviour patterns of different types of solar bursts may easily be recognised.

This account of highly specialised instrumentation and techniques for solar observing could be extended almost indefinitely, and might encompass all the equipment designed for parallax observations, solar radius determinations, oblateness determinations, eclipse work, tree rings and C^{14} determination, the designs of individual solar satellites and so on. However, at least in the author's opinion, these are becoming too specialised for inclusion in a general book like this, and so the reader is referred to more restricted texts and the scientific journals for further information upon them.

Exercise

5.3.1 Calculate the maximum and minimum thicknesses of the elements required for an Hα birefringent filter based upon calcite, if its whole bandwidth is to be 0.05 nm, and it is to be used in conjunction with an interference filter whose whole bandwidth is 3 nm. The birefringence of calcite is -0.172.

5.4   MAGNETOMETRY

Background

The measurement of astronomical magnetic fields is accomplished in two quite separate ways. The first is direct measurement by means of apparatus carried by spacecraft, while the second is indirect and is based upon the Zeeman effect of a magnetic field upon spectrum lines (or more strictly upon the inverse Zeeman effect, since it is usually applied to absorption lines).

Zeeman effect

The Zeeman effect describes the change in the structure of the emission lines in a spectrum when the emitting object is in a magnetic field. The simplest change arises for singlet lines, that is lines arising from transitions between levels with

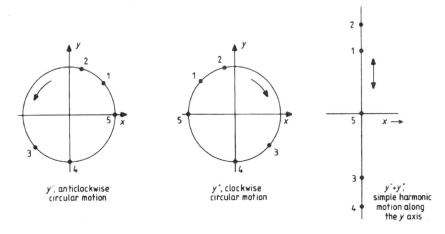

Figure 5.4.1. Resolution of simple harmonic motion along the y axis into two equal but opposite circular motions.

a multiplicity of one, or a total spin quantum number, M_s of zero for each level. For these lines the effect is called the normal Zeeman effect. The line splits into two or three components depending on whether the line of sight is along, or perpendicular to the magnetic field lines. An appreciation of the basis of the normal Zeeman effect may be obtained from a classical approach to the problem. If we imagine an electron in an orbit around an atom, then its motion may be resolved into three simple harmonic motions along the three coordinate axes. Each of these in turn we may imagine to be the sum of two equal but opposite circular motions (figure 5.4.1). If we now imagine a magnetic field applied along the z axis, then these various motions may be modified by it. Firstly, the simple harmonic motion along the z axis will be unchanged since it lies along the direction of the magnetic field. The simple harmonic motions along the x and y axes, however, are cutting across the magnetic field and so will be altered. We may best see how their motion changes by considering the effect upon their circular components. When the magnetic field is applied, the radii of the circular motions remain unchanged, but their frequencies alter. If ν is the original frequency of the circular motion, and H is the magnetic field strength, then the new frequencies of the two resolved components ν^+ and ν^- are

$$\nu^+ = \nu + \Delta\nu \tag{5.4.1}$$

$$\nu^- = \nu - \Delta\nu \tag{5.4.2}$$

where

$$\Delta\nu = \frac{eH}{4\pi m_e c} \tag{5.4.3}$$

$$= 1.40 \times 10^{10} H \qquad (\text{Hz T}^{-1}). \tag{5.4.4}$$

Thus we may combine the two higher frequency components arising from the x and y simple harmonic motions to give a single elliptical motion in the xy plane at a frequency of $\nu + \Delta\nu$. Similarly we may combine the lower frequency components to give another elliptical motion in the xy plane at a frequency of $\nu - \Delta\nu$. Thus the electron's motion may be resolved, when it is in the presence of a magnetic field along the z axis, into two elliptical motions in the xy plane, plus simple harmonic motion along the z axis (figure 5.4.2), the frequencies being $\nu + \Delta\nu$, $\nu - \Delta\nu$ and ν, respectively. Now if we imagine looking at such a system, then only those components which have some motion across the line of sight will be able to emit light towards the observer, since light propagates in a direction perpendicular to its electric vector. Hence looking along the z axis (i.e. along the magnetic field lines) only emission from the two elliptical components of the electron's motion will be visible. Since the final spectrum contains contributions from many atoms, these will average out to two circularly polarised emissions, shifted by $\Delta\nu$ from the normal frequency (figure 5.4.3), one with clockwise polarisation, and the other with anticlockwise polarisation. Lines of sight perpendicular to the magnetic field direction, that is within the xy plane, will in general view the two elliptical motions as two collinear simple harmonic motions, while the z axis motion will remain as simple harmonic motion orthogonal to the first two motions. Again the spectrum is the average of many atoms, and so will therefore comprise three linearly polarised lines. The first of these is at the normal frequency of the line and is polarised

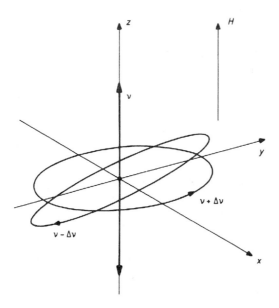

Figure 5.4.2. Components of electron orbital motion in the presence of a magnetic field.

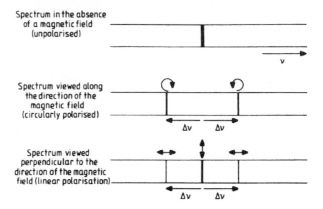

Figure 5.4.3. The normal Zeeman effect.

parallel to the field direction. It arises from the z axis motion. The other two lines are polarised at right angles to the first and are shifted in frequency by Δv (figure 5.4.3) from the normal position of the line. When observing the source along the line of the magnetic field, the two spectrum lines have equal intensities, while when observing perpendicular to the magnetic field, the central line has twice the intensity of either of the other components. Thus imagining the magnetic field progressively reducing then, as the components re-mix, an unpolarised line results, as one would expect. This pattern of behaviour for a spectrum line originating in a magnetic field is termed the normal Zeeman effect.

In astronomy, absorption lines, rather than emission lines, are the main area of interest, and their behaviour is described by the inverse Zeeman effect. This, however, is precisely the same as the Zeeman effect except that emission processes are replaced by their inverse absorption processes. The above analysis may therefore be equally well applied to describe the behaviour of absorption lines. The one major difference from the emission line case is that the *observed* radiation remaining at the wavelength of one of the lines is preferentially polarised in the opposite sense to that of the Zeeman component, since the Zeeman component is being subtracted from unpolarised radiation.

If the spectrum line does not originate from a transition between singlet levels (i.e. $M_s \neq 0$), then the effect of a magnetic field is more complex. The resulting behaviour is known as the anomalous Zeeman effect, but this is something of a misnomer since it is anomalous only in the sense that it does not have a classical explanation. Quantum mechanics describes the effect completely. The orientation of an atom in a magnetic field is quantised. The angular momentum of the atom is given by

$$[J(J+1)]^{1/2}h/2\pi \tag{5.4.5}$$

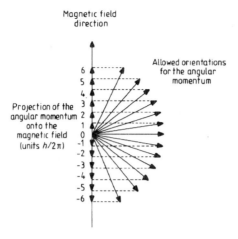

Figure 5.4.4. Space quantisation for an atom with $J = 6$.

where J is the inner quantum number, and its space quantisation is such that the projection of the angular momentum onto the magnetic field direction must be an integer multiple of $h/2\pi$ when J is an integer, or a half-integer multiple of $h/2\pi$ when J is a half-integer. Thus there are always $(2J + 1)$ possible quantised states (figure 5.4.4). Each state may be described by a total magnetic quantum number, M, which for a given level can take all integer values from $-J$ to $+J$ when J is an integer, or all half-integer values over the same range when J is a half-integer. In the absence of a magnetic field, electrons in these states all have the same energy (i.e. the states are degenerate), and the set of states forms a level. Once a magnetic field is present, however, electrons in different states have different energies, with the energy change from the normal energy of the level, ΔE, being given by

$$\Delta E = \frac{eh}{4\pi m_e c} M g H \tag{5.4.6}$$

where g is the Landé factor, given by

$$g = 1 + \frac{J(J + 1) + M_s(M_s + 1) - L(L + 1)}{2J(J + 1)} \tag{5.4.7}$$

where L is the total azimuthal quantum number. Thus the change in the frequency, Δv, for a transition to or from the state is

$$\Delta v = \frac{e}{4\pi m_e c} M g H \tag{5.4.8}$$

$$= 1.40 \times 10^{10} M g H \quad \left(\text{Hz T}^{-1}\right). \tag{5.4.9}$$

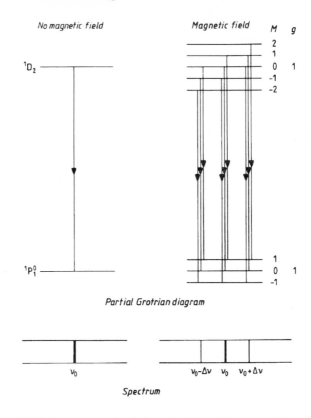

Figure 5.4.5. Quantum mechanical explanation of the normal Zeeman effect.

Now for transitions between such states we have the selection rule that M can change by 0 or ± 1 only. Thus we may understand the normal Zeeman effect in the quantum mechanical case simply from the allowed transitions (figure 5.4.5) plus the fact that the splitting of each level is by the same amount since the singlet levels

$$M_s = 0 \tag{5.4.10}$$

so that

$$J = L \tag{5.4.11}$$

and so

$$g = 1. \tag{5.4.12}$$

Each of the normal Zeeman components for the example shown in figure 5.4.5 is therefore triply degenerate, and only three lines result from the nine possible transitions. When M_s is not zero, g will in general be different for the two levels and the degeneracy will cease. All transitions will then produce separate lines, and hence we get the anomalous Zeeman effect (figure 5.4.6). Only one of many

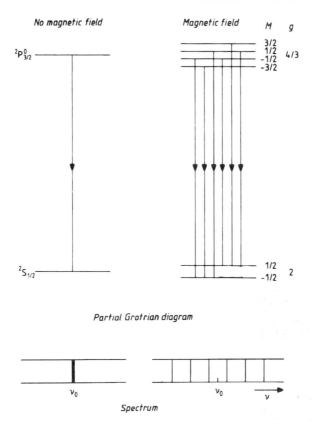

No magnetic field   Magnetic field   M   g

Partial Grotrian diagram

Spectrum

Figure 5.4.6. Quantum mechanical explanation of the anomalous Zeeman effect.

possible different patterns is shown in the figure; the details of any individual pattern will depend upon the individual properties of the levels involved.

As the magnetic field strength increases, the pattern changes almost back to that of the normal Zeeman effect. This change is known as the Paschen–Back effect. It arises as the magnetic field becomes strong enough to decouple L and M_s from each other. They then no longer couple together to form J, which then couples with the magnetic field, as just described, but couple separately and independently to the magnetic field. The pattern of spectrum lines is then that of the normal Zeeman effect, but with each component of the pattern formed from a narrow doublet, triplet, etc, accordingly as the original transition was between doublet, triplet, etc levels. Field strengths of around 0.5 T or more are usually necessary for the complete development of the Paschen–Back effect.

At very strong magnetic field strengths ($> 10^3$ T), the quadratic Zeeman effect will predominate. This displaces the spectrum lines to higher frequencies by an amount $\Delta \nu$, given by

$$\Delta \nu = \frac{\epsilon_0 h^3}{8\pi^2 m_e^3 c^2 e^2 \mu_0} n^4 (1 + M^2) H^2 \tag{5.4.13}$$

$$= 1.489 \times 10^4 n^4 (1 + M^2) H^2 \quad \text{(Hz)} \tag{5.4.14}$$

where n is the principal quantum number.

Magnetometers

Amongst the direct measuring devices, the commonest type is the flux gate magnetometer illustrated in figure 5.4.7. Two magnetically soft cores have windings coiled around them as shown. Winding A is driven by an alternating current. Winding B has equal and opposite currents induced in its two coils by the alternating magnetic fields of the cores so long as there is no external magnetic field, and so there is no output. The presence of an external magnetic field introduces an imbalance into the currents, so that there is a net alternating current output. This may then easily be detected and calibrated in terms of the external field strength. Spacecraft usually carry three such magnetometers oriented orthogonally to each other so that all the components of the external magnetic field may be measured.

Another type of magnetometer which is often used on spacecraft, is based upon atoms in a gas oscillating at their Larmor frequency. It is often used for calibrating flux gate magnetometers in orbit. The vector helium magnetometer operates by detecting the effect of a magnetic field upon the population of a metastable state of helium. A cell filled with helium is illuminated by 1.08 μm radiation which pumps the electrons into the metastable state. The efficiency of the optical pump is affected by the magnetic field, and the population of

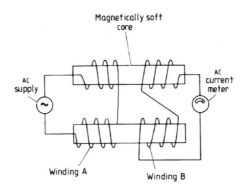

Figure 5.4.7. Flux gate magnetometer.

that state is monitored by observing the absorption of the radiation. A variable artificially generated magnetic field is swept through the cell until the external magnetic field is nullified. The strength and direction of the artificial field are then equal and opposite to the external field.

Any magnetometer on board a spacecraft usually has to be deployed at the end of a long boom after launch in order to remove it from the magnetic effects of the spacecraft itself. An alternative technique altogether, is to look at the electron flux. Since electron paths are modified by the magnetic field, their distribution can be interpreted in terms of the local magnetic field strength. This enables the magnetic field to be detected with moderate accuracy over a large volume, in comparison with the direct methods which give very high accuracies, but only in the immediate vicinity of the spacecraft.

Most of the successful work in magnetometry based upon the Zeeman effect has been undertaken by the Babcocks or uses instruments based upon their designs. For stars, the magnetic field strength along the line of sight may be determined via the longitudinal Zeeman effect. The procedure is made very much more difficult than in the laboratory by the widths of the spectrum lines. For a magnetic field of 1 T, the change in the frequency of the line is 1.4×10^{10} Hz (equation (5.4.4)), which for lines near 500 nm corresponds to a separation for the components in wavelength terms of only 0.02 nm. This is smaller than the normal linewidth for most stars. Thus even strong magnetic fields do not cause the spectrum lines to split into two separate components, but merely to become somewhat broader than normal. The technique is saved by the opposite senses of circular polarisation of the components, which enables them to be separated from each other. Babcock's differential analyser therefore consists of a quarter-wave

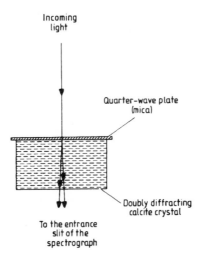

Figure 5.4.8. Babcock's differential analyser.

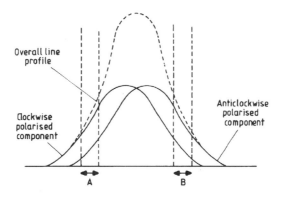

Figure 5.4.9. Longitudinal Zeeman components.

plate (section 5.2) followed by a doubly refracting calcite crystal. This is placed immediately before the entrance aperture of a high dispersion spectroscope (figure 5.4.8). The quarter-wave plate converts the two circularly polarised components into two mutually perpendicular linearly polarised components. The calcite is suitably oriented so that one of these components is the ordinary ray and the other the extraordinary ray. The components are therefore separated by the calcite into two beams. These are both accepted by the spectroscope slit and so two spectra are formed in close juxtaposition, with the lines of each slightly shifted with respect to the other due to the Zeeman splitting. This shift may then be measured and translated back to give the longitudinal magnetic field intensity. A lower limit on the magnetic field intensity of about 0.02 T is detectable by this method, while the strongest stellar fields found have strengths of a few tesla. By convention the magnetic field is positive when it is directed towards the observer.

On the Sun much greater sensitivity is possible, and fields as weak as 10^{-5} T can be studied. Hale first detected magnetic fields in sunspots in 1908, but most of the modern methods and apparatus are again due to the Babcocks. Their method relies upon the slight shift in the line position between the two circularly polarised components, which causes an exaggerated difference in their relative intensities in the wings of the lines (figure 5.4.9). The solar light is passed through a differential analyser as before. This time, however, the quarter-wave plate is composed of ammonium dihydrogen phosphate, and this requires an electric potential of some 9 kV across it, in order to make it sufficiently birefringent to work. Furthermore, only one of the beams emerging from the differential analyser is fed to the spectroscope. The line is then detected by a pair of photomultipliers which accept the spectral regions labelled A and B in figure 5.4.9. Thus the wing intensities of, say, the clockwise polarised component are detected when a positive voltage is applied to the quarter-wave plate, while those of the anticlockwise component are detected when a negative voltage is applied.

The voltage is switched rapidly between the two states, and the photomultiplier outputs are detected in phase with the switching. Since all noise except that in phase with the switching is automatically eliminated, and the latter may be reduced by integration and/or the use of a phase-sensitive detector (section 3.2), the technique is very sensitive. The entrance aperture to the apparatus can be scanned across the solar disc so that a magnetogram can be built up with a typical resolution of a few seconds of arc and of a few microtesla. Specialised CCD detectors can be used which have every alternate row of pixels covered over. The accumulating charges in the pixels are switched between an exposed row and a covered row in phase with the switching between the polarised components. Thus, at the end of the exposure, alternate rows of pixels contain the intensities of the clockwise and anti-clockwise components. Similar instruments have been devised for stellar observations, but so far are not as sensitive as the previously mentioned photographic system.

Several other types of magnetometer have been devised for solar work. For example, at the higher field strengths, Leighton's method provides an interesting pictorial representation of the magnetic patterns. Two spectroheliograms are obtained in opposite wings of a suitable absorption line, and through a differential analyser which is arranged so that it only allows the passage into the system of the stronger circularly polarised component in each case. A composite print is then made of both spectroheliograms, with one as a negative, and the other as a positive. Areas with magnetic fields less than about 2 mT are seen as grey, while areas with stronger magnetic fields show up light or dark according to their polarity. More modern devices use video recording but the basic technique remains unaltered. Vector-imaging magnetographs image the Sun over a very narrow wavelength range, and then rapidly step that image through a magnetically sensitive spectrum line. This enables both the direction and strength of the magnetic field to be determined over the selected region. Usually this region is just a small part of the whole solar disk because of the enormous amounts of data that such instruments can generate.

In the radio region, lines such as the 1.42 GHz emission from hydrogen are also affected by the Zeeman effect. Their Zeeman splitting is found in a similar manner to Babcock's solar method—by observing in the steepest part of the wings. Interstellar magnetic fields of about 10^{-9} T are detectable in this way.

The very strongest fields, up to 10^8 T, are expected to exist in white dwarfs and neutron stars. However, there is usually little spectral structure in such objects from which the fields might be detected. The quadratic Zeeman effect, however, may then be sufficient for circular polarisation of the continuum to be distinguished at points where it is varying rapidly in intensity. Field strengths of upwards of 1000 T are found in this way.

A few devices have been constructed to detect electric fields via the Stark effect upon the Paschen hydrogen lines. They are basically similar to magnetometers and for solar studies can reach sensitivities of 500 V m^{-1}.

Data reduction and analysis

Once a flux gate magnetometer has been calibrated, there is little involved in the analysis of its results except to convert them to field strength and direction. However it may be necessary to correct the readings for influence from the spacecraft, and for the effects of the solar wind and the solar or interplanetary magnetic fields. Measurements with accuracies as high as 10^{-11} T are possible when these corrections are reliably known.

With the first of Babcock's methods, in which two spectra are produced side by side, the spectra are simply measured as if for radial velocity. The longitudinal magnetic field strength can then be obtained via equation (5.4.9). The photoelectric method's data are rather more complex to analyse. No general method can be given since it will depend in detail upon the experimental apparatus and method and upon the properties of the individual line (or continuum) being observed.

With strong solar magnetic fields, such as those found in sunspots, information may be obtainable in addition to the longitudinal field strength. The line in such a case may be split into two or three components, and when these are viewed through a circular analyser (i.e. an analyser which passes only one circularly polarised component) the relative intensities of the three lines of the normal Zeeman effect are given by Seare's equations

$$I_v = \tfrac{1}{4}(1 \pm \cos\theta)^2 I \qquad\qquad (5.4.15)$$

$$I_c = \tfrac{1}{2}(\sin^2\theta) I \qquad\qquad (5.4.16)$$

$$I_r = \tfrac{1}{4}(1 \mp \cos\theta)^2 I \qquad\qquad (5.4.17)$$

where I_v, I_c and I_r are the intensities of the high frequency, central and low frequency components of the triplet, I is the total intensity of all the components and θ is the angle of the magnetic field's axis to the line of sight. Thus the direction of the magnetic field as well as its total strength may be found.

Appendix I

Magnitudes and spectral types of bright stars

Star	V	$B - V$	Spectral type	Star	V	$B - V$	Spectral type
α And	2.02	-0.10	B9p	γ Ori	1.64v?	-0.23	B2 III
β Cas	2.25	$+0.35$	F2 IV	β Tau	1.65	-0.13	B7 III
γ Peg	2.83v	-0.23	B2 IV	δ Ori	2.20v	-0.21	O9.5 II
β Hyi	2.79v?	$+0.62$	G2 IV	α Lep	2.59	$+0.22$	F0 Ib
α Phe	2.39	$+1.08$	K0 III	ι Ori	2.77	-0.25	O9 III
α Cas	2.24v	$+1.18$	K0 II–III	ε Ori	1.70	-0.19	B0 Ia
β Cet	2.04v?	$+1.02$	K1 III	α Col	2.63v?	-0.12	B8e V
γ Cas	2.65v	-0.22	B0e IV?	ζ Ori	2.05v?	*	O9.5 lb
β And	2.03v?	$+1.63$	M0 III	$\varkappa$ Ori	2.04v?	-0.18	B0.5e I
δ Cas	2.68v	$+0.13$	A5 V	α Ori	0.80v	$+1.86$	M2 Iab
α Eri	0.47	-0.19	B5 IV	β Aur	1.90v	$+0.03$	A2 V
α UMi	2.5v	*	F8 Ib	θ Aur	2.69v?	-0.08	B9.5pv
β Ari	2.65v?	$+0.13$	A5 V	β CMa	1.98v?	-0.24	B1 II
γ' And	2.28	*	K3 II	α Car	-0.73	$+0.16$	F0 Ib
α Ari	2.00v	$+1.15$	K2 III	γ Gem	1.93	0.00	A0 IV
o Cet	2.00v	*	gM6e	α CMa	-1.47	$+0.01$	A1 V
α Cet	2.52	$+1.64$	M2 III	τ Pup	2.92	$+1.18$	K0 III
β Per	2.2v	*	B8 V	ε CMa	1.50	-0.22	B2 II
α Per	1.79v?	$+0.48$	F5 Ib	δ CMa	1.84	$+0.68$	F8 Ia
ζ Per	2.83v?	$+0.13$	B1 Ib	π Pup	2.70v?	$+1.63$	K5 III
α Tau	0.86v?	$+1.53$	K5 III	η CMa	2.40	-0.07	B5 Ia
ι Aur	2.66v?	$+1.57$	K3 II	α Gem	1.99	*	A1 V
β Eri	2.80v?	$+0.13$	A3 III	α CMi	0.34v?	$+0.40$	F5 IV
β Ori	0.08v?	-0.03	B8 Ia	β Gem	1.15v?	$+1.00$	K0 III
α Aur	0.09v	$+0.80$	G8 III? + F	ζ Pup	2.25	-0.26	O5f

Appendix I (continued)

Star	V	$B - V$	Spectral type	Star	V	$B - V$	Spectral type
ρ Pup	2.88v	*	F6p II	α Lup	2.30	−0.22	B2 II
γ^2 Vel	1.82v?	−0.26	WC7 + O7?	ε Boo	2.60	*	K0 II–III + A2 V
ε Car	1.85v?	+1.30	K0 II + B	α^2 Lib	2.75	+0.15	Am
δ Vel	1.95	+0.04	A0 V	β UMi	2.08v?	*	K4 III
λ Vel	2.30v	+1.70	K5 Ib	β Lup	2.67	−0.23	B2 IV
β Car	1.67	0.00	A1 IV	β Lib	2.61	−0.11	B8 V
ι Car	2.24v?	+0.18	F0 I	γ Lup	2.77	−0.22	B2n V
$\varkappa$ Vel	2.49	−0.20	B2 IV	α CrB	2.23v	−0.02	A0 V
α Hya	1.99v?	+1.41	K4 III	α Ser	2.65	+1.17	K2 III
α Leo	1.36v?	−0.11	B7 V	δ Sco	2.32	−0.11	B0 V
γ^1 Leo	2.61	*	K0 III	β' Sco	2.63v?	−0.08	B0.5 V
μ Vel	2.68	+0.90	G5 III	η Dra	2.77v?	*	G8 III
β UMa	2.36v?	+0.02	A1 V	α Sco	1.08v	+1.80	M1 Ib
α UMa	1.79v?	+1.06	K0 II–III	β Her	2.83v?	*	G8 III
δ Leo	2.55v?	+0.13	A4 V	τ Sco	2.82	+0.26	B0 V
β Leo	2.14v?	+0.09	A3 V	ζ Oph	2.56	+0.02	O9.5 V
γ UMa	2.44v?	0.00	A0 V	α TrA	1.91	+1.44	K4 III
δ Cen	2.88v?	*	B2pe? V?	ε Sco	2.28v?	+1.15	K2 III–IV
γ Crv	2.60v?	−0.11	B8 III	η Oph	2.44	+0.05	A2.5 V
α' Cru	1.58	*	B1 IV	β Ara	2.84	+1.46	K3 Ib
γ Cru	1.62v?	+1.60	M3 II	ν Sco	2.70	−0.22	B3 Ib
β Crv	2.66v?	+0.89	G5 III	λ Sco	1.62	−0.24	B1 V
α Mus	2.71v?	−0.20	B3 IV	α Oph	2.08v?	+0.15	A5 III
γ Cen	2.16	−0.02	A0 III	θ Sco	1.88	+0.40	F0 Ib
γ Vir	2.90	*	F0 V + F0 V	$\varkappa$ Sco	2.41	−0.23	B2 IV
β Cru	1.24v	−0.24	B0.5 IV	β Oph	2.77	+1.16	K2 III
ε UMa	1.76v	−0.02	A0pv	γ Dra	2.22	+1.52	K5 III
α^2 CVn	2.89v	−0.12	B9.5pv	δ Sgr	2.70	+1.38	K2 III
ε Vir	2.81v?	*	G9 II–III	ε Sgr	1.84	−0.03	B9 IV
ι Cen	2.76	+0.04	A2 V	λ Sgr	2.84	+1.04	K2 III
ζ UMa	2.40v?	*	A2 V	α Lyr	0.04v?	0.00	A0 V
α Vir	0.96v	−0.23	B1 V	σ Sgr	2.10	−0.20	B2 V
ε Cen	2.30	−0.23	B1 V	ζ Sgr	2.60	+0.08	A2 III
η UMa	1.86v?	−0.20	B3 V	γ Aql	2.62	*	K3 II
η Boo	2.69	+0.58	G0 IV	α Aql	0.77	+0.22	A7 V
β Cen	0.59	−0.22	B1 II	γ Cyg	2.24v?	*	F8 Ib
θ Cen	2.05	+1.02	K0 III–IV	α Pav	1.93v?	−0.20	B3 IV
α Boo	0.06v?	+1.23	K2p III	α Cyg	1.26v?	+0.09	A2 Ia
η Cen	2.35v?	−0.20	B1.5ne? V	ε Cyg	2.45	+1.03	K0 III
α Cen	0.06	*	G2V + dK1	α Cep	2.41	+0.23	A7 IV–V

Appendix I (continued)

Star	V	$B - V$	Spectral type	Star	V	$B - V$	Spectral type
ε Peg	2.42v?	+1.56	K2 Ib	α PsA	1.16	+0.09	A3 V
α Gru	1.73v?	−0.13	B5 V	β Peg	2.56v	+1.66	M2 II–III
α Tuc	2.85	+1.39	K3 III	α Peg	2.49v?	−0.05	B9.5 III
β Gru	2.24v?	+1.8	M3 II				

The data in this table were taken from *Catalogue of Bright Stars* by D Hoffleit (Yale University Observatory, 1964).

Appendix II

North polar sequence

Star number	Position						Photographic magnitude
	RA(2000)			Dec(2000)			
	h	m	s	°	′	″	
1	17	32	14	86	35	22	4.48
2	10	20	12	85	51	20	5.28
3	11	30	09	87	01	41	5.81
4	17	30	40	86	58	19	5.99
5	01	32	07	89	00	56	6.49
6	09	22	30	88	34	27	7.11
7	11	21	38	87	38	25	7.31
8	03	38	26	89	06	32	8.23
9	04	04	56	88	55	35	8.83
10	10	02	19	89	34	58	9.02
11	11	51	33	88	55	55	9.55
12	12	55	56	88	57	41	9.86
13	13	15	16	89	09	59	10.30
14	11	51	55	89	35	23	10.65
15	10	52	47	89	18	09	11.08
16	13	38	38	89	36	04	11.40
17	13	59	18	89	21	58	11.63
18	13	45	57	89	26	23	12.06
19	12	48	23	89	25	09	12.42
20	12	30	10	89	28	46	12.73

Appendix II (continued)

| Star number | RA(2000) | | | Dec(2000) | | | Photographic magnitude |
	h	m	s	°	′	″	
21	13	28	36	89	21	28	12.96
22	13	27	05	89	25	18	13.19
23	12	10	16	89	26	55	13.34
24	13	05	08	89	25	18	13.64
25	12	50	05	89	24	23	13.87
26	12	35	37	89	26	49	14.33
27	12	02	46	89	30	20	14.69
28	11	51	17	89	19	58	15.11
29	11	48	31	89	21	17	15.66
30	11	54	21	89	19	57	16.04
31	12	22	44	89	24	34	16.22
32	11	47	51	89	27	07	16.62
33	12	44	01	89	26	12	16.96
34	12	20	20	89	24	31	17.25
35	11	59	34	89	26	23	17.50
36	12	03	00	89	26	44	17.75
37	12	17	13	89	29	03	17.94
38	12	20	21	89	27	26	18.31
39	12	10	35	89	28	47	18.67
40	12	00	45	89	27	37	19.03
41	12	08	02	89	27	51	19.28
42	11	54	30	89	24	39	19.68
43	12	12	48	89	28	01	20.22
44	12	11	33	89	27	57	20.44
45	12	13	21	89	28	24	20.74
46	11	55	42	89	26	16	21.10

Figure AII.1 gives a finder chart for the fainter stars in this sequence. The data for the table and the finder chart come from Leavitt H S *Annals of the Harvard College Observatory* (**71** (3), 49–52).

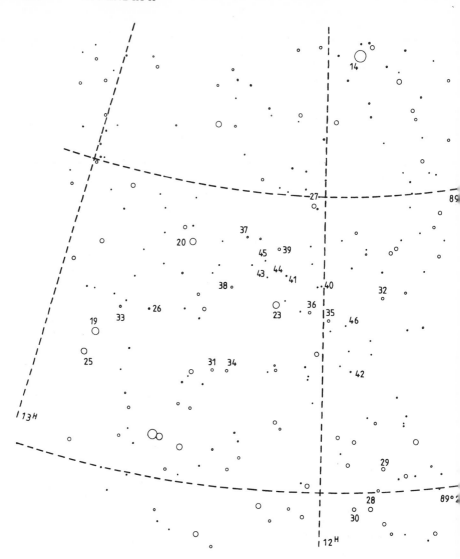

Figure II.0.1. Finder chart for faint stars in the north polar sequence (epoch 2000.0).

Appendix III

Standard stars for the UBV photometric system

PRIMARY STANDARD STARS FOR THE UBV PHOTOMETRIC SYSTEM

Star	Spectral type	Magnitudes		
		U	B	V
10 Lac	O9 V	3.64	4.68	4.88
η Hya	B3 V	3.36	4.10	4.30
τ Her	B5 IV	3.18	3.74	3.89
β Lib	B8 V	2.13	2.50	2.61
HD 18331	A1 V	5.30	5.25	5.17
α Ari	K2 III	4.27	3.15	2.00
α Ser	K2 III	5.06	3.82	2.65
ε CrB	K3 III	6.64	5.38	4.15
HD 219134	K3 V	7.47	6.58	5.57
β Cnc	K4 III	6.78	5.00	3.52

The data in this table were taken from Johnson H L *Basic Astronomical Data* p. 208 (ed W A Hiltner, University of Chicago Press, 1963).

SECONDARY STANDARD STARS FOR THE UBV PHOTOMETRIC SYSTEM

Star	Spectral type	Magnitudes			Star	Spectral type	Magnitudes		
		U	B	V			U	B	V
γ Peg	B2 IV	1.73	2.60	2.83	η Psc	G8 III	5.35	4.59	3.62
θ And	A2 V	4.71	4.67	4.61	107 Psc	K1 V	6.56	6.06	5.23
ν And	B5 V	3.80	4.38	4.53	τ Cet	G8p V	4.42	4.22	3.50
θ Cas	A7 V	4.61	4.50	4.33	β Ari	A5 V	2.88	2.78	2.65
δ Cas	A5 V	2.93	2.81	2.68	$-18°$ 359	*	12.87	11.71	10.18

Appendix III (continued)

Star	Spectral type	U	B	V	Star	Spectral type	U	B	V
$+2^\circ$ 348	*	12.55	11.47	10.03	β Vir	F8 V	4.26	4.16	3.61
ξ^2 Cet	B9 III	4.09	4.22	4.28	HR 4550	G8p V	7.37	7.20	6.45
HR 753 B	*	14.38	13.26	11.65	γ UMa	A0 V	2.45	2.44	2.44
$\varkappa$ Cet	G5 V	5.68	5.50	4.82	δ UMa	A3 V	3.46	3.39	3.31
o Tau	G8 III	5.10	4.48	3.59	γ Crv	B8 III	2.14	2.49	2.60
HR 1046	A1 V	5.16	5.13	5.08	$+0^\circ$ 2989	M0.5 V	11.16	9.90	8.49
ε Eri	K2 V	5.19	4.62	3.73	78 UMa	F2 V	5.30	5.29	4.93
ε Tau	K0 III	5.44	4.56	3.54	β Com	G0 V	4.92	4.85	4.28
π^3 Ori	F6 V	3.63	3.64	3.19	61 Vir	G6 V	5.71	5.46	4.75
π^4 Ori	B2 III	2.72	3.52	3.69	α Vir	B1 V	-0.21	0.73	0.96
η Aur	B3 V	2.32	2.99	3.17	80 UMa	A5 V	4.25	4.17	4.01
β Eri	A3 III	3.03	2.93	2.80	70 Vir	G5 V	5.95	5.69	4.98
HD 35299	B2 V	4.61	5.48	5.70	η Boo	G0 IV	3.46	3.27	2.69
γ Ori	B2 III	0.54	1.41	1.64	109 Vir	A0 V	3.71	3.74	3.74
β Tau	B7 III	1.03	1.52	1.65	α^1 Lib	*	5.53	5.57	5.16
-3° 1123	*	10.65	9.44	7.97	α^2 Lib	*	2.98	2.90	2.75
ν Ori	B0 V	3.30	4.37	4.63	-7° 4003	*	13.37	12.17	10.56
HD 36591	B1 V	4.21	5.15	5.35	β Ser A	A2 IV	3.80	3.73	3.67
ι Ori	O9 III	1.44	2.52	2.77	λ Ser	G0 V	5.13	5.03	4.43
ε Ori	B0 Ia	0.47	1.51	1.70	γ Ser	F6 V	4.30	4.33	3.85
ζ Lep	*	3.71	3.65	3.55	-12° 4523	*	12.91	11.73	10.13
134 Tau	B9 IV	4.65	4.83	4.90	ζ Oph	O9.5 V	1.72	2.58	2.56
$+17^\circ$ 1320	*	12.31	11.13	9.63	-4° 4225	K5 V	9.94	8.89	7.73
γ Gem	A0 IV	1.96	1.93	1.93	-4° 4226	M3.5 V	12.59	11.50	10.07
$+5^\circ$ 1668	*	12.50	11.38	9.82	-2° 3312	K7 V	10.16	8.90	7.54
λ Gem	A3 V	3.79	3.69	3.58	α Oph	A5 III	2.33	2.23	2.08
ρ Gem	F0 V	4.45	4.48	4.16	β Oph	K2 III	5.17	3.93	2.77
$\varkappa$ Gem	G8 III	5.18	4.50	3.57	γ Oph	A0 V	3.83	3.79	3.75
HR 3314	A0 V	3.86	3.88	3.90	$+4^\circ$ 3561	M5 V	12.57	11.28	9.54
ι UMa	A7 V	3.39	3.32	3.14	-3° 4233	*	12.11	10.90	9.38
θ Hya	B9.5 V	3.69	3.82	3.88	γ Lyr	B9 III	3.11	3.20	3.25
-12° 2918	*	12.74	11.59	10.06	ζ Aql	*	2.98	2.99	2.99
11 LMi	G8 IV–V	6.63	6.18	5.41	$+4^\circ$ 4048	M3.5 V	11.78	10.62	9.13
21 LMi	A7 V	4.74	4.66	4.48	$+3^\circ$ 4065	*	6.01	6.84	6.82
α Leo	B7 V	0.89	1.25	1.36	$\varkappa$ Aql	B0.5 III	4.08	4.95	4.96
λ UMa	A2 IV	3.54	3.48	3.45	α Aql	A7 IV–V	1.07	0.99	0.77
$+1^\circ$ 2447	*	12.34	11.15	9.63	β Aql	G8 IV	5.05	4.57	3.71
ρ Leo	B1 Ib	2.76	3.71	3.85	α Del	B9 V	3.49	3.71	3.77
90 Leo AB	*	5.15	5.79	5.95	ε Aqr	A1 V	3.82	3.78	3.77
β Leo	A3 V	2.30	2.23	2.14	-15° 6290	*	12.92	11.77	10.17

Appendix III (continued)

Star	Spectral type	Magnitudes			Star	Spectral type	Magnitudes		
		U	B	V			U	B	V
74 Aqr	*	5.41	5.73	5.81	ι Psc	F7 V	4.64	4.64	4.13
α Peg	B9 V	2.38	2.44	2.49	+1° 4774	M2 V	11.55	10.46	8.98

The data in this table were taken from Johnson H L *Basic Astronomical Data* p. 209 (ed W A Hiltner, University of Chicago Press, 1963).

Appendix IV

Standard stars for the *uvby* photometric system

Star	Spectral type	$(b - y)$	m_1	c_1
α And	B9p	− 0.046	0.120	0.520
β Cas	F2IV	0.216	0.177	0.785
22 And	F2II	0.273	0.123	1.082
γ Peg	B2IV	− 0.106	0.093	0.116
θ And	A2V	0.026	0.180	1.050
σ And	A2V	0.026	0.187	1.040
x Phe	A7nV	0.100	0.192	0.915
28 And	Am	0.169	0.165	0.869
ζ Cas	B2V	− 0.090	0.087	0.134
π Cas	A5	0.087	0.221	0.902
o Cas	B2V	0.007	0.076	0.479
HD 4775	A + dF	0.355	0.127	0.696
μ And	A5V	0.068	0.194	1.056
θ Cas	A7V	0.087	0.213	0.997
39 Cet	gG5	0.567	0.291	0.328
ρ Psc	dF1	0.256	0.148	0.485
ν And	F8V	0.344	0.179	0.409
107 Psc	K1V	0.492	0.364	0.294
χ Cet	F2IV	0.209	0.184	0.648
β Ari	A5V	0.064	0.211	0.983
α Ari	K2III	0.696	0.526	0.395
β Tri	A5III	0.071	0.191	1.065

Appendix IV (continued)

Star	Spectral type	$(b - y)$	m_1	c_1
14 Ari	F2III	0.210	0.185	0.874
δ Tri	G0V	0.386	0.191	0.254
10 Tri	A2V	0.011	0.161	1.145
9 Per	A2Ia	0.321	-0.038	0.753
12 Tri	sgA7	0.178	0.211	0.780
ν Ari	A7V	0.092	0.182	1.095
35 Ari	B3V	-0.052	0.097	0.333
π Cet	B7V	-0.048	0.096	0.607
38 Ari	A7IV	0.135	0.188	0.837
μ Cet	F0IV	0.189	0.187	0.762
ι Per	G0V	0.376	0.201	0.376
α Per	F5Ib	0.302	0.195	1.074
o Tau	G8III	0.547	0.335	0.424
16 Tau	B7IV	-0.001	0.105	0.647
18 Tau	B8V	-0.022	0.109	0.637
27 Tau	B8III	-0.019	0.092	0.708
HD 24357	dF1	0.221	0.166	0.610
ψ Tau	dF1	0.226	0.159	0.588
45 Tau	dF4	0.231	0.163	0.592
μ Per	G0Ib	0.614	0.268	0.551
HD 27022	G5III	0.510	0.289	0.405
ω Tau	Am	0.146	0.235	0.745
51 Tau	dA8	0.175	0.185	0.787
γ Tau	K0III	0.596	0.427	0.388
δ Tau	K0III	0.597	0.430	0.409
63 Tau	Am	0.180	0.237	0.738
64 Tau	A7V	0.081	0.210	0.981
$\varkappa$ Tau	A7V	0.070	0.200	1.054
67 Tau	A5n	0.149	0.193	0.840
68 Tau	A2IV	0.020	0.193	1.046
71 Tau	F0V	0.150	0.188	0.934
ε Tau	K0III	0.616	0.448	0.422
θ^1 Tau	K0III	0.580	0.390	0.389
θ^2 Tau	A7III	0.099	0.202	1.013
79 Tau	A6n	0.114	0.225	0.912
83 Tau	dF1	0.154	0.201	0.814
ρ Tau	F0V	0.144	0.205	0.823
α Tau	K5III	0.955	0.814	0.373

Appendix IV (continued)

Star	Spectral type	$(b - y)$	m_1	c_1
π^3 Ori	F6V	0.298	0.164	0.412
π^4 Ori	B2III	-0.055	0.070	0.131
ι Aur	K3II	0.937	0.775	0.307
ι Tau	A7V	0.080	0.203	1.031
η Aur	B3V	-0.085	0.104	0.318
104 Tau	G4V	0.415	0.197	0.332
16 Ori	Am	0.138	0.245	0.840
λ Aur	G0V	0.389	0.206	0.363
HD 36591	B1V	-0.077	0.074	-0.002
α Lep	F0Ib	0.139	0.148	1.504
122 Tau	A5	0.133	0.203	0.850
α Col	B8eV	-0.046	0.086	0.650
136 Tau	A0III	0.000	0.136	1.153
χ^1 Ori	G0V	0.378	0.194	0.307
λ Col	B5V	-0.076	0.121	0.408
γ Col	B2.5IV	-0.076	0.092	0.363
40 Aur	Am	0.139	0.222	0.923
71 Ori	F6V	0.293	0.163	0.448
74 Ori	F5IV–V	0.284	0.153	0.438
45 Aur	F5III	0.284	0.171	0.632
β CMa	B1II	-0.090	0.052	-0.002
HD 46089	A4V	0.101	0.172	0.999
γ Gem	A0IV	0.007	0.149	1.186
ε Gem	G8Ib	0.869	0.654	0.283
ψ^5 Aur	G0V	0.357	0.185	0.371
ξ Gem	F5IV	0.288	0.167	0.552
16 Lyn	A2V	0.011	0.163	1.101
π CMa	gF2	0.248	0.150	0.649
γ CMa	B8II	0.045	0.097	0.560
21 Mon	A8n	0.184	0.187	0.878
λ Gem	A3V	0.048	0.198	1.055
δ Gem	F0IV	0.221	0.156	0.696
β CMi	B7V	0.038	0.113	0.799
ρ Gem	F0V	0.214	0.155	0.613
64 Gem	A6V	0.062	0.202	1.013
δ^1 CMi	A8	0.131	0.173	1.203
68 Gem	A1V	0.036	0.149	1.178
25 Mon	F5III	0.289	0.169	0.653

Appendix IV (continued)

Star	Spectral type	$(b - y)$	m_1	c_1
o Gem	F3III	0.266	0.176	0.660
HD 61555/6	B8, B5IV	−0.074	0.114	0.402
HD 61831	B2.5V	−0.082	0.100	0.304
ϰ Gem	G8III	0.573	0.379	0.398
β Gem	K0III	0.611	0.427	0.420
81 Gem	K5III	0.893	0.743	0.444
HD 64503	B2.5V	−0.083	0.097	0.247
HD 65810	A3V	0.049	0.158	1.128
27 Lyn	A2V	0.021	0.150	1.099
β Cnc	K4III	0.911	0.765	0.367
χ Cnc	F6V	0.314	0.146	0.384
HD 71155	A0V	−0.005	0.158	1.024
δ Hya	A0V	0.008	0.153	1.091
η Hya	B3V	−0.088	0.092	0.240
HD 74395	G2Ib	0.519	0.289	0.476
σ^2 Cnc	A3	0.084	0.205	0.968
15 UMa	Am	0.169	0.233	0.776
τ UMa	Am	0.217	0.238	0.723
18 UMa	A5V	0.113	0.196	0.892
23 UMa	F0IV	0.211	0.180	0.752
τ^1 Hya	F6V	0.296	0.164	0.448
24 UMa	G2IV	0.488	0.254	0.347
θ UMa	F6IV	0.314	0.153	0.463
10 LMi	G8III	0.562	0.346	0.389
11 LMi	G8IV−V	0.473	0.304	0.372
ϰ Hya	B5V	−0.069	0.107	0.405
o Leo	A2 + F6II	0.306	0.234	0.615
HD 83944	B9V	−0.042	0.143	0.818
HD 84737	G1V	0.390	0.203	0.382
ν UMa	F2IV	0.196	0.162	0.830
19 LMi	F5V	0.300	0.165	0.457
20 LMi	G4V	0.416	0.234	0.388
21 LMi	A7V	0.111	0.196	0.870
α Leo	B7V	−0.041	0.102	0.712
ζ Leo	F0III	0.196	0.169	0.986
40 Leo	F6IV	0.297	0.171	0.459
γ Leo	K0IIIp	0.689	0.457	0.373

Appendix IV (continued)

Star	Spectral type	$(b-y)$	m_1	c_1
30 LMi	F0V	0.151	0.195	0.956
36 UMa	F8V	0.341	0.172	0.331
β Sex	B6V	-0.064	0.116	0.466
ρ Leo	B1Ib	-0.025	0.033	-0.039
37 UMa	F1V	0.228	0.159	0.574
37 LMi	G0II	0.512	0.297	0.477
θ Car	B0.5pV	-0.102	0.071	-0.075
47 UMa	G0V	0.392	0.203	0.337
49 UMa	F0m	0.145	0.194	1.007
HD 95370	A2IV	0.060	0.175	1.121
α UMa	K0II–III + F	0.660	0.437	0.394
ψ UMa	K1III	0.703	0.524	0.396
β Crt	A2IV	0.010	0.161	1.201
56 UMa	G8II	0.609	0.405	0.401
γ Crt	A7IV	0.117	0.193	0.899
90 Leo	B3V	-0.070	0.104	0.319
ξ Vir	A4	0.091	0.198	0.919
93 Leo	A + G5III–IV	0.352	0.186	0.725
β Leo	A3V	0.044	0.210	0.975
β Vir	F8V	0.354	0.187	0.414
HD 103095	G8Vp	0.484	0.224	0.166
γ UMa	A0V	0.006	0.153	1.113
HD 105382	B2IIIne	-0.079	0.102	0.264
16 Vir	K0III	0.720	0.485	0.513
12 Com	G0III–IV + A3V	0.322	0.175	0.779
18 Com	F5IV	0.289	0.172	0.611
η Crv	F0IV	0.247	0.158	0.550
β CVn	G0V	0.385	0.182	0.296
τ Cen	A2V	0.021	0.159	1.087
31 Com	G0III	0.435	0.193	0.411
HD 111968	A7III	0.128	0.176	0.977
78 UMa	F2V	0.244	0.168	0.578
β Com	G0V	0.370	0.191	0.337
59 Vir	F8V	0.376	0.191	0.383
20 CVn	F0pII–III	0.180	0.231	0.913
81 UMa	A5V	0.097	0.192	0.928
1 Cen	F2III	0.247	0.155	0.562
η UMa	B3V	-0.080	0.106	0.296

Appendix IV (continued)

Star	Spectral type	$(b - y)$	m_1	c_1
η Boo	G0IV	0.376	0.203	0.476
HD 122563	sdG0	0.633	0.090	0.550
χ Cen	B2V	-0.095	0.089	0.176
12 Boo	F8IV	0.348	0.175	0.441
α Boo	K2IIIp	0.755	0.526	0.491
γ Boo	A7III	0.116	0.191	1.008
σ Boo	F2V	0.253	0.132	0.488
109 Vir	A0V	0.007	0.134	1.081
α^1 Lib	F2V	0.262	0.155	0.497
α^2 Lib	Am	0.074	0.192	0.996
λ Lup	B3V	-0.080	0.099	0.282
45 Boo	F5V	0.285	0.165	0.449
1 Lup	F0I	0.244	0.136	1.381
δ Boo	G8III	0.587	0.346	0.410
β Lib	B8V	-0.040	0.100	0.750
ι Dra	K2III	0.711	0.567	0.429
α CrB	A0V	0.000	0.144	1.060
HD 139664	F5IV–V	0.264	0.150	0.470
α Ser	K2IIIp	0.715	0.572	0.445
λ Ser	G0V	0.385	0.199	0.354
1 Sco	B1.5Vn	0.005	0.068	0.127
γ Ser	F6V	0.319	0.151	0.401
λ CrB	F2	0.233	0.161	0.662
π Sco	B1V + B2	-0.066	0.058	0.028
ε CrB	K3III	0.751	0.570	0.414
δ Sco	B0.5IV	-0.019	0.038	-0.017
ρ CrB	G2V	0.394	0.183	0.322
θ Dra	F8IV–V	0.354	0.174	0.460
ω^1 Sco	B1V	0.033	0.041	0.022
ω^2 Sco	gG2	0.521	0.284	0.448
ν Sco	B2IV–V	0.072	0.059	0.150
τ Her	B5IV	-0.056	0.089	0.440
γ Her	A9III	0.168	0.192	1.008
22 Sco	B2V	-0.046	0.085	0.202
τ Sco	B0V	-0.100	0.051	-0.065
ζ Oph	O9.5V	0.085	0.012	-0.061
ζ Her	G0IV	0.415	0.207	0.408

Appendix IV (continued)

Star	Spectral type	$(b-y)$	m_1	c_1
20 Oph	F5IV–V	0.307	0.161	0.536
59 Her	A3III	0.002	0.180	1.094
60 Her	A3IV	0.070	0.203	0.988
η Oph	A2.5V	0.026	0.184	1.084
72 Her	G2V	0.409	0.182	0.309
β Dra	G2II	0.610	0.323	0.423
o Ser	A2V	0.047	0.166	1.115
ι Her	B3V	-0.064	0.078	0.294
58 Oph	F3V	0.301	0.150	0.413
β Oph	K2III	0.721	0.549	0.453
γ Oph	A0V	0.026	0.165	1.054
γ Dra	K5III	0.943	0.811	0.374
67 Oph	B5Ib	0.079	0.023	0.303
68 Oph	A1V	0.033	0.136	1.094
θ Ara	B2Ib	0.004	0.035	0.027
γ Sct	A3V	0.044	0.141	1.219
α Lyr	A0V	0.004	0.157	1.089
111 Her	A3V	0.061	0.216	0.942
ε CrA	F0V	0.255	0.149	0.633
γ Lyr	B9III	0.001	0.093	1.219
ζ Aql	A0V	0.012	0.147	1.080
HD 178233	A5	0.176	0.190	0.752
α CrA	A2	0.018	0.185	1.062
$\varkappa$ Cyg	K0III	0.579	0.390	0.430
ρ^1 Sgr	F0IV	0.129	0.188	0.955
δ Aql	F0IV	0.204	0.168	0.715
$\varkappa$ Aql	B0.5III	0.079	-0.014	-0.022
ι Aql	B5III	-0.017	0.086	0.577
σ Dra	K0V	0.472	0.320	0.261
θ Cyg	F4V	0.261	0.158	0.506
α Sge	G0II	0.497	0.260	0.466
16 Cyg	G2V	0.410	0.214	0.375
HD 186427	G5V	0.416	0.226	0.354
γ Aql	K3II	0.936	0.760	0.294
17 Cyg	F5V	0.316	0.155	0.435
α Aql	A7V	0.137	0.178	0.880
o Aql	F8V	0.356	0.188	0.404
β Aql	G8IV	0.521	0.306	0.341

Appendix IV (continued)

Star	Spectral type	$(b-y)$	m_1	c_1
φ Aql	A1V	0.002	0.174	1.020
30 Cyg	A3III	0.068	0.150	1.310
α^1 Cap	G3Ib	0.571	0.327	0.389
ν Cap	B9V	-0.022	0.135	1.025
α Pav	B2.5V	-0.092	0.087	0.271
γ Cyg	F8Ib	0.396	0.296	0.885
η Del	A2V	0.038	0.196	0.975
ζ Del	A3V	0.066	0.176	1.107
α Del	B9V	-0.019	0.125	0.893
ψ Cap	F5V	0.271	0.157	0.481
ε Cyg	K0III	0.627	0.415	0.425
55 Cyg	B3Ia	0.356	-0.067	0.153
56 Cyg	A4m?	0.108	0.209	0.897
η Cap	A3m?	0.088	0.186	0.949
61 CygA	K5V	0.656	0.677	0.134
61 CygB	K7V	0.791	0.676	0.067
ζ Cyg	G8II	0.591	0.446	0.296
σ Cyg	B9Iab	0.138	0.027	0.571
α Cep	A7IV–V	0.125	0.190	0.936
γ Pav	F8V	0.321	0.124	0.323
5 Peg	F0IV	0.203	0.170	0.892
9 Cep	B2Ib	0.275	-0.051	0.135
9 Peg	G5Ib	0.709	0.468	0.354
13 Peg	F2III	0.263	0.156	0.545
γ Gru	B8III	-0.048	0.104	0.734
α Gru	B7IV	-0.061	0.105	0.576
ι Peg	F5V	0.296	0.159	0.446
μ PsA	A2V	0.024	0.172	1.084
π Peg	F5II–III	0.304	0.177	0.778
ε Cep	F0IV	0.169	0.192	0.787
35 Peg	K0III–IV	0.638	0.426	0.404
α Lac	A2V	0.001	0.173	1.030
9 Lac	A7IV	0.142	0.174	0.948
10 Lac	O9V	-0.070	0.042	-0.110
β Oct	dA9	0.110	0.198	0.908
ζ Peg	B8V	-0.035	0.114	0.867
η Peg	G8II? + F?	0.535	0.296	0.499

Appendix IV (continued)

Star	Spectral type	$(b-y)$	m_1	c_1
ξ Peg	F7V	0.330	0.147	0.407
ε Gru	A2V	0.041	0.168	1.154
δ Aqr	A3V	0.034	0.162	1.176
α PsA	A3V	0.036	0.206	0.991
51 Peg	G5V	0.416	0.232	0.364
α Peg	B9.5III	−0.012	0.130	1.128
59 Peg	A2V	0.075	0.165	1.090
7 And	F0V	0.181	0.172	0.723
γ Tuc	F0III	0.262	0.146	0.579
τ Peg	A5IV	0.104	0.166	1.013
ν Peg	F8IV	0.390	0.186	0.461
ι And	B8V	−0.031	0.100	0.784
ι Psc	F7V	0.329	0.164	0.395
x And	B8V	−0.035	0.131	0.831
82 Peg	A3	0.099	0.181	0.967
ω Psc	F4IV	0.266	0.150	0.630
ε Tuc	B9IV	−0.032	0.104	0.894
85 Peg	G2V	0.428	0.189	0.215
ζ Scl	B5V	−0.067	0.107	0.461

The data in this table were taken from Crawford D J and Barnes J V 1970 *Astron. J.* **75** 978.

Appendix V

Standard stars for MK spectral types

Star	HD number	Spectral type	Star	HD number	Spectral type
(i) Luminosity class V			β^2 Sco	144218	B2 V
			22 Sco	148605	B2 V
	46202	O9 V		191746	B2 V
	52266	O9 V		208947	B2 V
	57682	O9 V		218440	B2 V
14 Cep	209481	O9 V	σ Sgr	175191	B2.5 V
10 Lac	214680	O9 V		6300	B3 V
	34078	O9.5 V	35 Ari	16908	B3 V
σ Ori	37468	O9.5 V	η Aur	32630	B3 V
ζ Oph	149757	O9.5 V	η Hya	74280	B3 V
ν Ori	36512	B0 V	η UMa	120315	B3 V
τ Sco	149438	B0 V		178849	B3 V
	206183	B0 V		191263	B3 V
	207538	B0 V	16 Peg	208057	B3 V
	8965	B0.5 V		218537	B3 V
40 Per	22951	B0.5 V		4142	B5 V
	7252	B1 V	ν And	4727	B5 V
	24131	B1 V		14372	B5 V
42 Ori	37018	B1 V	ρ Aur	34759	B5 V
ω Sco	144470	B1 V	$\varkappa$ Hya	83754	B5 V
ξ Cas	3901	B2 V	λ Cyg	198183	B5 V
o Cas	4180	B2 V	ψ^2 Aqr	219688	B5 V

Appendix V (continued)

Star	HD number	Spectral type	Star	HD number	Spectral type
β Sex	90994	B6 V	46 Tau	26690	F3 V
α Leo	87901	B7 V		27524	F5 V
18 Tau	23324	B8 V	45 Boo	134083	F5 V
21 Tau	23432	B8 V	ι Peg	210027	F5 V
ζ Peg	214923	B8 V	π³ Ori	30652	F6 V
ι And	222173	B8 V	γ Ser	142860	F6 V
ω² Aqr	222661	B9.5 V	110 Her	173667	F6 V
4 Aur	31647	A0 V	θ Per	16895	F7 V
	71155	A0 V	θ Boo	126660	F7 V
γ UMa	103287	A0 V	χ Dra	170153	F7 V
109 Vir	130109	A0 V	ι Psc	222368	F7 V
	140775	A0 V	γ And	9826	F8 V
	146624	A0 V		27808	F8 V
γ Oph	161868	A0 V		27383	F9 V
α Lyr	172167	A0 V	β Vir	102870	F9 V
α CrB	193006	A0 V	η Cas	4614	G0 V
	18331	A1 V	β CVn	109358	G0 V
	21447	A1 V	β Com	114710	G0 V
ι Ser	140159	A1 V	λ Ser	141004	G0 V
39 Dra	170073	A1 V		27836	G1 V
	193702	A1 V		115043	G1 V
ε Aqr	198001	A1 V		10307	G2 V
θ And	1280	A2 V	16 Cyg	186408	G2 V
θ Leo	97633	A2 V	Sun		G2 V
λ Gem	56537	A3 V	κ Cet	20630	G5 V
β Leo	102647	A3 V		186427	G5 V
α PsA	216956	A3 V	61 UMa	101501	G8 V
δ Leo	97603	A4 V	ξ Boo	131156	G8 V
δ Cas	8538	A5 V	54 Psc	3651	K0 V
β Ari	11636	A5 V		124752	K0 V
80 UMa	116842	A5 V	70 Oph	165341	K0 V
θ Cas	6961	A7 V	σ Dra	185144	K0 V
21 LMi	87696	A7 V	ε Eri	20320	K2 V
ρ Gem	58946	F0 V		109011	K2 V
γ Vir	110379/80	F0 V		110463	K3 V
78 UMa	113139	F2 V		128165	K3 V
σ Boo	128167	F2 V		219134	K3 V
	26015	F3 V	61 Cyg A	201091	K5 V
			61 Cyg B	201092	K7 V

Appendix V (continued)

Star	HD number	Spectral type	Star	HD number	Spectral type
$+56°\ 1458$		K7 V	o Per	23180	B1 III
	147379	M0 V	σ Sco	147165	B1 III
	95735	M2 V	π^4 Ori	30836	B2 III
			γ Ori	35468	B2 III
(ii) Luminosity class IV			12 Lac	214993	B2 III
δ Sco	143275	B0.3 IV		21483	B3 III
γ Peg	886	B2 IV	δ Per	22928	B5 III
ζ Cas	3360	B2 IV	τ Ori	34503	B5 III
τ Her	147394	B5 IV	ι Aql	184930	B5 III
19 Tau	23338	B6 IV			
			17 Tau	23302	B6 III
16 Tau	23288	B7 IV	η Tau	23630	B7 III
α Del	196867	B9 IV	β Tau	35497	B7 III
γ Gem	47105	A0 IV	27 Tau	23850	B8 III
ι UMa	76644	A7 IV	γ Lyr	176437	B9 III
μ Cet	17094	F0 IV			
			δ Cyg	186882	B9.5 III
57 Tau	27397	F0 IV	α Dra	123299	A0 III
ε Cep	211336	F0 IV	β Eri	33111	A3 III
1 Mon	40535	F2 IV	θ Gem	50019	A3 III
α Tri	11443	F6 IV	β Tri	13161	A5 III
θ UMa	82328	F6 IV			
			α Oph	159561	A5 III
40 Leo	89449	F6 IV	θ^2 Tau	28319	A7 III
σ Peg	216385	F7 IV	γ Boo	127762	A7 III
ν Peg	209747	F8 IV	γ Her	147547	A9 III
η Boo	121370	G0 IV	ζ Leo	89025	F0 III
ζ Her	150680	G1 IV			
			β Cas	432	F2 III–IV
μ Her	161797	G5 IV	14 Ari	13174	F2 III
β Aql	188512	G8 IV	16 Per	17584	F2 III
δ Eri	23249	K0 IV	36 Per	21770	F4 III
η Cep	198149	K0 IV		17918	F5 III
χ CrB	142091	K1 IVa			
				160365	F6 III
γ Cep	222404	K1 IV	ψ^3 Psc	6903	G0 III
			31 Com	111812	G0 III
(iii) Luminosity class III			24 UMa	82210	G4 III–IV
ι Ori	37043	O9 III		27022	G4 III
	193443	O9 III	ε Dra	188119	G7 IIIb
1 Cam	28446	B0 III	χ Gem	62345	G8 IIIa
	48434	B0 III		88009	G8 IIIa
ε Per	24760	B0.5 III	ε Vir	113226	G8 IIIab
χ Aql	184915	B0.5 III	φ^2 Ori	37160	G8 IIIb

Appendix V (continued)

Star	HD number	Spectral type	Star	HD number	Spectral type
η Psc	9270	G8 III	ε CrB	143107	K2 IIIab
o Psc	10761	G8 III		112127	K2 III
o Tau	21120	G8 III	ι Dra	137759	K2 III
δ Boo	135722	G8 III	α Ser	140573	K2 III
γ Lib	138905	G8 III	ϰ Oph	153210	K2 III
η Dra	148387	G8 III	β Oph	161096	K2 III
β Her	148856	G8 III	ξ Dra	163588	K2 III
ξ Her	163993	G8 III	ϰ Lyr	168775	K2 III
71 Oph	165760	G8 III	109 Her	169414	K2.5 IIIab
ρ Cyg	205435	G8 III	θ Dor	34649	K2.5 III
μ Peg	216131	G8 III		130705	K3 IIIb
10 LMi	82635	G8.5 III	μ Aql	184406	K3 IIIb
β Sge	185958	G9 IIIa	δ And	3627	K3 III
β LMi	90537	G9 IIIab	51 And	9927	K3 III
ι Gem	58207	G9 IIIb	ν UMa	98262	K3 III
δ Dra	180711	G9 III	ρ Boo	127665	K3 III
ϰ Cyg	181276	G9 III	39 Cyg	194317	K3 III
η Ser	168723	K0 III–IV	ω Boo	133124	K4 IIIab
α Cas	3712	K0 IIIa	β Cnc	69267	K4 III
α UMa	95689	K0 IIIa		114873	K4 III
ν Oph	163917	K0 IIIa		125332	K4 III
γ Tau	27371	K0 IIIab	β UMi	131873	K4 III
θ Cet	8512	K0 IIIb	76 Gem	62285	K4.5 III
β Gem	62509	K0 IIIb	ρ Ser	141992	K4.5 III
ϰ Per	19476	K0 III	γ Sge	189319	K5–M0 III
δ Tau	27697	K0 III	α Tau	29139	K5 III
ε Tau	28305	K0 III		82668	K5 III
77 Tau	28307	K0 III	41 Com	113996	K5 III
δ Aur	40035	K0 III	ν Boo	120477	K5 III
τ CrB	145328	K0 III	γ Dra	164058	K5 III
ε Oph	146791	K0 III	α Lyn	80493	K7 IIIab
η Cyg	188947	K0 III	31 Lyn	70272	K7 III
ε Cyg	197989	K0 III	29 Her	149161	K7 III
ι Cep	216228	K0 III	17 Per	17709	K7 III
ψ UMa	96833	K1 III	β And	6860	M0 IIIa
1 Peg	203504	K1 III	ν Gem	60522	M0 III
ι Cet	1522	K1.5 III	μ UMa	89758	M0 III
	37192	K2 IIIa	λ Dra	100029	M0 III
α Ari	12929	K2 IIIab			

Appendix V (continued)

Star	HD number	Spectral type	Star	HD number	Spectral type
	142574	M0 III	*(iv) Luminosity class II*		
δ Oph	146051	M0.5 III	16 Sgr	167263	O9 II
ν Vir	102212	M1 IIIab	δ Ori	36486	O9.5 II
x Ser	141477	M1 IIIab		43818	B0 II
75 Cyg	206330	M1 IIIab		1383	B1 II
55 Peg	218329	M1 IIIab		199216	B1 II
36 Com	112769	M1 IIIb	ε CMa	52089	B2 II
106 Her	168720	M1 IIIb	ι CMa	51309	B3 II
α Vul	183439	M1 IIIb		194779	B3 II
2 Peg	204724	M1 III	γ CMa	53244	B8 II
α Cet	18884	M1.5 III		43836	B9 II
φ Aqr	219215	M1.5 III	19 Aur	34578	A5 II
π Leo	86663	M2 IIIab	o Sco	147084	A5 II
87 Vir	120052	M2 IIIab		196379	A9 II
83 UMa	119228	M2 IIIab		6130	F0 II
χ Peg	1013	M2 III		25291	F0 II
8 And	219734	M2 III	22 And	571	F2 II
λ Aqr	216386	M2.5 IIIa	ν Her	144206	F2 II
η² Dor	43455	M2.5 III	ν Per	23230	F5 II
	148349	M2.5 III		186155	F5 II
	186776	M3 IIIa	41 Cyg	195295	F5 II
μ Gem	44478	M3 IIIab	ε Leo	84441	G1 II
ψ Vir	112142	M3 III	α Sge	185758	G1 II
δ Vir	112300	M3 III		176123	G3 II
	154143	M3 III	β Sct	173764	G4 II
104 Her	167006	M3 III	ω Gem	52497	G5 II
3 Aqr	198026	M3 III		20894	G7 IIb
ψ Peg	224427	M3 III		139862	G8 II–III
30 Psc	224935	M3 III	56 UMa	98839	G8 IIb
	214665	M4 III	ξ' Cet	13611	G8 II
TU CVn	112264	M5 III	ζ Cyg	202109	G8 II
	184313	M5–M5.5 III	ζ Hya	76294	G9 II–III
VY Leo	94705	M5.5 III		4404	G9 II
RZ Ari	18191	M6 III	θ Lyr	180809	K0 II
30 Her	148783	M6 III		26311	K1 II–III
EU Del	196610	M6 III	θ Her	163770	K1 IIa
θ Aps	122250	M6.5 III		45416	K1 II
	207076	M7 III	56 Ori	39400	K1.5 IIb
			π⁶ Ori	31767	K2 II

Appendix V (continued)

Star	HD number	Spectral type	Star	HD number	Spectral type
	75022	K2 II		190919	B1 Ib
σ Oph	157999	K2 II		190603	B1.5 Ia
α Hya	81797	K3 II–III		194279	B1.5 Ia
π Her	156283	K3 IIab		193183	B1.5 Ib
γ And A	12533	K3 IIb		14143	B2 Ia
ι Aur	31398	K3 II	χ² Ori	41117	B2 Ia
	167818	K3 II		13841	B2 Ib
γ Aql	186791	K3 II		13866	B2 Ib
	223173	K3 II	9 Cep	206165	B2 Ib
	193092	K3.5 IIab–IIb	3 Gem	42087	B2.5 Ib
	52938	K3.5 IIb		14134	B3 Ia
	168815	K5 II	o² CMa	53138	B3 Ia
	181475	M0 II	55 Cyg	198478	B3 Ia
	66342	M1 IIa		13267	B5 Ia
	23475	M2 IIab	η CMa	58350	B5 Ia
β Peg	217906	M2.5 II–III		167838	B5 Ia
	10465	M2.5 II	χ Aur	36371	B5 Iab
π Aur	40239	M3 II		9311	B5 Ib
ρ Per	19058	M4 IIb–IIIa	67 Oph	164353	B5 Ib
δ² Lyr	175588	M4 II		15497	B6 Ia
XY Lyr	172380	M4–M5 II		183143	B7 Ia
				14542	B8 Ia
(v) *Luminosity class I*			β Ori	34085	B8 Ia
τ CMa	57061	O9 Ib		199478	B8 Ia
	210809	O9 Ib	53 Cas	12301	B8 Ib
α Cam	30614	O9.5 Ia	13 Cep	208501	B8 Ib
	195592	O9.5 Ia		17088	B9 Ia
ζ Ori	37742/3	O9.5 Ib		21291	B9 Ia
19 Cep	209975	O9.5 Ib	σ Cyg	202850	B9 Iab
ε Ori	37128	B0 Ia	4 Lac	212593	B9 Iab
15 Sgr	167264	B0 Ia		35600	B9 Ib
69 Cyg	204172	B0 Ib		21389	A0 Ia
χ Ori	38771	B0.5 Ia		223960	A0 Ia
	194839	B0.5 Ia	13 Mon	46300	A0 Ib
	192422	B0.5 Ib	η Leo	87737	A0 Ib
χ Cas	2905	B1 Ia		12953	A1 Ia
	216411	B1 Ia		14433	A1 Ia
ζ Per	24398	B1 Ib	9 Per	14489	A2 Ia
ρ Leo	91316	B1 Ib			

Appendix V (continued)

Star	HD number	Spectral type	Star	HD number	Spectral type
α Cyg	197345	A2 Ia	9 Peg	206859	G5 Ib
ν Cep	207260	A2 Ia		77912	G7 Ib−II
	207673	A2 Ib	AX Sgr	165782	G8 Ia
	210221	A3 Ib	ε Gem	48329	G8 Ib
	17378	A5 Ia		208606	G8 Ib
	164514	A5 Ia	12 Peg	207089	K0 Ib
	59612	A5 Ib		63302	K1 Ia−Iab
φ Cas	7927	F0 Ia	ζ Cep	210745	K1.5 Ib
α Lep	36673	F0 Ib		4817	K2 Ib−II
89 Her	163506	F2 Ia		73884	K2 Ib
ν Aql	186689	F2 Ib	ε Peg	206778	K2 Ib
	10494	F5 Ia	o′ CMa	50877	K2.5 Iab
	17971	F5 Ia	BM Sco	160371	K2.5 Ib
	231195	F5 Ia	η Per A	17506	K3 Ib−IIa
44 Cyg	195593	F5 Iab	41 Gem	52005	K3 Ib
α Per	20902	F5 Ib	145 CMa	56577	K3 Ib
35 Cyg	193370	F5 Ib		11092	K4 Ib−IIa
δ CMa	54605	F8 Ia	63 Cyg	201251	K4 Ib−IIa
+31° 3907		F8 Ia		185622A	K4 Ib
γ Cyg	194093	F8 Ib	ξ Cyg	200905	K4.5 Ib−II
	18391	G0 Ia	ψ′ Aur	44537	K5−M0 Iab−Ib
	91629	G0 Iab	TV Gem	42475	M0−M1 Iab
μ Per	26630	G0 Ib		98817	M1 Iab−Ib
β Cam	31910	G0 Ib	BU Gem	42543	M1−M2 Ia−Ib
β Aqr	204867	G0 Ib	α Ori	39801	M1−M2 Ia−Ib
	57146	G1 Iab−Ib	α Sco	148478	M1.5 Iab
	52220	G1 Ib	CPD-57° 3502		M1.5 Iab−Ib
31 Mon	74395	G1 Ib	μ Cep	206936	M2 Ia
o′ Cen	100261	G2 Ia	CPD-59° 4547		M2 Iab
β Dra	159181	G2 Ib−IIa	119 Tau	36389	M2 Iab−Ib
	96746	G2 Iab		202380	M2 Ib
	44362	G2 Ib	AZ Cyg		M2−M4 Iab
ζ Mon	67594	G2 Ib		100930	M2.5 Iab−Ib
α Aqr	209750	G2 Ib	CPD-31° 1790		M3 Iab−Ib
	161664	G3 Iab−Ib	BZ Car	94613	M3 Ib
37 LMi	92125	G3 Ib−II		101007	M3 Ib + B
22 Vul	192713	G3p Ib−II		190788	M3 Ib
25 Gem	47731	G5 Ib	RW Cyg		M3−M4 Ia−Iab
	58134	G5 Ib			

Appendix V (continued)

Star	HD number	Spectral type	Star	HD number	Spectral type
SU Per	14469	M3–M4 Iab	σ CMa	52877	K7 Ib
BO Car	93420	M4 Ib		216946	M0 Ib
EV Car	89845	M4.5 Ia			
			RT Car		M2 0–Ia
α' Her	156014	M5 Ib–II			
			(*vii*) *No luminosity class*		
(*vi*) *Luminosity class 0*				46233	O4
	271182	F8 0		46150	O5
	269953	G0 0	ζ Pup	66811	O5f
R Pup	62058	G1 0–Ia	15 Mon	47839	O7
	269723	G4 0	μ Leo	85503	K1.5
	217476	G5 0			
			BK Vir	108849	M7
	268757	G7 0	R Dor	29712	M8e
	119796	G8 0–Ia	RX Boo	126327	M8
RW Cep	212466	K0 0–Ia			

The data in this table were taken from Johnson H L and Morgan W W 1953 *Astrophys. J.* **117** 313 and Morgan W W and Keenan P C 1973 *Ann. Rev. Astron. Astrophys.* **11** 29. (**N.B.** Recent versions of the MK system have rescaled some of the subclasses. Thus, O9.5 and B0.5 are regarded as full subtypes, while at lower temperatures some subtypes are regarded as less than a decimal subdivision in order to keep to an approximate equal change in the spectra between adjacent subtypes. The bottom of the classification now runs:

G0 G2 G5 G8 K0 K2 K3 K4 K5 M0 M1 M2 M3 M4 M5 M6 M7 M8

(Keenan P C 1985 *IAU Symposium* No 111, 'Calibration of Fundamental Stellar Quantities' p. 121 (eds Hayes D S, Pasinetti L E and Davis Philip A G); Keenan P C and McNeil R C 1976 *Atlas of Spectra of the Cooler Stars* (Publ. Perkins Observatory).))

Appendix VI

Standard stars for polarimetry

Star	Spectral type	V (mag)	π_L† (%)	θ† (deg)	P or U‡
HD 204827	B0 V	7.9	5.6	60	P
HD 154445	B1 V	5.7	3.7	90	P
9 Gem	B3 Ia	6.2	3.0	170	P
55 Cyg	B3 Ia	4.9	2.8	3	P
HD 80558	B7 Iab	5.9	3.3	162	P
HD 183143	B7 Ia	6.9	6.1	0	P
HD 21291	B9 Ia	4.2	3.5	115	P
HD 23512	A0 V	8.1	2.3	30	P
HD 14433	A1 Ia	6.4	3.9	112	P
HD 111613	A1 Ia	5.7	3.2	81	P
HD 160529	A2 Ia	6.7	7.2	20	P
α PsA	A3 V	1.2	0.006	89	U
o Sco	A5 II–III	4.6	4.3	32	P
φ Cas	F0 Ia	5.0	3.4	94	P
HD 84810	F0 – K0 Ib	3.3–4.0	1.6	100	P
β Cas	F2 IV	2.2	0.009	32	U
θ Cyg	F4 V	4.5	0.003	139	U
α CMi	F5 IV	0.3	0.005	145	U
ι Peg	F5 V	3.8	0.002	45	U
η Aql	F6 – G2 Ib	3.5–4.3	1.8	93	P
γ Lep A	F6 V	3.6	0.005	130	U
HD 165908	F7 V	5.0	0.002	39	U

Appendix VI (continued)

Star	Spectral type	V (mag)	π_L† (%)	θ† (deg)	P or U‡
β Vir	F8 V	3.6	0.017	162	U
χ Her	F9 V	4.6	0.012	31	U
χ' Ori	G0 V	4.4	0.013	20	U
β Com	G0 V	4.3	0.018	116	U
$\varkappa$ Cet	G5 V	4.8	0.006	135	U
α Men	G5 V	5.1	0.009	142	U
61 Vir	G6 V	4.8	0.010	132	U
β Aql	G8 IV	3.7	0.012	154	U
η Cep	K0 IV	3.4	0.006	101	U
HD 100623	K0 V	6.0	0.016	57	U
107 Psc	K1 V	5.2	0.116	175	U
HD 155885/6	K1 V	4.3	0.005	61	U
ε Ind	K5 V	4.7	0.006	88	U

† The errors in π_L are about $\pm 0.008\%$ for unpolarised standards, and $\pm 0.1\%$ for polarised standards. The errors in θ are about $\pm 1°$.
‡ The final column denotes whether the star is an unpolarised standard (U), or a polarised standard (P).

The data in this table were taken from Serkowski K *Methods of Experimental Physics* volume 12 part A, p. 361 (ed N Carlton, Academic Press, 1974).

Appendix VII

Julian date

The Julian date is the number of days elapsed since noon on 25 November 4714 BC (on the Gregorian calendar) or since noon on 1 January 4713 BC (on the Julian calendar).

Date (January 1 at noon Gregorian reckoning)	Julian day number	Date (January 1 at noon Gregorian reckoning)	Julian day number
2050	2469 808.0	1600	2305 448.0
2025	2460 677.0	1200	2159 351.0
2000	2451 545.0	800	2013 254.0
1975	2442 414.0	400	1867 157.0
1950	2433 283.0	0	1721 060.0
1925	2424 152.0	400 BC	1574 963.0
1900	2415 021.0	800 BC	1428 866.0
1875	2405 890.0	1200 BC	1282 769.0
1850	2396 759.0	1600 BC	1136 672.0
1825	2387 628.0	2000 BC	990 575.0
1800	2378 497.0	2400 BC	844 478.0
1775	2369 366.0	2800 BC	698 381.0
1750	2360 234.0	3200 BC	552 284.0
1725	2351 103.0	3600 BC	406 187.0
1700	2341 972.0	4000 BC	260 090.0
1675	2332 841.0	4400 BC	113 993.0
1650	2323 710.0	4714 BC (25 Nov)	0.0
1625	2314 579.0		

A cycle of the Gregorian calendar is 400 years long and has 146 097 days in it. For days subsequent to midday on 1 January the following number of days should be added to the Julian day number.

Date (noon)	Number of days (non-leap years)	Number of days (leap years)
1 Feb	31	31
1 Mar	59	60
1 Apr	90	91
1 May	120	121
1 Jun	151	152
1 Jul	181	182
1 Aug	212	213
1 Sep	243	244
1 Oct	273	274
1 Nov	304	305
1 Dec	334	335
1 Jan	365	366

Appendix VIII

Catalogues

There are many hundreds, if not thousands, of catalogues which are used by astronomers. Some may only amount to a short list of stars or spectrum lines etc in some obscure scientific journal. Others may contain data on millions of individual items and take the form of many and vast tomes occupying many feet of library shelving. A major 'catalogue of catalogues' has been published and the reader is referred to that for further information. That publication is:

Collins M 1977 *Astronomical Catalogues* 1951–75 (INSPEC Bibliography Series No 2)

Other large lists of catalogues may be found, amongst other sources, in:

Strand K Aa (ed) 1963 *Basic Astronomical Data* (vol III of *Stars and Stellar Systems*) (Chicago: University of Chicago Press)
Sidgwick J B 1971 *Amateur Astronomer's Handbook* (London: Faber and Faber)

Those catalogues specifically referred to within this work are listed below:

Spectral Type Atlases
Morgan W W, Keenan P C and Kellman E 1943 *An Atlas of Stellar Spectra* (Chicago: University of Chicago Press)
Yamashita Y. Nariai K and Norimoto Y 1977 *An Atlas of Representative Stellar Spectra* (Chicago: University of Chicago Press).

Spectrum Line Lists
Moore C E *Atomic Energy Levels* vols I, II, III NBS Circular No 467 (Washington, DC: US Govt. Printing Office). Reprinted in 1971 as NSRD-NBS 35
Moore C E 1959 *A Multiplet Table of Astrophysical Interest* NBS Technical Note No 36 (Washington, DC: US Govt. Printing Office)

Reader J and Corliss C H 1980 *Wavelengths and Transition Probabilities for Atoms and Atomic Ions* pt I, NSRDS-NBS 68 (Washington, DC: US Govt. Printing Office)

Wiese W L and Martin G A 1980 *Wavelengths and Transition Probabilities for Atoms and Atomic Ions* pt II, NSRDS-NBS 68 (Washington, DC: US Govt. Printing Office)

Striganov A R and Sventitskii N S 1968 *Tables of Spectral Lines of Neutral and Ionised Atoms* (New York: IFI/Plenum). Translated from the Russian

Moore C E 1950, 1952, 1962 *An Ultra-Violet Multiplet Table* NBS Circular No 488 pts 1 (1950), 2 (1952), 3, 4, 5 (1962) (Washington, DC: US Govt. Printing Office)

Bibliographies of atomic energy levels and spectra may be found in:

Hagan L and Martin W C 1972 *Bibliography on Atomic Energy Levels and Spectra, July 1968 through June 1971* NBS Special Publication 363 and its supplements (Washington, DC: US Govt. Printing Office)

Appendix IX

Answers to the exercises

1.1.1 69.21″ and 21.4″ (don't forget the Purkinje effect)
1.1.2 1.834 m and -3.387 m
1.1.3 0.31 mm
1.1.4 94 mm, 850× (for the exit pupil to be smaller than the eye's pupil)
1.1.5 0.014″
1.2.1 84 m along the length of its arms
1.4.1 0.002 m^3
1.4.3 250 per second
1.5.2 1.6×10^{-23}
1.5.3 About 44 $^{37}_{18}$Ar atoms
1.6.2 The calculated data are tabulated below:

Planet	Mass ($M_p/M_\odot$)	Period (days)	L_G (W)
Mercury	1.6×10^{-7}	88	62
Venus	2.4×10^{-6}	225	600
Earth	3.0×10^{-6}	365	180
Mars	3.2×10^{-7}	687	0.25
Jupiter	9.6×10^{-4}	4 330	5300
Saturn	2.9×10^{-4}	10 760	20
Uranus	4.3×10^{-5}	30 700	0.015
Neptune	5.3×10^{-5}	60 200	0.002
Pluto	2.5×10^{-6}	90 700	10^{-6}

1.6.3 400 pc
2.4.1 (a) 35 mm
         (b) 52 mm
2.4.2 3.1×10^{16} m (or 1 pc)
2.4.3 5×10^{17} m (or 16 pc)
2.7.1 28 mW, 72 km
3.1.1 $+25.9$, -20.6
3.1.2 3050 pc

3.1.3 $(U - B) = -0.84$, $(B - V) = -0.16$
$Q = -0.72$
$(B - V)_0 = -0.24$, $E_{B-V} = 0.08$
$E_{U-B} = 0.06$, $(U - B)_0 = -0.90$
Spectral type B3 V
Temperature 20 500 K
Distance (average) 230 pc
$U_0 = 2.82$, $B_0 = 3.72$
$V_0 = 3.96$, $M_v = -2.8$

4.1.1 -80.75 km s^{-1}

4.1.2   One prism,   0.79 nm mm^{-1}

4.2.1 $f' = 25$, $W_\lambda = 10^{-12}$ m, $\lambda = 625$ nm
$R = 6.3 \times 10^5$, $L = 0.42$ m, $D = 0.38$ m
$d\theta/d\lambda = 1.66 \times 10^6$ rad m^{-1}
$f_1 = 13.3$ m, $f_2 = 12.1$ m
$D_1 = 0.533$ m, $D_2 = 0.533$ m
$S = 22$ μm
Problems to be solved if possible in the next iteration through the
exercise. Grating is too large: a normal upper limit on L is 0.1 m.
Slit width is too small. The overall size of the instrument will lead
to major thermal control problems.

4.2.2 $+5.3$ (taking $q = 0.002$ and $\alpha = 5 \times 10^{-6}$)

4.2.3   64.96°

5.1.2   0.26 mm

5.3.1   50.6 mm,   0.79 mm

Bibliography

Some books and articles which provide further reading for selected aspects of this book, or from which further information may be sought, are listed below together with the section of the text to which the book or article applies. A cross-index of references relevant to each section appears on pages 456 and 457.

[1] Adams D J 1980 *Cosmic X-ray Astronomy* (Bristol: Adam Hilger)   section 1.3
[2] Adelman S J *et al* (ed) 1992 *Automated Telescopes for Photometry and Imaging (Astronomical Society of the Pacific Conference Series Vol 28)*   sections 1.1, 3.1 and 3.2
[3] Allen C W 1973 *Astrophysical Quantities* (London: Athlone)   General
[4] Allen D A 1975 *Infrared; the New Astronomy* (Keith Reid Ltd)   section 1.1
[5] Alloin D M and Mariotti J-M 1994 *Adaptive Optics for Astronomy* (Dordrecht: Kluwer)   sections 1.1 and 2.5
[6] Alloin D M Mariotti J-M 1989 *Diffraction-Limited Imaging with Very Large Telescopes* (Dordrecht: Kluwer)   sections 1.1, 1.2, 2.5, 2.6 and 2.7
[7] Bahcall J N 1979 Solar neutrinos—theory versus observation *Space Sci. Rev.* **24** 227   section 1.5
[8] Barden S C (ed) 1988 *Fiber Optics in Astronomy (Astronomical Society of the Pacific Conference Series Vol 3)*   sections 2.5, 4.1 and 4.2
[9] Barford N C 1985 *Experimental Measurements: Precision, Error and Truth* (New York: Wiley)   section 2.1
[10] Barlow B V 1975 *The Astronomical Telescope* (London: Wykeham)   section 1.1
[11] Bell R J 1972 *Introductory Fourier Transform Spectroscopy* (New York: Academic Press)   sections 4.1 and 4.2
[12] Bell-Burnell S J *et al* (ed) *Next Generation Space Observatory* (Dordrecht: Kluwer)   section 1.1
[13] Beynon J D E and Lamb D R (eds) 1980 *Charge Coupled Devices and Their Applications* (New York: McGraw-Hill)   section 1.1
[14] Birney D S 1991 *Observational Astronomy* (Cambridge: Cambridge University Press)   sections 1.1, 1.2, 2.2, 3.1, 4.1 and 5.3
[15] Bode M F 1995 *Robotic Observatories* (New York: Wiley)   section 1.1

451

[16] Bracewell R N 1978 *The Fourier Transform and its Applications* (New York: McGraw-Hill)   sections 2.1, 2.5 and 4.1

[17] Bradt H and Giacconi R (ed) 1973 *X- and Gamma-Ray Astronomy (IAU Symposium No 55)* (Dordrecht: Reidel)   section 1.3

[18] Brancazio P J and Cameron A G W 1968 *Infrared Astronomy* (London: Gordon and Breach)   section 1.1

[19] Bray D F 1992 *The Observation and Analysis of Stellar Photospheres* (Cambridge: Cambridge University Press)   sections 4.1 and 4.2

[20] Budding E 1993 *Astronomical Photometry* (Cambridge: Cambridge University Press)   sections 3.1 and 3.2

[21] Burbidge G and Hewitt A 1981 *Telescopes for the 1980's* (New York: Annual Reviews)   section 1.1

[22] Butler C J and Elliott I 1993 *Stellar Photometry—Current Techniques and Future Developments* (Cambridge: Cambridge University Press)   sections 3.1 and 3.2

[23] Carleton N (ed) 1974 *Methods of Experimental Physics* vol 12, pt A *Astrophysics* (New York: Academic)   sections 3.1, 3.2 and 5.2

[24] Christiansen W M and Högbom J A 1985 *Radiotelescopes* (Cambridge: Cambridge University Press)   sections 1.2, 2.5 and 2.8

[25] Clarke D and Grainger J F 1971 *Polarized Light and Optical Measurement* (Oxford: Pergamon)   section 5.2

[26] Comte G and Marcelin M (ed) 1995 *Tridimensional Optical Spectroscopic Methods in Astrophysics (Astronomical Society of the Pacific Conference Series Vol 71)*   sections 4.1 and 4.2

[27] Cooper W A and Walker E N 1989 *Getting the Measure of Stars* (Bristol: Adam Hilger)   sections 3.1 and 3.2

[28] Cornwell T J and Perley R A (ed) 1991 *Radio Interferometry: Theory, Techniques and Applications (Astronomical Society of the Pacific Conference Series Vol 19)*   sections 1.2 and 2.5

[29] Craig I J D and Brown J C 1986 *Inverse Problems in Astronomy* (Bristol: Adam Hilger)   section 2.1

[30] Davies P C W 1980 *The Search for Gravity Waves* (Cambridge: Cambridge University Press)   section 1.6

[31] Davis Phillip A G *et al* (ed) 1990 *CCDs in Astronomy II* (L Davis Press)   sections 1.1, 3.2, 4.1 and 4.2

[32] Dennis P N J 1986 *Photodetectors: An Introduction to Current Technology* (New York: Plenum)   section 1.1

[33] di Gesù V, Scarsi L, Crane P, Friedman J H and Levialdi S (ed) 1985 *Data Analysis in Astronomy* (New York: Plenum)   sections 1.1, 1.2, 3.1, 3.2, 4.1, 4.2, 5.1, 5.2, 5.3 and 5.4

[34] di Gesù V, Scarsi L and Crane P (ed) 1987 *Selected Topics on Data Analysis in Astronomy* (Singapore: World Scientific)   sections 1.1, 1.2, 3.1, 3.2, 4.1, 4.2, 5.1, 5.2, 5.3 and 5.4

[35] di Gesù V *et al* (ed) 1992 *Data Analysis in Astronomy IV* (New York: Plenum)   section 2.1

[36] Eccles M J Sim M E and Tritton K P 1983 *Low Light Level Detectors in Astronomy* (Cambridge: Cambridge University Press)   section 1.1

[37] Edwards A L 1984 *Introduction to Linear Regression and Correlatic̣n* (San Francisco: Freeman)   sections 2.1 and 2.5

[38] Emerson D Tand Payne J M (ed) 1995 *Multi-Feed Systems for Radio Telescopes (Astronomical Society of the Pacific Conference Series Vol 75)*   section 1.2

[39] Evans J V and Hagfors T 1968 *Radar Astronomy* (New York: McGraw-Hill)   section 2.8

[40] Fazio G G (ed) 1976 *Infrared and Submillimeter Astronomy* (Dordrecht: Reidel)   sections 1.1 and 1.2

[41] Fehrenbach C and Westerlund B E (ed) 1973 *Spectral classification and multicolour photometry (IAU Symposium No 50)* (Dordrecht: Reidel)   sections 3.1, 4.1

[42] Filippenko A V (ed) 1992 *Robotic Telescopes in the 1990s (Astronomical Society of the Pacific Conference Series Vol 34)*   sections 1.1, 3.1 and 3.2

[43] Genet R M and Hayes D S 1989 *Robotic Observatories* (Autoscope)   section 1.1

[44] Goddard D E and Milne D K (ed) 1994 *Parkes: Thirty Years of Radio Astronomy* (CSIRO)   sections 1.2 and 2.5

[45] Golay M 1974 *Introduction to Astronomical Photometry* (Dordrecht: Reidel)   sections 3.1 and 3.2

[46] Grandy W T Jr and Schick L H 1991 *Maximum Entropy and Bayesian Methods* (Dordrecht: Kluwer)   section 2.1

[47] Grey P M (ed) 1993 *Fiber Optics in Astronomy II (Astronomical Society of the Pacific Conference Series Vol 37)*   sections 4.1 and 4.2

[48] Guyenne T (ed) 1995 *Future Possibilities for Astrometry in Space* (European Space Agency)   section 5.1

[49] Hall D S *et al* (ed) 1986 *Automatic Photoelectric Telescopes* (Fairborn Press)   sections 1.1, 3.1 and 3.2

[50] Harwit M 1973 *Astrophysical Concepts* (New York: Wiley)   General

[51] Hearnshaw J B 1986 *The Analysis of Starlight* (Cambridge: Cambridge University Press)   sections 1.1, 4.1 and 4.2

[52] Hearnshaw J B 1996 *Measurement of Starlight* (Cambridge: Cambridge University Press)   sections 3.1 and 3.2

[53] Henry G W and Eaton J A (ed) 1995 *Robotic Telescopes: Current Capabilities, Present Developments, and Future Prospects for Automated Astronomy (Astronomical Society of the Pacific Conference Series Vol 79)*   sections 1.1, 3.1, 3.2, 4.1 and 4.2

[54] Herzberg G 1944 *Atomic Spectra and Atomic Structure* (New York: Dover)   section 4.1

[55] Hillas M 1972 *Cosmic Rays* (Oxford: Pergamon)   section 1.4

[56] Hiltner W A (ed) 1962 *Astronomical Techniques (Stars and Stellar Systems vol II)* (Chicago: University of Chicago Press)   sections 1.1, 2.2, 2.3, 3.1, 3.2, 4.1, 4.2, 5.1, 5.2, 5.4

[57] Hopkins J 1976 *Glossary of Astronomy and Astrophysics* (Chicago: University of Chicago Press)   General

[58] Jaschek C and Murtagh F (ed) 1990 *Errors, Bias and Uncertainties in Astronomy* (Cambridge: Cambridge University Press)   section 2.1

[59] Jenkins T E 1987 *Optical Sensing Techniques and Signal Processing* (Englewood Cliffs, NJ: Prentice-Hall)   sections 1.1 and 2.1

[60] Kane S R (ed) 1975 *Solar, Gamma, X and EUV radiation (IAU Symposium No 68)* (Dordrecht: Reidel)   sections 1.3 and 5.3

[61] Kenneth Mees C E and James T H (ed) 1966 *The Theory of the Photographic Process* (London: Macmillan)   section 2.2

[62] Kitchin C R 1995 *Optical Astronomical Spectroscopy* (Bristol: Institute of Physics Publishing)   sections 4.1 and 4.2

[63] Kitchin C R 1995 *Telescopes and Techniques* (Berlin: Springer)   sections 1.1, 3.1 and 4.1

[64] Kitchin C R and Forrest R F 1997 *Seeing Stars* (Berlin: Springer)   section 1.1

[65] Korsch D 1991 *Reflective Optics* (New York: Academic Press)   sections 1.1 and 1.3

[66] Krauss J D 1966 *Radio Astronomy* (New York: McGraw-Hill)   sections 1.2, 2.5 and 2.8

[67] Kruger A 1979 *Introduction to Solar Radio Astronomy and Radio Physics* (Dordrecht: Reidel)   section 5.3

[68] Kuiper G P and Middlehurst B M (ed) 1960 *Telescopes (Stars and Stellar Systems Vol I)* (Chicago: University of Chicago Press)   sections 1.1 and 1.2

[69] Laikin M 1995 *Lens Design* (New York: Dekker)   section 1.1

[70] Lang K R 1978 *Astrophysical Formulae* (Berlin: Springer)   General

[71] Longair M S 1981 *High Energy Astrophysics* (Cambridge: Cambridge University Press)   sections 1.3 and 1.4

[72] Mackintosh A 1986 *Advanced Telescope Making Techniques* Vols 1 and 2 (Willman Bell)   sections 1.1, 2.5 and 5.3

[73] Manly P L 1991 *Unusual Telescopes* (Cambridge: Cambridge University Press)   section 1.1

[74] Manly P L 1994 *The 20-cm Schmidt–Cassegrain Telescope* (Cambridge: Cambridge University Press)   sections 1.1, 2.2 and 3.2

[75] Manno V and Ring J (ed) 1972 *Infrared Detection Techniques for Space Research* (Dordrecht: Reidel)   section 1.1

[76] Mar J W and Leibowitz H (ed) 1969 *Structures Technology for Large Radio and Radar Telescope systems* (Cambridge, MA: MIT Press)   sections 1.2 and 2.8

[77] Mattok C (ed) 1993 *Targets for Space-Based Interferometry* (European Space Agency)   section 2.5

[78] McClean I S 1994 *Electronic and Computer Aided Astronomy* (Chichester: Ellis Horwood)   sections 1.1, 1.3, 2.3, 4.2 and 5.2

[79] Meeks M L (ed) 1976 *Methods of Experimental Physics* vol 12, part B *Radio Telescopes* and part C *Radio Observations* (New York: Academic)   sections 1.2 and 2.5

[80] Minnaert M G J 1969 *Practical Work in Elementary Astronomy* (Dordrecht: Reidel)   General

[81] Morrison L V and Gilmore G F (ed) *Galactic and Solar System Optical Astrometry* (Cambridge: Cambridge University Press)   section 5.1

[82] Motz L and Duveen A 1977 *Essentials of Astronomy* (New York: Columbia University Press)   General

[83] Murray C A 1983 *Vectorial Astrometry* (Bristol: Adam Hilger)   section 5.1

[84] Perley R A *et al* (ed) 1989 *Synthesis Imaging in Radio Astronomy (Astronomical Society of the Pacific Conference Series Vol 6)*   sections 1.2 and 2.5

[85] Pomerantz M A 1971 *Cosmic Rays* (New York: Van Nostrand) section 1.4

[86] Press W H, Flannery B P, Teukolsky S A and Vetterling W T 1986 Numerical Recipes (Cambridge: Cambridge University Press) sections 2.1, 2.5 and 4.1

[87] Rieke G H 1994 *Detection of Light: From the Ultraviolet to the Submillimeter* (Cambridge: Cambridge University Press) sections 1.1 and 1.2

[88] Rohlfs K 1986 *Tools of Radio Astronomy* (Berlin: Springer) sections 1.2, 2.5 and 2.8

[89] Rowan-Robinson M (ed) 1976 *Far Infrared Astronomy* (Oxford: Pergamon) section 1.1

[90] Roy A and Clarke D 1982 *Astronomy: Principles and Practice* 2nd edn (Bristol: Adam Hilger) General

[91] Rutten H and van Vemrooij M 1989 *Telescope Optics—Evaluation and Design* (Willman-Bell) section 1.1

[92] Schroeder D J 1987 *Astronomical Optics* (New York: Academic) sections 1.1 and 4.1

[93] Setti G and Fazio G G (ed) 1978 *Infrared Astronomy* (Dordrecht: Reidel) section 1.1

[94] Shurcliffe W A 1962 *Polarized Light* (Oxford: Oxford University Press) section 5.2

[95] Shurcliffe W A and Ballard S S 1964 *Polarized Light* (New York: Van Nostrand) section 5.2

[96] Sidgwick J B 1955 *Amateur Astronomer's Handbook* (London: Faber and Faber) sections 1.1, 3.1, 3.2, 4.1, 4.2 and 5.1

[97] Sterken C and Manfroid 1992 *Astronomical Photometry* (Dordrecht: Kluwer) sections 1.1, 2.2, 3.1 and 3.2

[98] Strand K Aa (ed) 1963 *Basic Astronomical Data (Stars and Stellar Systems Vol III)* (Chicago: University of Chicago Press) sections 3.1, 4.1, 5.1 and 5.2

[99] Struve O Lynds B and Pillans H 1959 *Elementary Astronomy* (Oxford: Oxford University Press) General

[100] Taylor H D 1983 *The Adjustment and Testing of Telescope Objectives* (Bristol: Adam Hilger) section 1.1

[101] Thorne A P 1974 *Spectrophysics* (London: Chapman and Hall) section 4.1

[102] Tyson R K 1991 *Principles of Adaptive Optics* (New York: Academic) section 1.1

[103] van de Kamp P 1967 *Principles of Astrometry* (San Francisco: Freeman) section 5.1

[104] van der Hucht K A and Vaiana G 1978 *New Instrumentation for Space Astronomy* (Oxford: Pergamon) sections 1.1, 1.3, 4.1 and 4.2

[105] Vaughan J M 1989 *The Fabry–Perot Interferometer* (Bristol: Adam Hilger) sections 4.1 and 4.2

[106] Wall J V 1993 *Optics in Astronomy* (Cambridge: Cambridge University Press) section 1.1

[107] Wall J V and Boksenberg A 1990 *Modern Technology and its Influence on Astronomy* (Cambridge: Cambridge University Press) sections 1.1, 1.2, 1.3, 2.5, 3.1, 3.2, 4.1 and 4.2

[108] Wallis B D and Provin R W 1988 *Manual of Advanced Celestial Photography* (Cambridge: Cambridge University Press)   section 2.2
[109] Warner B 1988 *High Speed Astronomical Photometry* (Cambridge: Cambridge University Press)   sections 2.7, 3.1 and 3.2
[110] West R M and Heudier J C (ed) 1978 *Modern Techniques in Astronomical Photography* (La Silla: European Southern Observatory)   section 2.2
[111] White H E 1934 *Introduction to Atomic Spectra* (New York: McGraw-Hill)   section 4.1
[112] Wilson J G 1976 *Cosmic Rays* (London: Wykeham)   section 1.4
[113] Zensus J A *et al* (ed) 1995 *Very Long Baseline Interferometry and the VLBA* (*Astronomical Society of the Pacific Conference Series Vol 82*)   sections 1.2 and 2.5

Cross-index for the Bibliography

Section   Book/Article

1.1   [4], [10], [13], [14], [15], [18], [21], [31], [32], [33], [34], [36] [40], [43], [49], [51], [56], [59], [63], [64], [68], [72], [74], [73] [75], [78], [87], [89], [91], [92], [93], [96], [97], [100], [102], [104] [107], [106]

1.2   [6], [14], [24], [28], [33], [34], [38], [40], [44], [66], [68], [76], [79] [84], [87], [88], [107], [113]

1.3   [1], [17], [60], [65], [71], [78], [104], [107]

1.4   [55], [71], [85], [112]

1.5   [7]

1.6   [30]

2.1   [9], [16], [29], [35], [37], [46], [58], [59], [86]

2.2   [32], [56], [61], [74], [97], [108], [110]

2.3   [5], [56], [78]

2.5   [6], [8], [16], [24], [28], [37], [44], [66], [72], [77], [79], [84], [86] [88], [107], [113]

2.6   [6]

2.7   [6], [109]

2.8   [24], [39], [66], [76], [88]

3.1   [2], [14], [20], [22], [23], [27], [31], [33], [34], [41], [42], [45], [49] [52], [53], [56], [63], [96], [97], [98], [107], [109]

3.2   [2], [20], [22], [23], [27], [33], [34], [42], [45], [49], [52], [53] [56], [74], [96], [97], [107], [109]

4.1   [8], [11], [14], [16], [19], [26], [31], [33], [34], [41], [47], [51], [53]
        [54], [56], [62], [63], [86], [92], [96], [98], [101], [105], [104], [107]
        [111]

4.2   [8], [11], [19], [26], [31], [33], [34], [47], [51], [53], [56], [62]
        [78], [96], [104], [107], [105]

5.1   [33], [34], [48], [56], [103], [81], [83], [96], [98]

5.2   [23], [25], [33], [34], [56], [78], [94], [95], [98]

5.3   [14], [33], [34], [60], [67], [72]

5.4   [33], [34], [56]

General   [3], [50], [57], [70], [80], [82], [90], [99]

Index